Case Studies in Food Microbiology for Food Safety and Quality

Rosa K. Pawsey

South Bank University, London, UK

ROYAL SOCIETY OF CHEMISTRY

ISBN 0-85404-626-7

A catalogue record for this book is available from the British Library

Published by The Royal Society of Chemistry,
Thomas Graham House, Science Park, Milton Road,
Cambridge CB4 0WF, UK

Registered Charity Number 207890

For further information see our web site at www.rsc.org

Typeset in Great Britain by Vision Typesetting, Manchester
Printed by TJ International Ltd, Padstow, Cornwall

Preface

It is obvious to any teacher that you cannot expect to provide students with all the 'facts' they need to go out into the world and become professional food scientists, food managers, safety officers, legislators, or enforcement officials. There are too many facts. It would be extremely dull to teach this way, even duller to be the recipient. Yet this is how much teaching has been implemented over the decades, and over the world.

What I have tried to do is to help the students to engage in the learning process, and in so doing learn techniques for learning, and subsequently for applying knowledge. This has been more interesting for me, the teacher, and I think for the students too. This book has grown from my ten years teaching food microbiology and hygiene to post-graduate students in Food Safety and Control at South Bank University, London. It was the decade during which, in the UK, the national policies behind the management of food safety changed rapidly, and this meant that managers in small and large companies alike had to learn new terminology and new ways. It also meant that food enforcement officers had to gain greater depth of specialist food knowledge in order to first understand, and secondly to ensure that new food law was being implemented. So I have tried out some ideas on the students, obviously some more successful than others, but what has been clear to me is that active learning is much more fun all round, and it's my hypothesis that it is more productive too.

I learnt alongside the students. Together we have explored ideas through working with real and imagined scenarios to identify what microbiological principles were at issue and how microbiological hazards should be managed, balancing theory with practicality. I learnt, as we all learnt (I like to hope) that we don't all think alike, that our priorities are different, and that different solutions can be offered for the same problem. Some solutions, of course, are better than others. But all solutions are driven by our own individual perspectives and priorities.

I learnt also that not all students have the same enthusiasm for food microbiology as I have. Their need was to place microbiology in the holistic context of their work – enforcement, management, training, hygiene.

So what was imperative was a sense of reality.

The case studies in this book are largely real. They really occurred as written.

One – the tomato paste – case study is imagined, but is based on my experience of how food is transported round the world, and how some of it (too much) is mishandled. There are touches of my travels (which I enjoy enormously) in several of the case studies whose inclusion is indeed there not because I was present at the scene of the event described, but because I have seen similar events round the world which I have recounted to my students. On other occasions the students have recounted horrors to me. One such example is the story of the blocked down pipe from a sink in which the waste water flowed upwards into the canned tuna which supposedly was draining ready for filling into sandwiches for the hospital contract . . . There is so often such a gulf between what we would like to imagine is the way food is handled and the awful reality. By exploring real events through case studies many issues arise. It is up to the teacher to steer the discussion and help the students gain understanding of specific topics the teacher wants the students to grasp. But there is a difference between face-to-face discussions in a classroom environment and sitting down alone and using a book of case studies like this for self managed learning – which I hope is what this book will also support.

What I have tried to do is to present a number of Chapters – 21 in fact – which address different issues through the exercises I put in the third section of each chapter. I have to trust the reader not to use Section 4 – which is my commentary on the exercises until they have thought things through themselves. That is their choice however. In the commentary I have tried to raise points for thinking about: I have not tried to provide answers. I have not tried to claim that my commentary is 'complete' or 'total', but hopefully does provide some guidance on the issues I ask the student to think about. But it may well be – I hope so – that there is sufficient information in the Case Study (Section 1 of each chapter) alone, or with the Background information (Section 2 of each chapter) for the material to be used for an entirely different purpose other than the one I propose. Equally it is not my intention that within the covers of the book 'all' the information needed to develop understanding of particular issues is there. But I have tried to provide key information for I am aware that text books are expensive to buy, and that access to good libraries and internet information is easier for some than others.

My intent is that the student needs to *use* the book – to search the index, to read the other chapters, to study Tables or Figures to find relevant information. I have included the well known food poisoning organisms – the ones which occur in the national statistics most frequently. Of course there are emerging pathogens whose significance may turn out to be greater than statistics currently show – that however is always the case as both science and technology advance. But since the purpose of this book is to help students to engage in active learning and to apply information already learned, I hope that the outcome for them will be personal development facilitating application of both principles and knowledge as pathogens emerge, and food safety priorities change. I have not tried, I did not intend, to provide text which 'teaches' fundamental microbiology. What I have tried to do is provide a series of vehicles within which principles can be taught, learned or developed.

There is a scheme to the book: it seeks to cover the key issues relating to the control and management of micro-organisms in food which compromise its safety. At the beginning of each chapter a number of key issues are listed: a type of food, an organism or organisms, a microbiological issue and a control issue. The 21 Case Studies work through a number of themes, but being real events they each potentially cover much more than I have addressed. Within these contexts my theme is simple and I have tried to communicate my emphases through the titles of the chapters. I have tried to address the incidence of organisms, their transfer from one place to another, when they grow in foods, strategies to prevent them growing and killing them, determining shelf life and post production history; finding out through surveillance the prevalence of organisms, global dissemination of organisms, addressing some of the problems of interpreting microbiological results and in interpreting the presence of organisms in foods, setting standards, managing hygiene, perceiving and managing risk, understanding of the changing concept of safe food.

I owe an enormous 'thank you' to all my students, for teaching them has been a pleasure. They have enriched my life and through their diversity (they have come from so many countries round the world) have helped me to understand that what may work in England, may not be so relevant in, for example, parts of Africa; or what works in a big company is difficult to implement in a small one. We all have so much to learn from each other. This is one of the pleasures of sitting round in a group sorting out a set of approaches to solve some problems. But we all, world wide, want and need safe food. I have wanted to help students work out for themselves that there is huge responsibility in determining how to improve food safety, that difficult choices sometimes have to be made, that raising standards requires enormous effort, and that solutions differ. I have enjoyed writing these case studies. I know that people's views will be different – I'd like to hear from you!

How to use this book

One way would be to start at the first chapter and work through the book to the end. I would hope that by then you have a comprehensive understanding of key areas of microbiological food control. Another would be to read a single Case Study, to work on it and do no more.

My suggestion is that you look at the Table of Contents and at Table A on p. x, which together give you a summary of the whole book. Each chapter has a set of Key Issues and a Challenge to focus you on the issues I am interested in you addressing. Table A will also facilitate you choosing the material you wish to study.

I suggest that you need to decide from where you will get your information (see Good supporting books on p. ix) to undertake the exercises in any chapter. One source is this book itself. You need to be clear that I have not provided within each chapter everything you may feel you need to know. But within the covers of the book there is lots of information. The book as a whole is a resource, but it cannot provide everything.

Active learning is about engaging in the issues and asking questions. You will need, like a detective, to read the text of the case study provided in section 1 of any chapter. There is information in there which may not have been emphasised, but is there just the same. There is information in the Tables of data, and in the Figures. There are clues to understanding, but you have to read closely, and think in terms of details. There is information elsewhere in the book, so use the Index as much as possible. There is information which was relevant at the time of the event described which may have been superseded by newer policies and newer knowledge. To understand the Case Study you may have to put aside more modern knowledge, but to develop yourself you will then have to consider what information is relevant today. Say to yourself – had this occurred today, 'here' in the country in which you live – what would or should be done, now? So when you start addressing the exercises ask yourself or your group all sorts of questions. You will need to sort the questions – some will be more central than others, some may prove irrelevant. Secondly you need to amass information – from this book, from other resources, from the internet, libraries, real experience and so on.

The stepwise approach outlined below provides a generic structure which could guide you:

Step 1. Read the Case Study again and ensure that you understand the meaning of all the terms used in it, for example:
What is the food? Have I eaten this? Where does is come from, how is packed, what is its shelf life . . .
What is the organism? What do I know about it? Is what I know enough to deal with this problem? What else do I need to find out about it?
Step 2. What aspect of the exercise – the problem – can be tackled first?
Step 3. Analyse the problem you have selected.
Note down all the ideas you have giving possible explanations, ideas, facts.
Step 4. Summarise and organise your ideas to clarify where you have got to.
Step 5. Identify what information you do not yet have but need to obtain. Be as specific as possible.
Step 6. Obtain the information listed at Step 5.
Step 7. Collate the information and apply it to the problem. Identify where you have clarified the problem and where lack of clarity still exists. Repeat the earlier steps until you feel confident that you are now able to provide an answer/approach to the problem set.

This stepwise structure is based on Sivik and Ljungquist (2001). It is a valuable approach which ensure you are actively engaged in developing your abilities in microbiological food safety.

Reference

Sivik, B., and Ljungquist, I. (2001). *Problem based learning in food science: experience at Lund University*, in *Personal skills: An integrated component of food science courses*. eds. Walters, M. and Pawsey, R., pp.195–211. Published for the Working Group 'Development of personal skills', Socrates Thematic Network Project in Food Studies – project number 55792-CP–3–00-FR-ERASMUS-ETN. Now published by the Royal Society of Chemistry, London.

Good supporting books

Adams, M. and Moss, M. (2000). *Food Microbiology*, 2nd Edition. Royal Society of Chemistry, Cambridge. This is an excellent, affordable text book.

The publications of the International Commission for Microbiological Specifications for Foods (ICMSF) – see the references at the end of many of the following chapters. These publications are more expensive, but should be available in every University library.

Table A *Structure of the book*

Case study	Primary food system	Key issues	Main organism(s)
A. DEVELOPING STRATEGIES FOR CONTROL			
1. Water as a vector of organisms	Water	Microbial survival and growth in waters	*Vibrio cholerae*
2. Expectations of food control systems – in the past and now	Meat paste	Food safety control and national infrastructure Sources of pathogens FP outbreak: identifying its cause Microbial growth conditions Toxin production in food	*Clostridium botulinum*
3. Zoonotic disease	Drinking milk	Control of food safety in manufacture – then and now MILK: a primary source of pathogens Pasteurised milk and salted telema cheese Zoonotic disease; tuberculosis	*Mycobacterium bovis*
4. Should pasteurisation of drinking milk be obligatory?	Drinking milk	The effect of political change on control efficiency The need for a holistic approach to control	*Campylobacter jejuni*
B. TECHNIQUES FOR CONTROL			
5. Surveillance and microbiological analyses	Kebabs	MEAT: a primary source of pathogens Analysis of foods for their microbial population Microbiological surveillance Hygiene	*Salmonella* spp.
6. Microbial hazards – what are they?	Fish – caviar and fish roes	FISH: a primary source of organisms Points of safety and quality loss Microbial hazards – what are they? Controls for microbial hazards Hazard analysis	Spoilage organisms and pathogens
7. Post-production product handling and acceptability	Canned tomato paste	VEGETABLES and fruit: primary sources of organisms Microbial death and commercial sterility Microbial growth conditions Control of food safety and quality during manufacture and after Product acceptability	Spoilage organisms and *Clostridium botulinum*
8. HACCP and the responsibilities of the food producer	Shell eggs and chocolate mousse	Cross contamination HACCP	*Salmonella* spp.
9. Product formulation for control	Yoghurt	The 'combined treatment' approach to control HACCP	*Clostridium botulinum*

Acknowledgements

A number of people have helped me in putting together this book. Some are those who have experienced very difficult episodes in food production, and who very generously have allowed me access to their papers and sat and talked with me. Some are colleagues in the UK and elsewhere in the world, others are librarians, former students, government officials, food producers, legislators, enforcement officers . . .. I am extremely grateful to everyone who has helped me both directly and indirectly, in big and small ways, in the writing of this book, and in helping me gain some of the experience which is embodied in it. But of course the mistakes are all mine:

Dr. Tony Burns, Mr. Dominic Man (South Bank University), Mrs. Sam Colley (Dorset District Council), Dr. Carmen Tudorica (University of Galatz, Romania; now of Exeter University), Prof. Stephane Desobry (ENSAIA, Nancy, France), Prof. Elisabeth Dumoulin (ENSIA, Massy, France), the late Mr. James Aldridge, Mr. David Wright (Meat and Livestock Commission), Dr. Chris Little (PHLS,UK), Mrs. Aine O'Brian (private consultant), Dr. Tamara Orlova (Sevrybtechcentr, Murmansk), Mr. Rolf Kummerlin (University of Concepcion), Mrs. Idiat Amusu (Yaba College of Food Technology, Lagos, Nigeria), Dr. Saverio Mannino (University of Milan), Dr. Mike Walters (Nestlé, Switzerland), Dr. Alan Varnam (University of North London), Dr. Maurice Moss (University of Surrey), Dr. Judith Hilton (Food Standards Agency, London), Mr. Robert Steenson (formerly of Clydesdale District Council), Dr. Duncan McDougall (South Lanarkshire Council), Mr. Humphrey Errington (cheesemaker, Lanarkshire), Sheriff Douglas Allan and Sheriff John McInnes (South Lanarkshire), Ms. Kathryn Callaghan, (FSA). Also a special thanks to my husband, John Pawsey.

Rosa Pawsey,
Shalford, Surrey, UK
randjpawsey@btinternet.com
December 2001.

Contents

A. Developing Strategies for Control

CHAPTER 1

Water as a Vector of Organisms 3

Key issues
- Water
- *Vibrio cholerae*
- Microbial survival and growth in water
- Food safety control, and the need for national infrastructure

Challenge

In 1974, cholera spread in Portugal. Not only did potable water supplies become polluted with cholera infected faeces, but so did shellfish, other foods and bottled water. This case study invites you to start thinking about how organisms become disseminated, whether they die in that process, and what affects their growth and survival. You can, of course, extend your thoughts beyond this case and wonder what control systems prevent water in all its various uses in the food industry both becoming contaminated and being a vector for a very wide range of pathogens.

CHAPTER 2

Expectations of Food Control Systems – in the Past, and Now   26

Key issues
- Meat – bottled wild duck paste
- *Clostridium botulinum*
- Sources of pathogens
- Food poisoning outbreak – identifying its cause
- Microbial growth conditions
- Toxin production in food
- Control of food safety in manufacture: then and now

Challenge
A very serious outbreak of food poisoning, which occurred in 1922, is described. It provides a means by which you can analyse how the cause of the food poisoning was identified, and see whether such an approach is relevant today. In understanding how this particular outbreak of food poisoning arose you are also encouraged to think about how food contamination arises, why and when micro-organisms grow in foods, what is needed to ensure their destruction, and what we expect of a microbiological control system in a food manufacturing process today.

CHAPTER 3

Zoonotic disease 56

Key issues
- Milk: a primary source of pathogens
- Pasteurised milk and salted telemea cheese
- *Mycobacterium bovis*
- Zoonotic diseases; tuberculosis
- The effect of political change on control efficiency

Challenge
This case study addresses the age-old problem of the transfer of disease from animals to man – in this instance *via* milk and milk products. The case study demands that you think about the elements needed in a control system – from where organisms originate, what controls microbial growth in a food system, what can be manipulated to control it, the use of heat for the destruction of organisms, and when we are confident in the safety of a food at local level. But the exercises also demand that you consider these things in order to evaluate the broader regional context in which food control systems are implemented.

CHAPTER 4

Should Pasteurisation of Drinking Milk Be Obligatory?   **82**

Key issues
- Drinking milk
- *Campylobacter jejuni*
- Developing strategies for control
- The need for a holistic approach to control

Challenge
Three outbreaks of *Campylobacter jejuni* infection contracted through milk are presented to you. While some of the issues raised in Chapter 3 are also relevant here, the main challenge to you is to consider the role pasteurisation has in managing drinking milk safety and whether that role is unique.

B. Techniques for Control

CHAPTER 5

Surveillance and Microbiological Analyses   **99**

Key issues
- Meat: grilled kebabs
- Meat: a primary source of pathogens
- *Salmonella* and other organisms
- Microbiological surveillance as an aid to control
- Microbiological analysis as a tool
- Hygiene

Challenge
Meats, like milk, carry zoonotic organisms. Three case studies challenge you to
think about the microbiology of grilled meat. They demand consideration of the
significance of the organisms in the raw and cooked meat. They also ask you to
think about how microbiological tests produce results, what the results mean,
and when surveillance is of value.

CHAPTER 6

Microbial Hazards 127

Key issues
- Fish – caviar and fish roes
- Microbial hazards – spoilage organisms and pathogens
- Fish: a primary source of organisms
- Points of safety and quality loss
- Microbial hazards – what are they?
- Controls for microbial hazards
- Hazard analysis

Challenge
Caviar is a delicate, rare, expensive commodity and as such you would expect that great care would be taken at every stage of its production. This case study should lead you to explore which microbial hazards need be controlled to assure its safety and quality. It should also cause you to consider how economic forces influence perceptions of safety and quality, and thus how control systems bend to those forces, whether in an ideal world they should or should not.

CHAPTER 7

Post-production Product Handling and Acceptability   **150**

Key issues
- Vegetables and fruit; canned tomato paste – a high acid food
- Spoilage organisms and *Clostridium botulinum*
- Canning
- Microbial death and commercial sterility
- Microbial growth conditions
- Control of food safety and quality during manufacture, and after
- Product acceptability

Challenge
Canned tomato paste, stored in tropical conditions, is known to be vulnerable to spoilage. The product is liable to deteriorate, the cans to swell and burst.

The case study presents you with some microbiological data for tomato paste and should cause you to think about the meaning of 'product acceptability'. It allows you the opportunity to identify for yourself the key microbiological issues in production which must be addressed to ensure that the product is microbiologically safe and stable in storage. It also enables you to consider whether the market for which a product is destined should be part of that process.

CHAPTER 8

HACCP and the Responsibilities of the Food Producer **174**

Key issues
• Shell eggs and chocolate mousse
• *Salmonella* spp.
• Cross contamination
• HACCP

Challenge
Shell eggs, often used in catering, can contain *Salmonella*. The main case study describes an outbreak of salmonellosis which arose from a chocolate mousse dessert made with raw shell eggs. The mousse was made by a caterer and supplied to a party. In reading this, and the other two case studies, you should be evaluating what controls are potentially available to reduce risk of salmonellosis and food poisoning in the catering production of lightly cooked, chilled foods.

But you should also be considering why a HACCP system properly implemented was believed, in the main case study, to lead to a level of control where risk of salmonellosis was negligible.

CHAPTER 9

Product Formulation and Control   **207**

Key issues
- Yoghurt
- *Clostridium botulinum*
- The 'combined treatment' approach to control
- HACCP

Challenge
Yoghurt, a fermented milk product, has widely been considered to be a 'safe' product. Yet this case study describes a situation which should lead you to an exploration of when the formulation of a product can, but may not, control the potential hazards within it.

C. Risk

CHAPTER 10

Views of Risk   235

Key issues
- Semi-soft cheeses, and raw milk
- *Escherichia coli* O157:H7
- Cross contamination
- Control of *Escherichia coli* O157:H7 in cheesemaking
- HACCP
- Risk perception

Challenge
The case study presents you with the opportunity to consider both how the
application of HACCP assists the production of safe cheese, and whether raw
milk cheese should be made. It further asks you to think about whether sampling
of made cheeses can demonstrate the safety of a batch.

CHAPTER 11

Hazards and risks   **258**

Key issues
- Cheese – blue cheese
- *Listeria monocytogenes*
- Virulence and pathogenicity: when is an organism a food hazard?
- Risk to public health and food safety policy
- Microbiological tests and their meaning
- Risk management

Challenge
The case study relates to a consignment of raw sheep's milk cheese in which *Listeria monocytogenes* was reportedly found 'at high level', and the authorities required its removal from the market. The small family cheese-making business, put at risk of bankruptcy, challenged this decision through the courts.

The case study allows you to think about when the identification of *Listeria* strains in foods represent a risk to public health.

CHAPTER 12
The Need for Food Hygiene 307

Key issues
- Cooked meat pies
- *Escherichia coli* O157:H7
- Infection and cross contamination in the food chain
- Food hygiene training and management

Challenge
The case relates to an outbreak of food poisoning caused by *E.coli* 0157:H7 in which of 496 known cases, 21 people died, with 17 as a direct result of the infection. A Government commissioned report by the Pennington Group made a number of recommendations for future strategies to minimise the risk to the public from this organism, and a Fatal Accident Enquiry identified the errors made and what safety measures could reasonably be expected in the production and retail sale of cooked products. Although the case study may enable you to consider a wide number of issues both in and outside the scope of this book, the focus of the exercises presented in Chapter 12 is to challenge you to consider how cross-contamination may arise, and the reasonable precautions, including food hygiene training, which may manage and reduce that risk.

CHAPTER 13

A Shelf Life Problem   333

Key issues
- Chilled desserts and bottled vegetables
- Psychrotrophic hazards and spoilage organisms
- Perception of risk
- Shelf life – safety and quality

Challenge
This case concerns products made on a small scale for commercial sale. Determining the shelf life of a product is a challenge.

CHAPTER 14

Airline Food and Control Failure   343

Key issues
- Cook–chill foods and airline foods
- *Staphylococcus aureus*
- Global dissemination of pathogens by air travel
- Risk factors and high and low risk foods
- HACCP

Challenge
This case study concerns the safety not only of the airline food, but also of the aircraft itself – for if the pilot and the aircrew succumb to food poisoning who will fly the plane? The challenge is therefore to consider the control systems necessary to ensure that neither this nightmare scenario nor illness in the passengers occurs.

D. Sampling, Criteria and Acceptance

CHAPTER 15

Global Dissemination of Organisms and Their Control 357

Key issues
- Fish meal and animal feeds; low a_w materials
- Globalisation of food supply and the spread of pathogens
- *Salmonella* spp.
- Contamination and cross contamination
- Microbial survival and significance
- Import acceptance – sampling criteria

Challenge
This case study asks you consider the movement of feeds and foods in international trade and the world wide dissemination of pathogens.

CHAPTER 16

Extending Shelf Life – Compromising Safety? 374

Key issues
- Raw fish and sushi
- *Vibrio parahaemolyticus*
- Shelf life
- HACCP in seafood management
- Microbiological criteria

Challenge
Extension of the shelf life of fresh fish offered the opportunity of cornering the quality fresh fish market in Australia, but could this also have compromised the safety of sushi – specialised preparations of raw sea-foods combined with glutinous rice?

CHAPTER 17

Acceptable, Unsatisfactory and Unacceptable Concentrations of Pathogens in Ready-to-eat Food **385**

Key issues
- Meat pâté
- *Clostridium perfringens*
- Food poisoning risk factors
- Developing microbiological criteria
- Microbiological quality guidelines

Challenge
This Case Study concerns commercially processed Belgian pâtés, made in 2000, retail samples of which were shown in January 2001 to have 'unsatisfactory' levels of *C. perfringens* present, resulting in the national withdrawal of the product range from the UK market. The challenge to you is to consider the circumstances under which *C. perfringens* presents an unacceptable human health risk, and how that relates to the presence of the organism in food.

E. Managing Risk

CHAPTER 18

Managing Risk 399

Key issues
- Sandwiches
- Food poisoning organisms
- Factors affecting shelf life.
- Risk ranking, and risk management
- HACCP training

Challenge
You are placed in the position of a manager who has to decide between two
systems of production of sandwiches.

CHAPTER 19

Changing a Risk Management Strategy 403

Key issues
- Raw fruit and vegetable salads
- *Shigella* spp.
- Risk management

Challenge
Pre-prepared raw fruit and vegetable salads for retail sale or catering use represent processed foods for which achieving acceptably low food safety risk is difficult. The outbreak described relates to an incident of Shigellosis associated with the retail sale of raw fruit salads. The challenge for you is to determine the procedures which should reduce risk to an acceptable level and give confidence in the suitability of raw fruit salads for retail sale.

CHAPTER 20

Hygiene Improvement at Source   **410**

Key issues
- Raw meat in the abattoir
- Microbial load and enteric pathogens
- *E. coli* O157:H7
- Hygiene
- Risk based scoring system for hygiene improvement
- Risk management

Challenge
In abattoirs traditional meat inspection techniques which are based on observation of visible defects in meat, such as tubercles, and worms, do not detect the presence of pathogens such as *E. coli* O157:H7 which cause no visible change. In managing the risk to public health from zoonotic disease spread through infected meat, approaches other than the traditional inspection techniques are being developed. Evaluation of hygienic practice through a risk based scoring system was introduced in the UK in 1994, and improvement encouraged through the regular publication of the scores associated with the named companies. The challenge to you is first to understand the system, and then to evaluate whether, on microbiological grounds, you agree with the system and its weightings.

CHAPTER 21

What is Safe Food? 429

Key issues
- Milk, rice and other foods
- *Bacillus cereus*
- Risk analysis: assessing and managing risk
- Safe food

Challenge
This case study considers foods in which the food poisoning organism *Bacillus cereus* occurs and asks you to assess the risk of food poisoning occurring in extended shelf life pasteurised milk. You should then be able to define for yourself what constitutes a safe food.

A. Developing Strategies for Control

Water as a Vector of Organisms

Key issues
- Water
- *Vibrio cholerae*
- Microbial growth and survival in water
- Food safety control, and the need for national infrastructure

Challenge

In 1974, cholera spread in Portugal. Not only did potable water supplies become polluted with cholera infected faeces, but so did shellfish and other foods and bottled water. This case study invites you to start thinking about how organisms become disseminated, whether they die in that process, and what affects their growth and survival. You can, of course, extend your thoughts beyond this case and wonder how control systems prevent water in all its various uses in the food industry both becoming contaminated and being a vector for a very wide range of pathogens.

1.1 THE CASE STUDY: CHOLERA IN PORTUGAL, 1974

(Most of the text below comes from Blake *et al.*, (1977a and 1977b, see 1.1.4). The text in the original papers has been slightly shortened and some linking sentences put in to allow textual flow. Paragraphs from the two papers have been mixed to provide a shorter version of the two texts which are used here with the permission of the copyright holders – The Johns Hopkins University School of Hygiene and Public Health.

From Blake et al., *1977a*:

On April 24, 1974, in Tavira, a town on the southern coast of Portugal, a 33-year old man developed diarrhoea and dehydration so severe that he suffered a cardiac arrest. An alert physician suspected cholera, and the Ricardo Jorge Institute in Lisbon isolated *Vibrio cholerae* biotype El Tor serotype Inaba from his stool. *V. cholerae* were subsequently isolated from a Lisbon woman who had contracted a diarrheal illness on May 10 and from a child in Porto on May 14. Within six weeks after the first case, cholera had been reported from eight of

Portugal's 18 districts. Eventually 17 of the 18 districts reported cholera, all due to El Tor Inaba organisms. The epidemic peaked in the last week in August, then declined rapidly; on November 29, 1974, Portugal was declared free of cholera. In this seven month period 2467 cases and 48 deaths were reported to the World Health Organisation (WHO). The case–fatality ratio was 1.9%.

Control measures included individual case investigations by public health nurses, distribution of two- or three-day supply of tetracycline to all contacts of cholera patients, a massive health education program, and an extensive effort to promote the chlorination of public water systems. These measures were supplemented by distribution of free bottles of chlorine for disinfection of drinking water from unsafe sources. Cholera vaccine was given only to those who demanded it.

Investigators from the Center for Disease Control [USA] and the Portuguese Directorate General of Health carried out two studies in Tavira and Faro, two cities on the southern coast, and in Lisbon to determine the important modes of transmission of cholera in Portugal.

1.1.1   Background to the Outbreak

Both Tavira and Faro are separated from the open sea by the Ria de Faro, a complex of islands and mud flats that extends approximately 50 km along the southern coast [of Portugal]. Sewage from the coastal towns emptied into the Ria, and water and shellfish in the Ria had high coliform bacteria counts. Most shellfish consumed throughout Portugal are caught in this area. After anecdotal information suggested that shellfish had caused some cases of cholera, the Maritime Biology Institute in Faro systemmatically cultured water and shellfish from the Ria from May to August 1974; *V. cholerae* was isolated from 24% of 121 seawater samples and 42% of 154 shellfish, including clams (44 of 114 positive), cockles (11 of 22), oysters (7 of 15), and mussels (2 of 3).

Tavira's municipal water supply was chlorinated, but some residents preferred to drink from two springs in the town. One of these, the Fonte do Bispo, emerged from a pipe in a populated area and had allegedly produced clear water until September 1973 when, following blasting of the rock above the spring and a heavy rain, the water was muddy for a few days. Both springs were closed on May 10 and 11, shortly after the cholera outbreak began, because health authorities suspected that some of the cases might have resulted from drinking from these springs. No water from either spring was cultured. Troops travelled back and forth between a military base uphill from the Fonte do Bispo and what were then Portugal's African colonies – Angola, Mozambique and Portuguese Guinea. Untreated sewage from the base and the town entered the Gilao river which bisects the city, and flowed with it into the Ria. *V. cholerae* was isolated from 9 of 20 water samples taken from the Gilao river May 14–22, 1974.

In the rural area north of Faro, outbreaks of cholera were attributed by the Health Authorities to consumption of water from wells, several of which were culture positive for *V. cholerae*. The cause of the 59 cholera cases reported in Faro itself in May–September 1974 was unclear, although shellfish were suspect.

In the area around Lisbon, where sewage was sometimes used to irrigate and fertilise vegetable gardens, health authorities initially felt that contaminated raw vegetables, as well as shellfish, might be responsible for most cases . . . But

[*From Blake* et al., *1977b*]

. . . in August, during the peak of the cholera epidemic, Lisbon Health Department noted that some cholera patients reported recent travel to a spa in Lisbon district. At the same time *Vibrio cholerae* El Tor Inaba was isolated from both the springs which supplied mineral water to the spa and to a commercial water bottling plant. Both the spa and the plant were closed at once . . .

The spa had 51 718 visitors in 1973, of whom about 20 000 came during August. . . . Most of the water used by the spa and all of the water for the bottling plant came from spring A, which entered a concrete covered underground reservoir excavated in the limestone from which the spring emerged. Untreated water from this reservoir was piped to the baths, a swimming pool, multiple drinking water sources, and the water bottling plant. A small river carrying sewage from upstream towns ran near the springs. Some of the characteristics of the water from spring A are given in Table 1.1.

At the bottling plant approximately half of the water was carbonated before bottling, and half was bottled without treatment . . . There was little change in the amount of this water (Brand 'A') distributed annually from 1969 to 1973 when 10.5 million litres were bottled, but in 1974 production increased by about 50% apparently because many people distrusted public water during the cholera outbreak. In August, the month of greatest demand, bottles moved from the production line directly to waiting trucks and could be on store shelves in Lisbon within four hours after production.

On August 13, after two cases of cholera were reported in residents of a village near the spa, a sanitarian cultured water from springs A and B as part of a sanitation inspection of the area. On August 22 the National Public Health Laboratory reported that *V. cholerae* had been isolated from water samples from both springs. The springs and the bottling plant were ordered closed on August 23, and the bottled water recalled. A press release on August 24 warned the public that consumption of brand A mineral water could be dangerous. *V. cholerae* was isolated from five or six water samples taken from spring A on August 13, 22, 26, and 28, and from two of six water samples collected from spring B on those dates; subsequent cultures were negative. Bottled water was not cultured for vibrios. Cultures of three of five water samples collected on August 13 and 26 from the river were also positive for *V. cholerae* . . .

Table 1.1 *The characteristics of water from spring A*[a]

Temperature	27 °C
pH	7.4
Sodium chloride content	428 mg l^{-1}
Calcium carbonate content	415 mg l^{-1}

[a]Data obtained by Blake *et al.* (1977b) from the label of the bottled water. Reproduced with permission.

1.1.2   The Investigation in Tavira and Faro, Southern Portugal

[*From Blake* et al., *1977a*]:

Tavira

In Tavira, an attempt was made in October 1974 to locate the 15 cholera patients whose cases were reported during the first four weeks of the outbreak and to question them about their basic demographic data, travel histories, frequency of consuming a variety of foods (including raw fruits and vegetables and seven types of shellfish), methods of cooking shellfish, and sources of drinking water . . .

Fourteen of the first 15 cholera patients were located; although their homes were scattered throughout the town, a matched pair analysis implicated water from one spring, the Fonte do Bispo, as the probable source of cholera in 11 cases ($p = 0.001$). The epidemic curve dropped sharply after the spring was closed on May 10. The other spring was not implicated. Although three patients had eaten raw or semi-cooked cockles in April or May, there was no statistically significant association between eating cockles and illness. The first cholera patient was a truck driver who three days before the onset of his illness ate cockles which he had gathered from the Ria de Faro near the mouth of the Gilao river. The cockles were heated until they opened and eaten immediately by three persons. The patient, who took antacids regularly, was the only one of the three who developed diarrhoea. He had not been out of Portugal during 1974.

Faro

In Faro, the 59 patients who had *V. cholerae* isolated from their stools between the beginning of the epidemic in May and the end of September were studied . . . 53 (90%) of the 59 patients were interviewed and matched [with case controls]. Two different exposures, consumption of Brand A commercially bottled mineral water and of raw or semi-cooked cockles were significantly associated with cholera in the initial analysis . . .

Sources of water other than bottled water (public water supplies, cisterns, wells, and springs) were not significantly associated with cholera . . .

1.1.3   Investigation of the Bottled Water

[*From Blake* et al., *1977b*]:

During August 1974 the attack rate for the 14 000 visitors to the spa from the other counties in Lisbon District was 2.57 : 1000 (36 : 14 000) while for residents of those counties who did not visit the spa it was only 0.25 : 1000 (382 : 1 530 831). . . . the relative risk for visitors to non-visitors was 10.3 . . .

To determine if Brand A bottled water had caused cases of cholera, a case-control study was performed in the city of Lisbon which showed that the risk for those who consumed the non-carbonated bottled water, rather than carbonated water, was 12. Consumption of carbonated brand A bottled water was not found to be associated with cholera.

1.1.4 The Case Study Source Materials

Blake, P.A., Rosenberg, M.L., Bandeira Costa, J., Ferreira, P.S., Guimaraes, C.L. and Gangarosa, E.J., (1977a), Cholera in Portugal, 1974. I. Modes of transmission, *American Journal of Epidemiology*, **105** (4), 337–343.

Blake, P.A., Rosenberg, M.L., Florencia, J., Bandeira Costa, J., Prado Quintino, L. and Gangarosa, E.J., (1977b), Cholera in Portugal, 1974. II. Transmission by bottled mineral water, *American Journal of Epidemiology*, **105** (4), 344–348.

1.2 BACKGROUND

1.2.1 Potable Water

In March 2002 the world population stood at 6.2 billion (6.2×10^9) people, all needing safe potable (drinking) water (www.world-gazetter.com/home/htm). Gastroenteritis and waterborne diseases constantly threaten the lives of children, the weak, the poor and the elderly, and rank high among the causes of mortality.

Human and animal faeces are the primary sources of the incriminated organisms, so secure separation of sewage and night soil from clean and treated water supplies is required. However, implementing the necessary actions takes commitment, time, and investment of huge sums of money. Many societies do not have safe water because, for political and economic reasons, those investments have not been made. Sometimes it takes a serious outbreak of waterborne disease to jolt the relevant authorities into action, until which many peoples in the poorer countries of the world, ill equipped to make investment in infrastructure, continue to suffer endemic waterborne disease.

Drinking water can be supplied variously from the mains, wells, boreholes and springs for use in homes, hotels, restaurants, canteens, cafes and factories. There it is used for drinking, cooking, for hand and body washing, and other tasks, while in food businesses it may not only become part of the foods but is also used for cleaning and rinsing the food processing equipment. Contaminated, it puts the health of users directly at risk. Polluted water used in aquaculture and in irrigation, particularly in horticulture, may contaminate the fruit, vegetables and shellfish, products often eaten without cooking.

In 1987 Galbraith and his colleagues published a review (Galbraith *et al.*, 1987) of the outbreaks of disease transmitted through water in the UK in the fifty year period from 1937. Table 1.2 summarises the incidence of waterborne disease in the UK in the years 1937 to 1986, and Table 1.3 breaks those statistics down indicating the types of water sources and the probable causes of contamination. In 1937, Croydon, to the south of London, suffered a waterborne outbreak of typhoid fever infecting 341 people of whom at least 43 died. It was a turning point. Following that outbreak, national surveillance of water quality was stepped up while increased attention was paid to the effectiveness of purification and chlorination procedures, as well as to maintenance of the integrity of water supply systems.

Galbraith and his colleagues demonstrated that the principal causative bacteria of waterborne disease in the UK in that half century were *Salmonella*

Table 1.2 *Waterborne disease: UK 1937–1986*

Disease	Public supply No. of outbreaks, contamination at:			Private supply		Total number of:		
	Source	Distribution	Cases	Outbreaks	Cases	Outbreaks	Cases	Deaths
Typhoid fever								
Croydon	1(1)		341			1	341	43 +
other outbreaks		2	72	3(3)	38 +	5	110 +	5
Paratyphoid fever	1(1)		90	2(2)	27	3	117	0
Amoebiasis				1(1)	17	1	17	0
Campylobacter enteritis		2	399	3(3)	520	5	919	0
Chemical gastroenteritis	1	3	531 +			4	531 +	1
Cryptosporidiosis	2		66 +			2	66 +	0
Dysentery: bacillary	3(3)	1	5088 +			4	5088 +	0
Gastroenteritis, ?viral	4(4)	1	3536 +	3(3)	998 +	8	4534 +	0
Giardiasis		1	108 +			1	108 +	0
Streptobacillary fever				1(1)	304	1	304	0
Total	11(8)[a]	10	9890 +[a]	13(13)	1904 +	34[a]	11 794 +[a]	6[a]

Number associated with unchlorinated or defectively chlorinated water in parentheses.

[a]Excluding Croydon outbreak.

Reproduced with permission from Galbraith, N.S., Barrett, N. and Stanwell-Smith, R., 1987. *Journal of the Institute of Water and Environmental Protection*, **1**, 7–21.

typhi (typhoid), *Salmonella paratyphi B* (paratyphoid), *Campylobacters* (gastro-enteritis), *Shigella sonnei* and other *Shigella* strains (bacillary dysentery). Additionally other organisms, amoebae such as *Entamoeba histolytica* (amoebic dysentery), *Cryptosporidia* (cryptosporidiosis), *Giardia lamblia* (giardiasis) and viruses (gastroenteritis) caused a significant number of cases of illness of varying degrees of severity both directly through infecting water, and indirectly through water contaminating food (see Table 1.4).

Although some outbreaks were associated with untreated water supplies, other water supplies, although chlorinated, were polluted by inward leakage of sewage and became unsafe. Note that other agents such as hepatitis, polio and Norwalk viruses, other salmonellae and *E. coli*, also associated with sewage and excreta from birds and animals, also cause waterborne illness.

1.2.2 Bottled Water

The origins of bottled water probably arise from the drinking of spa water for health reasons followed by taking a few bottles away after the visit to prolong the perceived health giving effects of the mineral water. In the UK sales of bottled water have been rising rapidly. It was estimated, for example, that the volume of cooled bottled waters sold in the UK was approximately 195 million litres in 1999, predicted to increase to 333 million litres by 2003 (Bottled Water Coolers Association, 2002).

Contrary to many people's expectation bottled water is neither sterile, nor incapable of supporting the growth of a microbial population. A review by Hunter (Hunter, 1993) shows that the microbial population (the TVC, 'total' viable count) of both still and carbonated water at the point of retail sale can range between $< 10^2$ and as high as 10^4 organisms per millilitre, while the range for still water can be wider, sometimes reaching as much as $> 10^6 \, \mathrm{ml}^{-1}$.

That population can be very mixed and, because of concerns that pathogens could be among that flora, several workers have looked for specific types of organisms. Table 1.5 shows the variety both in types of organisms and numbers of samples of water in which they were present.

Many factors associated with the water – its source, how it is produced, and its chemical composition and pH, combine together to affect the numbers and types of organisms present. Later in the time period in which the bottled water is distributed and sold other factors influence whether the population dies out or increases. The major concern is of course whether the population of organisms it contains represent a health risk. Table 1.6 summarises some of the factors influencing the microbiology of bottled water.

Hunter says, in relation to the microbial flora of bottled mineral waters (many species of which are also commonly present in foods) "it is unlikely that [the flora] will be a significant additional source of these organisms in the diet". But he also says "there is no hard evidence that all mineral waters are free from adverse health effects". The case study described in Section 1.1 concerns an outbreak of cholera in which bottled water, contaminated with the organisms *V. cholerae*, contributed to the dissemination of the organism in the epidemic. So

Table 1.3 *Waterborne diseases (UK) 1937–1986: water sources and probable causes*

Year	Location	Disease	Cases (deaths)	Water source	Probable cause
Public water supply					
1942	Biggleswade RAF station	Typhoid fever	22	Deep bore hole	Sewage leaking into the bore hole from a blocked drain
1959	North Kerry, NI	Typhoid fever	50	Water in distribution, C	Faecal pollution from a carrier into a rising main
1970	Yorkshire	Paratyphoid fever (B)	90	Spring water, C	Sewage from nearby farm cottages; probable carrier identified; C, but at inadequate level at times of peak demand
1983	Sussex, school	Campylobacter enteritis	142+	Mains, C	Defective sewer leaking into defective mains water supply
1986	Hertfordshire, school	Campylobacter enteritis	257	Mains, C	Engineering works on water main allowed infection
1983	Surrey, Cobham	Cryptosporidiosis	16	Spring, C	Spring water, C to 0.5 ppm, softened, filtered
1985	Surrey, Cobham	Cryptosporidiosis	50		As above, after heavy rain
1942	Somerset	Bacillary dysentery	400	Deep wells, C	Sewage pollution, plus defective chlorination
			a few	Raw milk (dairy)	Bottles infected with the polluted water in the washing process
?	Scottish village	Bacillary dysentery	500	Spring, UC	Sewage pollution
1966	Montrose, Scotland	Bacillary dysentery	4000	River water, C	Defectively C
1950	Leicester	Bacillary dysentery plus GE (*Shigella flexneri*) plus GE (*Salmonella typhimurium*)	188	River water for a factory	Accidental connection of the pipe to the public water mains
1937	Kilmarnock, Scotland	Viral GE	162	Surface water, UC	Contaminated with human and bird excreta
1938	Surrey, Haslemere	Viral GE	200	Well, C	Heavy rain, plus failure of chlorination
1980	Bramham, NE Leeds	Viral GE	3000	Borehole, C	Sewage polluted stream affected bore hole; and failure of chlorination
?1980s	Cumbria, camp site	Viral GE	60	Upland stream water, C	Sewage from camp site affected stream; plus chlorination failure
1974	Lancashire, Rochdale, factory canteen	Viral GE	114	Orange juice machine	Mistaken connection to raw river water supply for the factory
1985	Bristol	Giardiasis	108	Mains	Engineering work may have allowed contamination
1965	Worcester	GE probably viral	30 000	Mains	Flooding by River Severn – but not confirmed
?	Cruise liner	GE possibly ETEC	301	Ship's water tank	Ship's sewage leaked into water tank; plus inadequate chlorination

Private water supply

Year	Location	Disease	Number	Water supply	Notes
1941–44	Cornwall, hotel; intermittent cases	Typhoid fever	6(5)	Well water, UC	Sewage (from a carrier) overflowed from a cesspit *via* a drain to the well head
1942	Norfolk, Hingham; group of cottages	Typhoid fever	3	Well water, UC	Sewage from pail closets seeped into the wells. A typhoid carrier, who had lived there since 1927, was identified
1959	Ballreagh, NI, holiday caravan site	Typhoid fever	29	Well water, UC	Faecal contamination of well water from buried contents of dry closets; source – a chronic carrier on the site
1941	Brixworth, Northants	Paratyphoid fever (B)	21	Well water, UC	Sewage from cottage drains
1975	Herefordshire	Paratyphoid fever (B)	6	Well water, UC	Leakage of sewage from domestic drain
1950	RAF station	Amoebiasis: (amoebic dysentery active amoebiasis, symptomless infections)	43 6 11 26	Borehole	Contaminated from broken sewer
1981	Essex, school	Campylobacter enteritis	257	Borehole	Borehole water infected after distribution. Epidemiological investigation indicated the uncovered cold water storage tank was infected with the faeces of birds and bats
1985	Buckinghamshire, school	Campylobacter enteritis	234	Private spring, C	Affected by melting snow making C ineffectual
1986	Wales, camping party	Campylobacter enteritis	29	Well water, UC	Possibly contaminated at source by grazing cattle
1980	Somerset, holiday camp	Viral GE	160	Surface water supply, C	Defective C
1982	Derbyshire, youth hostel	Viral GE	138	Surface water supply, UC	Not identified
1986	Scotland, skiing lodge	Viral GE	700	Surface water supply, C	Defective C
1983	Essex, school	Streptobacillary fever	304	Spring	Rat infestation infected the hot water supply – the hot water could enter the potable water supply

Imported disease

Year	Location	Disease	Number	Water supply	Notes
1963	Zermatt, Switzerland; locals and tourists	Typhoid fever	437 (all) (78 UK)	Mains, C	Infected sewage leaked at source into the chlorination tank, and subsequent inadequate C
1976	Salou, Spain	Typhoid fever	8 UK	Mains, C	Infected public water supply
1978	Russian cruise liner	Illnesses recorded: typhoid fever GE bacillary dysentery	5 (4 UK) 150 10	Ships water supply, C	Replenishing ships tanks with contaminated water at one port; plus inadequate C

NI = Northern Ireland; GE = gastroenteritis; UK = United Kingdom; ETEC = enterotoxigenic *Escherichia coli*; C = chlorinated; UC = unchlorinated; ORHA-UR = Oxfordshire Regional Health Authority unpublished report; PHLS-UR = Public Health Laboratory Service unpublished report.
From data in Galbraith *et al.*, 1987.

Table 1.4 *Food contaminated by polluted water: 1937–1986*

Affected food	Date	Place	Illness	Outbreaks	Cases	Probable cause
Milk	1937–83	England and Scotland	Typhoid fever	4	44 (2 deaths)	Raw milk from dairies which had sewage polluted water supplies; or cows grazing on sewage polluted pasture, or had access to polluted stream water; or polluted water was used to wash the bottles at the dairy
			Paratyphoid fever	6	199	
			Dysentery	1	30	
Canned products						
Imported corned beef	1964	Scotland, Aberdeen	Typhoid fever	1	500	Sewage polluted river water, UC or ineffectively C, used for can cooling, together with leaking cans
Imported meats	1937–84	UK	Typhoid fever	11	approx. 400	
Imported ham (France)	1948	England, Herts.	Salmonellosis (*S. wein*)	1	49	
Imported corned beef (Argentina)	?	UK?	Salmonellosis (*S. newport*)	1	25	
Vegetables						
Miscellaneous vegetables	Several years	England, Devon	Typhoid and paratyphoid		a number of cases	Sewage pollution of local river by carriers, and this water used to irrigated commercial vegetable production
Tomato salad	1983	Greece, Kos, hotel	Typhoid fever	1	32 (UK tourists)	Contaminated either directly by a carrier, or the well water became polluted by sewage from the carrier

Crops from contaminated water					
Shellfish	1941–44	England, Devon	Typhoid fever	sporadic	Sewage polluted seawater, or estuarine water
Oysters	1952–58	England, Essex	Typhoid fever	sporadic	
Shellfish	1941–83	England and Wales	Norwalk virus	22	Sewage polluted seawater, or estuarine water
			Hepatitis A	10	
			Unknown cause	65	2300+
Watercress	1941–43	England, Devon	Typhoid fever	sporadic	Infected sheep or cattle having access to the water supplying the watercress beds
	1959–86	England and Wales	Liver fluke	sporadic	11

From data in Galbraith *et al.*, 1987.
Key: C, UC – see Table 1.3.

Table 1.5 *Bottled water: types of organisms found in it*

Group	Type	Data from surveys
Protozoa and algae		
Gram negative bacilli	*Aeromonas hydrophila*	
	Enterobacteraceae	
	Non-faecal coliforms	in 3% of 73 lots of mineral water
	Non-faecal coliforms	in 2% of 41 lots purified water
	Coliforms	in 0.5–3.7% of samples
	E. coli	in 0.2–1.7% of samples
	Pseudomonas strains	in varying percentages of samples
	Pseudomonas aeruginosa	in 4% of 104 samples
	Pseudomonas aeruginosa	in 1.2–10.2% of samples
	Pseudomonas aeruginosa	in 11% of 18 lots of mineral water
Gram positive cocci	Micrococci	
	strains of Staphylococci	
Gram positive bacilli		

Surveys reported in Hunter (1993) indicate that a very wide range of organisms have been shown to survive or grow in bottled waters. The table indicates some of the results reported.
Source: Hunter, 1993.

it was ironic that at the height of that outbreak people were electing to drink bottled water as a means of reducing their risk of exposure, in the expectation that bottled water would be safe.

Clearly for the cholera organisms to be disseminated they had at least to survive in the water. Little has been known about the duration of survival of organisms in mineral and bottled water, and where research has been conducted the results have been confusing. A general conclusion which can be drawn from both Table 1.7 and Table 1.8 is that survival depends very much on the type of water, its storage and the organisms of concern. Nothing can be assumed. That has, of course, serious implications for the production of safe bottled water.

1.2.3   Ice and Drinking Ice

At one time ice for cooling food was a luxury. Historically in richer houses in Britain and Europe, blocks of pond ice or winter snow were preserved packed together in subterranean ice houses for summer use. Elizabeth David, the cookery and food writer, has written an interesting book on this subject (David, 1994). In the twentieth century refrigeration has made possible not only the refrigerated storage of food materials but also the manufacture and conservation of ice in its own right. Ice is now widely used for three general purposes – for putting around foods to cool and maintain them at a low temperature, for putting into drinks, and, in some food manufacturing processes, as a food ingredient.

Ice is made from water in ice-making machines. Thus the quality of the water

Table 1.6 *Factors influencing the range and numbers of organisms found in bottled water*

The geology of the aquifer
The potential of the aquifer for contamination by surface water or sewage
The chemical characteristics of the water
The extraction technology
The bottling system

and

Time of sampling	The flora increases in the first week after bottling
Period of storage	The flora remains stable for about 6 months, after which it declines
The type of water	Non carbonated water counts reach higher levels than do carbonated waters (in the latter case due to the antimicrobial action of carbon dioxide)
The type of bottle	PVC bottles tend to have higher counts than glass bottles
	Very thoroughly cleaned glass bottles have higher counts than those less well flushed free of cleaning agents
Temperature for plate culture	Culture plates incubated at higher temperatures show higher counts
Temperature of water storage	Water stored at 6 °C was shown to have higher counts than the same water stored at 22 °C
Bottle size	Smaller bottles with low volume to surface ratio show higher counts
Mineral composition of the water	Minerals act as nutrients and support the growth of some of the flora

Source: Hunter, 1993.

and the cleanliness of the machines affect the microbial quality of the ice. Two outbreaks of gastroenteritis due to infected drinking ice and caused by Norwalk virus and *Giardia lamblia* have been well documented (Khan *et al.*, 1994; Quick *et al.*, 1992). In the first case, associated with a cruise ship, the virus was present in the ship's water supplies, but in the second the evidence suggests there was direct transfer from the unwashed hands of a *Giardia* carrier who, after defaecating, scooped up ice for restaurant customers with her unwashed bare hands. Surveys from many countries, for example the USA (Moyer, 1993) and the UK (LACOTS, 1996, 1997) have shown that the microbiological quality of ice manufactured for use in foods and drinks can be poor and not meet WHO guidelines for drinking water (WHO, 1996) or legal standards for the drinking water from which they were made (Anon, 1989). Other surveys have shown the intermittent presence in commercial ice samples of members of the genera *Streptococcus*, *Shigella*, and *Listeria* – pathogens capable of causing gastrointestinal disease, as well as the presence of faecal indicators such as coliforms, again demonstrating either the poor quality of the water used or a severe lack of hygiene in production (Alvarez-Seoane, 1980; Tsuno, 1984; Moyer, 1993).

Table 1.7 *Microbial survival in mineral waters*

Organism	Conditions of investigation	Survival
E. coli[3]	Initial concentration $10^7\,ml^{-1}$ – gas free mineral water	Undetectable in 4 days
E. coli[4,5]	Initial concentration $1.2 \times 10^5\,ml^{-1}$ – sterile water	One log decline in 3 months
E. coli[4]	Initial concentration $10^5\,ml^{-1}$ – mineral water	Undetectable in 1–2 months
E. coli[2]	Bottled mineral water	Detectable to about 42 days
Salmonella typhimurium,[2] Ps. aeruginosa,[2] Aeromonas hydrophila[2]	Still mineral water	Detectable to about 70 days
Campylobacter jejuni[2]	Still mineral water,	Detectable for only 2–4 days;
	Carbonated water	Period of detectability reduced to 1-2 days
Hepatitis A virus and polio virus[1]	Bottled mineral waters	Survival greater than 120 days

Source: Hunter, P.R., 1993.
Key: [1]Biziagos, E., Passagot, J., Crance, J.-M., and Deloine, R., 1988. *Applied and Environmental Microbiology*, **54**, 2705–2710. [2]Burge, S.H., and Hunter, P.R., 1990. *Rivista Italiana D'Igiene*, **50**, 401–406. [3]Ducluzeau, R., Bochand, J.M., and Dufresne, S., 1976. *European Journal of Applied Microbiology*, **2**, 127–134. [4]Lucas, F. and Ducluzeau, R., 1990. *Science des Aliments*, **10**, 62–73. [5]Lucas F. and Ducluzeau, R.,1990. *Rivista Italiana D'Igiene*, **50**, 383–393.

Table 1.8 *E. coli: survival in still mineral water bottled in polyvinyl chloride (PVC) and glass*

Days	PVC bottles with autochthonous flora ($Log_{10}\,cfu\,ml^{-1}$)	PVC bottles sterile mineral water ($Log_{10}\,cfu\,ml^{-1}$)	Glass bottles sterile mineral water ($Log_{10}\,cfu\,ml^{-1}$)	Glass bottles sterile tap water ($Log_{10}\,cfu\,ml^{-1}$)
0	4.5	4.5	4.5	4.5
5	3	3.4	4	3.5
10	2.1	2.3	3.5	3.4
15	1.4	1.2	2.8	3.3
20	0.5	0.5	1.8	3.2

Adapted from: Moreira, L. *et al.*, 1994. With permission.

On occasion other organisms have been found to be present in ice where control in production was poor. Laussucq *et al.* (1988) reported that a form of pulmonary tuberculosis was transmitted to immuno-compromised patients through the ice produced in an ice machine in the hospital ward. The machine was shown to be colonised by the causative organisms *Mycobacterium fortuitum*.

Freezing at very low temperature, such as −80 °C in nitrogen, is a technique used to preserve organisms. Water ice on the other hand is normally stored in the temperature range −25 to −20 °C. But since ice only melts at 0 °C, it can have a temperature up to that value, dependent on the conditions of storage. If organisms are present in the original water, the process of freezing does not destroy them. In fact many micro-organisms can survive in ice, although their numbers gradually decline with time. When ice is thawed the organisms remaining can be injured and may, as a result, be more difficult to detect. But research has shown that the organisms in the ice tend to recover their viability so that, when the ice melts into drinks for example, they may be robust enough to survive there too, even in tequila where the alcohol can be as strong as 47% by volume (Dickens *et al.*, 1985). This means that if organisms are present in the original water from which the ice is formed, they may also be viable in the ice when it is used, and capable of causing infection in the consumer. Equally organisms added to the wet surfaces of melting ice through careless hygienic practices or by other means can survive too, and thus also represent a risk to the consumer.

1.2.4 Surveillance

In the UK potable water, bottled water and ice are all subject to regular surveillance and the minimum microbial quality standards they must meet are defined in the following boxes.

The Water Supply (Water Quality) Regulations, 1989. SI (1989) No. 1147 (Schedule 2, Table C)

Total coliforms	Maximum concentration 0 per 100 ml
Faecal coliforms	Maximum concentration 0 per 100 ml
Faecal streptococci	Maximum concentration 0 per 100 ml
Sulfite reducing clostridia	Maximum concentration <1 per 20 ml
Colony count (TVC) at 22 °C	No significant increase over that normally observed per ml
Colony count (TVC) at 37 °C	No significant increase over that normally observed per ml

---

The Drinking Water in Containers Regulations, 1994. SI (1994) No. 743

Total coliforms	Maximum concentration 0 per 100 ml
E. coli	Maximum concentration 0 per 100 ml
Enterococci	Maximum concentration 0 per 100 ml
Sulfite reducing clostridia	Maximum concentration <1 per 20 ml
Colony count (TVC) at 22 °C	Maximum concentration 100 per ml
Colony count (TVC) at 37 °C	Maximum concentration 20 per ml

The Natural Mineral Waters Regulations, 1985. SI (1985) No. 71

These are more stringent, requiring the tests to be performed at the mineral water source and immediately after bottling:

Total coliforms	Maximum concentration 0 per 250 ml
E. coli	Maximum concentration 0 per 250 ml
Enterococci	Maximum concentration 0 per 250 ml
Sulfite reducing clostridia	Maximum concentration 0 per 50 ml
Pseudomonas aeruginosa	Maximum concentration 0 per 250 ml

Ice intended to be used in drinks falls within the definition of 'food' under the Food Safety Act, 1990 in UK legislation (Anon, 1990), and ice quality used in drinks or in food, or around food is controlled through Regulations made under the Act: the Food Safety (General Food Hygiene) Regulations 1995 (Anon, 1995), which require it to be made from potable water. With the exception of *Pseudomonas aeruginosa*, the organisms listed above are 'indicator organisms' which, if they exceed the numbers shown, are taken to indicate faecal pollution inferring the potential presence of enteric pathogens. This system is universally recognised as a means of judging water quality and safety, but requires the use of standardised methods of micro-organism detection. It does not reliably provide inferences for organisms which survive differently from the indicator organisms – including organisms such as viruses, protozoans or algae. Specific tests for these have to be undertaken if there is a significant risk of their presence. Definitions of 'coliforms' and '*Escherichia coli*' are given in an official UK publication 'The bacteriological examination of water supplies' (Anon, 1983).

1.3   EXERCISES

Exercise 1. Systematically work out the factors which came together to cause the early single cases of cholera, and which built up into a 6 month epidemic affecting the whole Portuguese community, and illness in over two thousand people.

Exercise 2. As you do Exercise 1 consider what had prevented such an outbreak occurring before, and what controls were really needed to have prevented it then, or in the future.

1.4 COMMENTARY

It is probably helpful in building your analysis to use the data provided and subsequently identify where you need additional information.

The types of questions you may need to ask are:

- What is the normal habitat of the cholera organisms?
- Where did the organism come from in this particular outbreak?
- How was that fact established?
- What are the characteristics of the organism? Does it survive in sewage, river water, seawater, bottled water, in chlorinated water?
- When and where does it grow? Does it grow in shellfish? does it grow in any types of waters? Does it grow in foods?
- What is the infective dose to cause illness? How does it cause illness? What are the symptoms of cholera?
- How is the cholera organism detected in food and water samples?
- How long do the tests take to achieve results?
- How did the organism spread?
- How did the shellfish beds and bottled waters become affected?
- How is the organism killed? Is cooking a preventative measure?
- What preventative measures were implemented in Portugal at the time?
- Are there better preventative measures today?
- How did this particular outbreak arise and spread?
- How was the outbreak brought under control?

After thinking out the questions and finding the answers to them an approach you could use in addressing the problems set is as follows: Read through the Case Study, Section 1.1, and Section 1.2, which provides background materials, and the data Tables 1.1 to 1.9 associated with waterborne disease. Read everything very carefully, and think about it all.

Create a timetable of events for the epidemic:

April 24	Southern Portugal: Tavira – 33 year old man – diarrhoea, cardiac arrest, death (shown to be due to cholera caused by *V. cholerae* El Tor Inaba)
May 10/11	Southern Portugal: Tavira – both springs at Fonte do Bispo closed
May 14 to 22	Southern Portugal: Tavira – nine out of 22 water samples from the River Gilao shown to be *V. cholerae* positive
May to August	Southern Portugal – shellfish from Ria frequently shown positive for *V. cholerae*
May 10	Western seaboard: Lisbon – a woman affected by cholera
May 14	North-western seaboard: Porto – a child affected by cholera
June 6	(six weeks from first case) 17 of 18 districts of Portugal now affected

August	Lisbon Health Department had noted some cholera patients reported recent travel to a spa (spa A) in county A to the north of Lisbon
August 13	Two cases reported in the vicinity of spa A
August 13 to 28	Water samples from the spring A, and two of six samples from spring B at spa A found positive for *V. cholerae*
August 15 to 25	Peak number of cases
August 13 and 26	Samples from the sewage carrying river adjacent to spa A proved positive for *V. cholerae*
August 22	The National Public Health Laboratory reported *V. cholerae* had been isolated from water samples from both springs at spa A
August 23	Spa A: Springs and bottling plant closed
August 24	Press release warned public that bottled water could be dangerous
September	Number of new cases declining
October	Southern Portugal: Tavira and Faro – case control study to identify modes of transmission
November 29	Portugal declared free of cholera: 2467 cases, 48 deaths

Work out the probable events associated with Faro and Tavira, and with Lisbon, which caused the dissemination of the organisms leading to the nation-wide outbreak of cholera.

Southern Portugal: Faro and Tavira
Portuguese soldiers returned from Mozambique, Angola and Portuguese Guinea and were billeted at the camp uphill of Fonte do Bispo. Untreated sewage from the camp flowed into the river Gilao, downhill and into the Ria de Faro, a complex of islands and mud flats adjacent to the open sea. This sewage contained *V. cholerae* from infected soldiers, and contaminated the seawater, and the shellfish beds. Shellfish filter feed and accumulate suspended organisms on their body mass within the shell. There the organisms can grow and increase in number.

Shellfish (mussels, cockles, oysters, clams) were eaten raw by many people. This partially accounts for the local infections both at Tavira and Faro, but also more widely in Portugal, for this area was the primary shellfish producing area for the whole country.

Sewage from the population, including infected people, also passed untreated into the river Gilao, and directly into the Ria, and further contaminated the shellfish beds. Sewage from the military base probably contaminated untreated spring water following blasting of the rock above the spring in Fonte do Bispo in 1993.

Lisbon: Spa producing Brand A waters, from springs A and B
Water from the springs A and B was used in the spas for drinking, and for bottling. The springs became contaminated with sewage and *V. cholerae*. Sewage

from upstream towns flowed in the river near the springs.

The production volume of bottled water was 50% carbonated, 50% non-carbonated; production volume in total was about 10.5 million litres. In 1994 production volume increased by 50% to meet demand. Spring water was shown to be *V. cholerae* positive as it emerged from the ground.

Contamination of the spring water must have occurred either by underground seepage of sewage from the polluted river nearby into the spring water (helped by the lowering of the water table due to high summer water consumption, increased bottled water production and lack of rain) or by sewage from nearby villages.

Bottled contaminated non-carbonated water, distributed far and wide in Portugal, disseminated the organism across the country in August. *V. cholerae* survived in the water and infected those people who were vulnerable. Their faeces carried *V. cholerae* which went into the local sewerage system.

Consider *Vibrio cholerae*
What would have happened to its population (increase or decrease) in the sewage, the seawater, the bottled water, mains water, sewage fertilised vegetables, shellfish or any other infected food material? What factors would influence these population changes?

V. cholerae is a Gram negative, polar flagellated, curved rod. It can be cultured readily in the laboratory under aerobic conditions on ordinary media, favoured by alkaline conditions. It is facultatively anaerobic and ferments sugars to produce a mixed range of acids. The culture time from water, stools or food to tentative identification using methods available in 1974 would have been about 48 hours (Mackie and McCartney, 1960) Today, in the early 2000s, detection of the presence of low numbers of organisms takes a similar time, but could be much quicker if high numbers were present in the sample.

It multiplies extensively in the small intestine, and produces a toxin which acts on the mucosal cells there; this disrupts normal absorption in the gut leading to water and electrolyte loss from the body causing sudden intense diarrhoea (often containing mucous and known as 'rice water stools') with consequent muscular cramps, shock, trauma and collapse.

Outside the body the organism is killed easily at temperatures above 56 °C; and it dies rapidly when dried, but it survives readily in alkaline waters for many days (ICMSF, 1996).

In the waters and foods affected in the Portugal outbreak the organism both survived for periods long enough to infect consumers and, depending on the precise circumstances, may have actually grown. Clearly in many samples the time period between infection and consumption could have been such that the organisms had diminished in numbers or died out completely. *V. cholerae* death rates would be particularly dependent on conditions of pH, salinity, nutrients and temperature. But at the height of the epidemic, however, bottled water evidently could be on the shelves of shops within a very short period, four hours, after processing.

Thus whether a sample of food or water actually caused cholera would have depended on the number of viable organisms present in the volume of material consumed. Many of the organisms would then be destroyed in the acid in the stomach. Once in the gut, disease would only be experienced if the organisms were able to multiply there – and that would depend on the susceptibility of the individual. It was noted in the investigation of the outbreak that, in a group of three men who ate shellfish, only the one who regularly took antacids suffered illness.

For the various reasons such as those instanced above, in an exposed population only a small number of people experience symptoms of the disease. In this outbreak the exposure must have been many millions of people; the confirmed cases were 2467, and with 48 deaths.

Original source of infection?

The cholera organism central to this outbreak was *V. cholerae* El Tor Inaba. *V. cholerae* includes a number of strains of the organism which can be differentiated on the combined basis of their somatic (O) antigens, and biotype. The true cholera organisms – the two biovars 'classic' and 'El Tor' – both possess the O1 somatic antigen and are capable of causing cholera epidemics, as is the more newly identified O139 strain. Non-O1 strains cause much less severe gastro-enteritis than 'true' strains and often fail to produce the complete cholera toxin molecule.

The two biotypes differ in the following way:

V. cholerae (classical type): non-haemolytic, Voges-Proskaur negative
V. cholerae El Tor: haemolytic,   Voges-Proskaur positive

Using polyvalent O antisera O1 Strains can be further differentiated –
V. cholerae Inaba, *V. cholerae* Ogawa and *V. cholerae* Hikojima.

These biochemical and serological characteristics enabled the epidemiologists to identify the cholera organisms whenever they were found in water and food samples as being of the same type, and therefore probably of the same origins. Furthermore, although it was never proven, it seemed likely that the cholera strain originated from Portugal's African colonies where *V. cholera* El Tor Inaba was known to occur, and was brought to Portugal by the troops. At that time (1974) there was much exchange between those colonies and Portugal.

How did the cholera spread from the soldiers to the people?

One reason is the potential for the carrier state. Infected people shed the organisms in their stools. This is how sewage becomes a vehicle for the organisms. But once an infected person has recovered from the symptoms of the illness they may still shed the organisms for a while. They are then symptomless carriers. However, in cholera, in contrast to typhoid and other *Salmonella* caused illnesses, it is rare for symptomless carriers to shed the organisms longer than three or four weeks. But occasionally people who have recovered from the illness carry the viable organisms in their intestine for longer periods (months or even years) and regularly shed them with their faeces.

The natural reservoir for the organisms is those who are ill and show symptoms, those recovering, and short and long term symptomless carriers.

Secondly, viable but non-culturable organisms are an issue in water microbiology. These are organisms which can be seen in the microscope to be intact, but which do not form colonies on laboratory culture media. Nevertheless these organisms may be capable of infecting and causing illness in susceptible people. This phenomen is now recognised to occur in many pathogens of significance in water safety. Their presence must therefore be considered even if conventional plate culture techniques fail to show colonies of organisms of concern (see Defives *et al.*, 1999).

Widening your thoughts, and perhaps also referring to UK experience 1937–1987, in the Portuguese outbreak sewage was clearly readily able to enter the drinking water systems, and infect the shellfish beds. It is worth considering whether some Portuguese people were permanently suffering from a whole range of waterborne diseases associated with sewage organisms and their consumption in the drinking water, and in the shellfish prior to the cholera outbreak.

What control measures were taken to bring the epidemic under control?

- investigations of individual cases,
- distribution of courses of tetracycline to those in contact with cases,
- health education,
- promotion of chlorination of public water supplies,
- free chlorine supplies for unsafe drinking water,
- cholera vaccine for those who demanded it.

Since the epidemic did come under control clearly the measures taken were effective, but it is worth re-reading the case study and seeing whether there are areas in the way the epidemic investigation was carried out which might have led to much speedier control of the outbreak.

Also consider what controls could have been implemented to prevent such an epidemic happening. Portugal needed much better control over its sewage, and much higher level of security of separation of its drinking water supplies from its sewage. An outcome of the epidemic was reported to be a commitment to such a programme. It also committed itself to monitor the water quality on the coast.

Clearly the shellfish beds needed protection from sewage too. Today Portugal, as a member of the European Union, must comply with EU Regulations concerning the quality of drinking water and management and microbial quality of shellfish.

Finally, today the production of bottled water would have to be run under a modern management system which included a HACCP system to anticipate microbiological hazards, and to control the risk of their occurrence.

1.5 SUMMARY

This case study illustrates that water is a vector of organisms, in this case the pathogen *Vibrio cholerae*, which was shown to survive in ground water and in sewage, and also multiply in the infected shellfish. It is not known whether the *V. cholerae* multiplied in the bottled water affected, but other research shows that some organisms do survive or multiply in bottled water, dependent on a number of environmental factors. Pathogens can be carried by water and infect consumers or contaminate foods. These examples together demonstrate that food processing waters can therefore also be carriers of organisms and could infect the processing plant and the foods processed. Thus the microbiological quality of food processing water is highly significant to both food safety and quality, and so in developing a strategy for control of food safety the quality of water in contact with food and food plant has to be taken into account.

- Water is a vector of organisms; faecal organisms survive in water often long enough and often in sufficent numbers to cause disease in consumers;
- *Vibrio cholerae* is a faecal organism which can survive in some waters, and can grow in shellfish;
- Drinking water can become polluted with faecal organisms;
- Microbial tests for potable water safety examine the water for organisms which indicate faecal pollution; drinking water quality should be the minimal standard for water used in food processing;
- Water can be a source of pathogens in foods and, being a vector, can be a source of other organisms which may affect the safety and quality of foods.

1.6   REFERENCES

Alvarez-Seoane, G., 1980. Ice cubes and the conditions under which they were used in Vigo 1977–79, *Alimentaria*, **114**, 53–58.

Anon, 1983. The bacteriological examination of water supplies, 1982. Report 71 (1983). HMSO, London, UK.

Anon, 1989. Water supply (Water Quality) Regulations. S.I. (1989) No. 1147. HMSO, London, UK.

Anon, 1990. Food Safety Act, 1995. HMSO, London, UK.

Anon, 1995. Food Safety (General Food Hygiene) Regulations, 1995. HMSO, London, UK.

Bottled Water Coolers Association website, www.bwca.org.uk – data dated 2000 sourced for them from Zenith International Ltd., www.zenint.co.uk

David, E., 1994. *Harvest of the cold months*. Published by Michael Joseph Ltd., London, UK.

Dickens, D.L., DuPont, H.L. and Johnson, P.C., 1985. Survival of bacterial enteropathogens in the ice of popular drinks. *Journal of the American Medical Association*, **253** (21), 4141–3143.

Galbraith, N.S., Barrett, N. and Stanwell-Smith, R., 1987. Water and disease after Croydon: a review of water borne and water associated disease in the UK 1937–86. *Journal of the Institute of Water and Environmental Protection*, **1**, 7–21.

Hunter, P.R., 1993. A review: The microbiology of bottled natural mineral waters. *Journal of Applied Bacteriology*, **74**, 345–352.

Khan, A.S., Moe, C.L., Glass, R.I., Monroe, S.S., Estes, M.K., Chapman, L.E., Jiang, X.,

Humphrey, C., Pon, E., Iskander, J.K. and Schonberger, L.B., 1994. Norwalk virus-associated gastroenteritis traced to ice consumption aboard a cruise ship in Hawaii: comparison and application of molecular method-based assays. *Journal of Clinical Microbiology*, **32** (2), 318–322.

LACOTS, 1996. Ice to be added to drinks or in direct contact with ready to eat food from public houses, clubs, catering establishments and retailers. Sampling and examination protocols. LAC 6 96 13.

LACOTS, 1997. Co-ordinated food liaison group sampling programmes – results of the survey of ice used to cool drinks and ready-to-eat food. LAC 3 97 13 and FG1 97 5.

Laussucq, S., Baltch, A.L., Smith, R.O., Smithwick, R.W., Davis, B.J., Desjardin, E.K., Silcox, V.A., Spellacy, A.B., Zeimis, R.T., Gruft, H.M., Good, R.C. and Cohen, M.L., 1988. Nosocomial *Mycobacterium fortuitum* colonization from a contaminated ice machine. *American Review of Respiratory Disease*, **138** (4), 891–894.

Mackie and MacCartney's *Handbook of Bacteriology*, 1960. 10th edition. Published by E. and S. Livingstone, UK.

Moreira, L., Agostinho, P., Morais, P.V. and da Costa, M.S., 1994. Survival of allochthonous bacteria in still mineral water bottled in polyvinyl chloride (PVC) and glass. *Journal of Applied Bacteriology*, **77**, 334–339.

Moyer, M.P., Breuer, G.M., Hall, N.H., Kempf, J.L., Friell, L.A., Ronald, G.W. and Hauser, Jr., W.J., 1993. Quality of packaged ice purchased at retail establishments in Iowa. *Journal of Food Protection*, **56** (5), 426–431.

Quick, R., Paugh, K., Aldiss, D., Kobayashi, J. and Baron, R., 1992. Restaurant associated outbreak of Giardiasis. *Journal of Infectious Diseases*, **166**, 673–676.

Tsuno, M., Thunghai, M., Bhanthumumkosol, D., Tonogai, T. and Jaengsawang, C., 1984. Bacteriological survey of water and ice for general users in Thailand. *Food Microbiology*, **1**, 123–128.

WHO, 1996. Guidelines for drinking water quality, Second Edition, Volume 2. Health criteria and other supporting information, World Health Organization, Geneva.

Other useful references

Defives, C., Guyard, S., Oulare, M.M., Mary, P. and Hornez, J.P., 1999. Total counts, culturable and viable, and non-culturable microflora of a French mineral water: a case study. *Journal of Applied Microbiology*, **86**, 1033–1038.

ICMSF (International Commission on Microbiological Specifications for Foods), 1996. *Micro-organisms in Foods*. Volume 5, *Characteristics of Microbial Pathogens*. Chapter 22, *Vibrio cholerae*. Published by Blackie Academic and Professional, London.

ICMSF (International Commission on Micro-biological Specifications for Foods), 1998. *Micro-organisms in Foods*. Volume 6, *Microbial Ecology of Food Commodities*. Chapter 14, Water. Published by Blackie Academic and Professional, London.

Wang, G. and Doyle, M.P., 1998. Survival of enterohaemorrhagic *Escherichia coli* O157:H7 in water. *Journal of Food Protection*, **61** (6), 662–667.

Expectations of Food Control Systems – in the Past, and Now

Key issues
- Meat – bottled wild duck paste
- *Clostridium botulinum*
- Sources of pathogens
- Food poisoning outbreak – identifying its cause
- Microbial growth conditions
- Toxin production in food
- Control of food safety in manufacture: then and now

Challenge

A very serious outbreak of food poisoning, which occurred in 1922, is described. It provides a means by which you can analyse how the cause of the food poisoning was identified, and see whether such an approach is relevant today. In understanding how this particular case of food poisoning arose you are also encouraged to think about how food contamination arises, why and when micro-organisms grow in foods, what is needed to ensure their destruction, and what we expect of a microbiological control system in a food manufacturing process today.

2.1 THE CASE STUDY: WILD DUCK PASTE, 1922

The tragic deaths of eight people through eating contaminated food. (The text below is quoted from Leighton, G., 1923. *Botulism and Food Preservation (The Loch Maree Tragedy)*, published by Collins, London, UK).*

2.1.1 How it Happened

"August, 1922, will always be a memorable date in the history of botulism on

---

* Despite every effort having been made to establish the identity of the copyright holder of this material, this has not proved to be possible. The Royal Society of Chemistry is publishing this material in good faith. If the rightful copyright holder identifies him/herself to The Royal Society of Chemistry, the matter of obtaining permission for use of the said material will be resolved.

account of the terrible tragedy which occurred at that time in a hotel at Loch Maree, in the Western Highlands of Scotland. The fact of this occurring in the height of the holiday season partly accounted for the extraordinary interest aroused by the tragedy, and the great publicity given to it in the Press made the whole English speaking community familiar with the tragic details. In one week, from 14th August to 21st August, eight adult persons died from botulism as the result of consuming sandwiches prepared from potted wild duck paste. No other persons but these eight partook of the sandwiches made from this particular container. Two were boatmen. The others were spending holidays in this romantic spot, famous for the beauty of its natural surroundings and for the fishing on the loch itself. Some of them had been in the habit of going to this same place for many years in succession. The hotel itself was well-known and extremely popular on account of its reputation for excellent management. The guests included some well-known people, and there were in the immediate neighbourhood at the same time, as it happened, quite a number of eminent medical men on holiday. All these facts together contributed to the sensation caused by the sudden announcement of the fatalities. It was some days before anything definite was known as to the cause of death. Very soon, however, the facts pointed unmistakably to the cause being botulism, and this to the popular mind, only made the case more mysterious as botulism was unknown in Great Britain. Indeed the Loch Maree tragedy is the very first authenticated instance of botulism which is known to have occurred in the British Isles; and this fact, together with the great detail and accuracy with which the whole occurrence was worked out, makes it of importance to place on record. This is how it happened.

On 14th August there were staying at the hotel some forty-four people. Every morning a number of visitors proceeded on the loch fishing, or went mountaineering, or took part in other excursions in the locality. Many of them stayed out for the day, returning only to dinner in the evening. For these people it was the custom at the hotel to provide lunch, principally in the form of various kinds of sandwiches. These were prepared for this purpose every morning according to the number required, and made into separate packets, either for one or more persons. On this particular day some twenty lots of sandwiches were thus prepared, and amongst them were sandwiches made from two containers of potted meat, one being wild duck paste, the other ham and tongue. These were distributed along with the other sandwiches made from cold beef, cold ham, and tongue, together with jam, butter, hard-boiled eggs, scones and cake. These were the ingredients which composed the lunches on that day.

Thirteen of the guests went fishing, taking with them, as usual their ghillies (boatmen), the latter numbering seventeen. Two of the guests who were fishing were accompanied by their wives. Another party of three went mountaineering. Of these thirty-five persons eight were seized with illness in the course of the next day or two, all of them having similar symptoms, and every single person of these eight died during the week.

On the evening of Monday 14th August some forty-eight people, including those who had been fishing, and those who had been mountaineering, dined at the hotel. The boatmen employed returned to their cottages. There was no sign nor complaint of illness on the part of anyone. About 3 a.m. on the morning of the 15th one visitor was taken ill. Later that same morning several others complained of illness and the local doctor was summoned. During the day the condition of at least one became serious, and the doctor summoned Professor T.K. Munro, M.D., of Glasgow University, who was in the district, in consultation. He arrived that night about 9 p.m. and found one of the first patients was already dead. About the same time a message was received saying that one of the boatmen had been seized with illness. While the doctors were visiting this case, and before they arrived back, a second death took place. Next day, the 16th, a second boatman began to be ill; and at noon on that day another death occurred. It was at this stage that the cases were reported to the Procurator-Fiscal, and from that time onwards the whole affair was in the hands of the legal authorities. Several other doctors from neighbouring hotels were also on the spot by this time, all of them giving the closest attention to the progress of the cases. The fourth death occurred that night, and the fifth and sixth on the following day (17th). There remained alive at this stage two of the eight persons who had been taken ill, one elderly visitor and one boatman. These last two cases lingered until the following Monday, 21st August, on which day they both died. No other case of illness of any kind had occurred amongst those in the hotel or in the few houses in the district.

Early in the week the doctors in attendance suspected that some form of food poisoning was the cause of the trouble, and on 18th August Dr. William Mac-Lean, Medical Officer of Health, in his report to the Procurator-Fiscal, stated that "the symptoms and course of the disease are identical in essentials with those described by Van Ermengem as his 'second type' in his investigations of sausage poisoning in the eighteen nineties." The full investigation which followed proved this diagnosis to be absolutely correct.

Suspecting food poisoning, the medical men in attendance set themselves to make a very careful scrutiny of all the food stuffs which had been used on the 13th and 14th August. One feature of the case narrowed down their enquiry very definitely and that was the occurrence of illness in the case of the two boatmen as well as in the six people from the hotel. If food poisoning were responsible then obviously it was some food which had been partaken of by these two boatmen in common with the six visitors. All of those in the hotel had eaten the same meals, but none were taken ill except these six people.

Obviously, therefore, the incriminating material had not been eaten at a meal in the house. Obviously, also, the only meal which the ghillies had shared with others was the lunch on the Monday, in which as usual, they partook of the lunch provided for the visitors. The inquiries of the doctors along these lines presented no difficulties, because the patients, although very ill, were perfectly conscious, and able to say what they had eaten, and it was not long before it was quite certain that the secret of the trouble was to be found in the sandwiches. A further scrutiny of what kind of sandwiches each person had had revealed the fact that all the eight patients, including the two ghillies, had partaken, amongst other things, of the potted paste sandwiches. These were the only food materials common to all cases. Inquiry was therefore directed very closely to this feature of the case, and the following facts were elicited:

On the morning of the 14th August two glass containers of potted meat were used in the making of the sandwiches for the day. Each of these potted meat containers were sufficient to make nine or twelve sandwiches according to size. There would thus be eighteen or twenty-four potted meat sandwiches distributed amongst the guests of that day. As the two containers were of different kinds of potted meat it was very unlikely that more than one was involved in the catastrophe, and if that were so there would not be more than twelve sandwiches at the most which could be held responsible for the disaster. On enquiring into the manner of making up the parcels of sandwiches it was shown that most likely two potted meat sandwiches would be put into each parcel. The twelve sand-wiches, therefore, would be found in not more than six or eight parcels. It is quite obvious that this is what actually occurred. There can be very little doubt that if twelve single potted meat sandwiches had been distributed in twelve packets and consumed by twelve different people the mortality would have been twelve instead of eight.

In all these eight fatal cases the symptoms exhibited by the patients were almost identical, differing slightly in degree of severity and time of onset, and in duration of the illness [. . .]. The absence of certain symptoms was as striking as the uniform presence of others. Thus, there was no fever, no headache, no pain, no disturbance of the sphincters, no facial paralysis, no deafness, no diarrhoea, no retention of urine, and no interference with consciousness or mental activity. It may be mentioned that in 4 of the 8 cases there was some vomiting; in only one was there any diarrhoea. In 2 cases pupils were noted as being distinctly dilated. The ages of the patients ranged from 22 years to 70 years. The cases consisted of 6 males and 2 females.

One or two features of interest in individual cases may be noted.

One patient (Mr. W., aged 66) on rising in the morning found that he was dizzy, and staggered on trying to walk. He also had diplopia (double vision). He nevertheless came down to breakfast, thinking nothing of his condition and even apologised to the doctor for troubling him. It was not until some 24 hours later that his symptoms became more marked, and from then onwards the progressive paralytic symptoms developed, and he died at 9.30 p.m. on 16th August.

In another case (Mr. S., aged 70) the first patient to be taken ill, the initial symptom was some vomiting, which continued more or less throughout the day of the 15th. In this case the symptoms became rapidly pronounced, and death ensued at 8 p.m. that night.

In the third case (Mrs. D., aged 56) the disease ran a very rapid course, death occurring at 11.30 p.m. on the 15th.

In the 4th case (Mr. T., aged 22) the first symptoms were slight, being some sickness and double vision. Loss of speech occurred on the 16th, but an hour before death, when asked by the doctor if he had eaten the paste sandwiches for lunch on the 14th the patient was able to nod his head in assent. Death occurred at 12.30 p.m. on the 16th August.

The fifth case (a ghillie, aged 35) first complained of illness when out in a boat on Tuesday the 15th, a day after the poisoned sandwiches had been eaten. That night this patient complained of some pain and tenderness in the abdomen, and his was the only case in which the symptom was present. During the night of the 16th paralytic symptoms supervened, with intense restlessness, and he died about midday on Thursday 17th.

In the 6th case (Mrs. A., aged 45) dizziness, double vision, and drowsiness were the first appearances. She became seriously ill, however, after 24 hours. Death took place at 4.05 a.m. on the 17th.

In the 7th case (Mr. D., aged 60) the patient noticed dizziness and double vision on rising in the morning of the 15th August. He however did not regard it as anything serious, and thinking that exercise would be beneficial he went out in his boat that day and actually rowed about four miles. He remarked to his boatman that if a fish were to rise he would see two. He was still going about on the following morning, but on the succeeding day, the 17th, he remained in bed, and the double vision, which had disappeared for a time reappeared. His speech also, which had recovered somewhat, became much worse. For a day or two there was but little change, and it was not until Sunday, 20th August, that paralytic symptoms set in, death occurring at noon on Monday, 21st August.

In the eighth case (the second boatman, aged 40) the patient first complained on Wednesday, 16th August, and the symptom mentioned was double vision on looking at objects at a distance. In his case the symptoms developed slowly, and he alternately seemed better, then worse, up till Thursday 17th. On the 18th speech was indistinct and difficulty in swallowing began. Paralytic symptoms came on markedly on the night of the 19th, and gradually increased until he died at 11 p.m. on 21st August [. . .].

Thus the only important difference in these eight cases was to be found in the duration of the illness, this possibly corresponding to the amount of the actual toxin absorbed by each patient on the Monday, on which date the sandwiches were eaten. From the subsequent symptoms the probability would seem to be that even if antitoxin had become available when the condition was diagnosed, treatment would have been unavailing. This supposition is borne out by the subsequent bacteriological investigation which proved that this particular meat paste contained a toxin of most extraordinary potency [. . .].

During the week in which these cases were running their fatal course, two official inquiries had begun. The first was that of the legal authorities under the Procurator-Fiscal. In this connection a large number of seizures of various kinds were made in order to determine their connection, if any, with the actual poisonous material. These were sent to a bacteriologist for examination. In addition, the usual police enquiries took place.

The second inquiry was that instigated by the Scottish Health Board, and carried out for them by Dr. Dittmar and the writer. This enquiry had in view the taking of any steps necessary to prevent any further danger to public health, and to allay anxiety.

At the time the tragedy occurred the whole affair seemed in the public mind a great mystery. That word was frequently used in the various Press accounts which appeared. When one considers the circumstances, it is not surprising that it was regarded in that way. Botulism had hardly been mentioned before as a possibility in this country. No-one had ever suspected its occurrence. Probably no physician in the country had ever seen a case. It was not one of the causes of death recognised as likely to happen in our midst. When it was given out officially by the Scottish Board of Health in the statement they issued to the Press on 25th August that this was the diagnosis which had been come to, coupled with the statement that there need be no general public alarm on that account, the mystery did not seem any the less. As a matter of fact, however, there was no mystery at all about it, although there was infinite pathos and tragedy. From the scientific and technical point of view it was simply a case of a pathogenic microbe which had the power of producing an immensely virulent poison, having found its way into a small container of preserved food under conditions which seemed ideal for its growth and multiplication. There it lived for some time until the toxin was generated and became spread throughout the small mass of meat contained, and this took place without leaving that material in such condition that its danger could easily be recognised. In some cases of food-stuffs which have caused death from botulism the material itself has been readily recognised as spoiled, but in others no obvious changes have taken place and this was one of the latter. If it be said that it is still a mystery where the microbe came from, one can only reply that it is equally mysterious where any microbe come from. The bacillus botulinus, like the organisms of tetanus, typhoid fever, diphtheria, and many other diseases, exists free in nature. Such organisms are carried about from place to place in many different ways. They find their way now and then directly into our bodies; or, as in the case in point, into a food-stuff and there they work their havoc. That is the only sense in which there is any mystery. The Loch Maree tragedy was one of those appalling and unfortunate events which no human foresight could readily have prevented, and for which no one could be blamed. If botulism were a common occurrence, doubtless such stringent regulations would have to be made for the control of the preparation of preserved foods as the authorities might see fit, but in a country where the condition was absolutely unknown it would be unreasonable to expect such stringent regulations to have been already laid down."

2.1.2 The Bacteriology

"In order to establish scientifically the proof of the diagnosis of botulism, a very careful and prolonged series of bacteriological investigations was necessary. This work was done by Mr. Bruce White, B.Sc., in the bacteriological laboratory of the Pathology Department of the University of Bristol. All specimens which seemed likely to have a bearing upon the case were sent to him, in accordance with an arrangement existing in this country for the investigation of outbreaks of food poisoning. Upon receipt of these specimens a long series of bacteriological cultures was instituted on different kinds of nutrient media, and these were grown at various temperatures. As usual in such an investigation, the organisms present were divided into the two groups of those which would grow in the presence of oxygen (aerobic); and, secondly, those which would only grow in the absence of oxygen (anaerobic). A very careful examination was, of course, made to ascertain the presence or absence of the organisms responsible for most of the ordinary outbreaks of food poisoning. In this case none of this group of organisms was found at all. Further, in this connection, an examination of a specimen of blood taken from one of the patients showed no sign which would associate the outbreak with this cause. From the aerobic organisms cultivated, a number of inoculation experiments were done on mice, with negative results. No poisonous element was found associated with the aerobic germs, and the conclusion was thus reached quite definitely that no organisms occurring in aerobic cultures could have been concerned in this outbreak.

The bacteriological investigation was then directed to the second group of organisms, namely, the anaerobic cultures. Immediately a very different state of things was discovered. Mr. Bruce White's report on this point says: "Of these cultures, all the sandwich meat cultures and all the wild duck cultures were found to be terribly pathogenic to mice when minute quantities were injected subcutaneously . . . The cultures giving positive results were now closely scrutinised . . . Microscopically the wild duck cultures have every appearance of purity, consisting entirely of large bacilli producing large egg shaped terminal spores . . . An anaerobic sporing bacillus had been isolated from the wild duck paste, which was highly pathogenic to mice . . . as soon as full cultures had been set up, experiments were initiated to test the toxicity or otherwise of the samples themselves."

A great number of very detailed experiments were made on these lines, which it is unnecessary to detail here. Many further particulars will be found in the official report on the Loch Maree tragedy issued by H.M. Stationery Office. It will be sufficient for our purpose if we state the conclusions which the bacteriologist arrived at:

"(1) The wild duck paste contained a potent toxin, the action of which is inhibited by botulinus (type A) antitoxin.
(2) The anaerobic spore-bearing bacillus isolated from the wild duck paste produces a similar toxin which is likewise counteracted by botulinus (type A) antitoxin.
(3) The identity of the wild duck bacillus with *Bacillus botulinus* (type A) seems established, as also the identity of botulinus (type A) toxin and that of the wild duck paste.
(4) It seems certain that the wild duck paste and the sandwich were the only toxic food stuffs submitted for examination."

The sandwich referred to in this last paragraph is one of the actual sandwiches made from the poisonous paste, which was ultimately recovered in dramatic circumstances to be related here-after.

The bacteriological investigation thus established the presence of *Bacillus botulinus* and its toxin in the suspected wild duck paste. It proved that this particular sample was of terrible virulence. Indeed, it was estimated that an amount of this paste in bulk equal to that of a pin's head contained as much toxin as would have been sufficient to kill 2000 mice. That calculation was readily made from the dilutions made for the experiments.

It should be remembered that this examination required the most extraordinary care and delicacy, because the total amount of the meat paste left in the glass container was extremely small. The glass jar – for such it was – in ordinary language would have been described as empty. That is to say, when opened the whole contents were taken out, leaving only tiny shreds of material which under those circumstances would adhere to the sides. Never the less this minute quantity was more than sufficient for the purpose in hand.

It so happened, however, that there was a further source of material for experiment, and the recovery of this forms one of the most dramatic incidents of this whole amazing affair. It may be referred to as the recovery of the buried sandwich, and the incident is so remarkable as to deserve full description.

At lunch time on Monday, 14th August, a boatman with one of the fishing parties was given some of the sandwiches prepared for their lunch. For some reason or other he put one of these sandwiches in his pocket and ate the rest. When he got to his home at night he placed this retained sandwich in a box in which he kept his food. On the following day he was taken ill, as were a number of others who had eaten similar sandwiches, and naturally it was not long before it became known that the inquiries of the doctors led to suspicion being attached to this particular food. This patient then remembered that he had one of these sandwiches left. It was taken out of the box, and those who were in the cottage noticed that it was composed of potted meat and that the sandwich was wrapped in white paper. One of the inmates took the sandwich and buried it in the garden, and the reason afterwards given for this procedure was that they thought it would be safer to dispose of it in this way in case it should be eaten by the hens and kill them. The sandwich thus buried was quite intact, and none of it had been eaten. Two days later, when the legal authorities were searching for all kinds of suspected materials for examination, the knowledge of this sandwich and its burial came to them, and they proceded in the darkness of the night of the 17th August to dig it up. It was found exactly as buried, still wrapped in its white paper. It was packed up and sent along with other material to the bacteriological laboratory, and on examination proved to be terribly toxic. This sandwich, recovered in this remarkable manner, proved to be one of the most important links in the bacteriological chain of evidence leading to the complete proof of the diagnosis of botulism."

2.1.3 The Public Enquiry

The public enquiry was held on 5th Deptember, 1922.
"The Sheriff directed the attention of the jury to the following points:

1) When and where the death or deaths took place.
2) The cause or causes of such death or deaths.
3) The person or persons, if any, to whose fault the accident was attributable.
4) Precautions, if any, by which it might have been avoided.

5) Any defects in the system or mode of working which contributed to the accident.

6) Any other facts that might be relevant to the enquiry."

The jury found:

"The consensus of medical and scientific opinion established the cause of death as botulism, due to consumption of the infected potted wild duck paste . . .

Throughout the whole of the evidence which had been given there was never any suggestion that anybody had been to blame. On the contrary, it was emphasised again and again by different witnesses that it was an accident which was quite unavoidable, and which could not have been foreseen . . .

They found they were not able to state any precautions by which this accident might have been avoided . . .

They also found that "it is desirable that every vessel containing preserved meat, fish, fruit or vegetables intended for human consumption should bear a distinct mark by which details of its manufacture can be traced" but carefully avoided using the word 'dating', suggesting instead 'a distinct mark' . . .

It was ascertained that the batch from which the poisonous container came had been manufactured not more than four months previously, not a long time for a preserved food on the market. The date of manufacture, the date of purchase by the grocer who stored it in the first place, and the date on which it was ordered from him by the proprietor of the hotel were all made clear by documents and invoices produced before the court.

The process of manufacture at the factory was described to the Court. It was shown that in the preparation of this food the materials composing it were first of all cooked in bulk, and were afterwards sterilised in open containers in retorts, and that subsequently, after it had been placed in the small glass containers in which it was ultimately issued, it was subjected to another cooking process at a temperature about boiling point. The manufacturers concerned, therefore, obviously took every reasonable precaution, and probably exercised a great deal more care than is very often the case. Indeed it was one of the most curious facts in the whole case that the manufacturers concerned were amongst the very forefront of those with a good reputation for this class of business. It was stated that they issued over a million such jars per year, and that as far as was known, this was the first time that anything had gone wrong. That fact in itself is sufficient to show that the methods adopted are quite adequate to provide against all ordinary risks."

Handling at the hotel:

"Two dozen potted meat containers had reached the hotel on 30th June, 1922 some six weeks before the tragedy. These were all kept locked up in the store and issued by the store keeper one by one as required. Six of them had been used prior to 14th August for making sandwiches in the usual way. On that day the storekeeper took out two glass containers himself and he gave them to the cook, and it was proved that these were still wrapped up in the way in which they had been received when ordered. The cook used these two pots, opening them in the usual way with a tin opener, piercing the lid. She was perfectly certain that the lids were not loose. There was no obvious indication to the senses of sight or smell that the meat paste was in any way abnormal, or unfit for food. It was perfectly clear that while the paste had been sealed up during the weeks prior to its use, the *Bacillus botulinus* which had found its way into the container had been growing and producing its deadly toxin, in this case without altering the appearance of the contents. Various witnesses spoke of the general excellence with which the hotel itself was managed, its cleanliness, and so forth, and here again was the strange feature that the tragedy should take place in a hotel famous for its excellent management . . ."

Finally

"The jury were clear upon the point that no fault could be imputed to anyone concerned . . .

They were also satisfied that reasonable precautions were taken, and did not see their way to make any suggestions as to how it could have been prevented."

2.2 BACKGROUND

2.2.1 The Organism: *Clostridium botulinum* and Botulism

Botulism is an illness of particular severity and high mortality because neurotoxins produced by the causative organism – *Clostridium botulinum* – severely affect the sympathetic nervous system. Rapid diagnosis of the condition, early administration of trivalent A, B, E antitoxin, and high quality of care of the patient can be very effective and save life. There are three forms of human botulism all associated with toxin produced during the growth and multiplication of the organism: food borne, which arises due to ingestion of the toxin present in food, infant botulism, in which the organism multiplies in the baby's intestine, and thirdly, rarely, wound infections.

Clostridium botulinum is an organism which has three important characteristics which have strongly influenced how canned and bottled foods are processed: it is normally considered to only grow in anaerobic conditions, it produces deadly toxins and it forms heat resistant spores. Today vacuum and modified atmosphere packs (MAP) may also provide conditions in which these organisms can grow, and because of various incidents of botulism the organism is known to be able to grow on occasions in conditions which may not be strictly anaerobic.

There are a number of strains differing in their origins, the foods they are likely to be found in, the conditions in which they can grow, the types of toxins formed, and the conditions in which toxin can be formed. They also differ in the resistance of spores and of toxins to heat and to other destructive forces such as irradiation, and pressure.

Table 2.1 *C. botulinum – toxigenic types and the animal species they affect*

Toxigenic type of C. botulinum	Species mainly affected						Commonest vehicles
	Man	Chickens	Horses	Cattle	Aquatic birds	Fish	
A	yes	yes					Home-canned vegetables, fruits, meat, fish
B	yes		yes	yes			Prepared meats, especially pork
E	yes					yes	Marine and fish products
F	yes						Meat products
C$_a$					yes		Rotting vegetation of alkaline marshes, invertebrates
C$_b$			yes	yes			Toxic food, carrion, pork liver
D				yes			Carrion
G	maybe						Soil

Source: ICMSF, 1996.

The group of organisms

All organisms producing botulinal neurotoxin are classified as *C. botulinum*. Generally the strains are Gram positive, motile, anaerobic rods, bearing oval sub-terminal spores, which distend the cells bearing them. They are widely distributed in nature being found in soils, fresh water and marine muds and silts, faeces, animal, fish and human intestines, and can occur on almost all foods of vegetable and animal origin. Although widespread, the levels of contamination are generally low but in specific food types can be high. The seven types (A to G) (Table 2.1), affecting different animal species, are classified on the antigenic specificity of their toxins.

The types fall into four groups (I to IV) based on metabolic and serological characteristics. Group I are proteolytic, causing in their growth breakdown of protein and the production of foul odours. Group II are non-proteolytic and affected foods may show no signs of change. Strains from both of these groups have been shown to be associated with outbreaks of botulism, whereas strains from Groups III and IV, producing toxins C, D and G, are very rarely implicated (Table 2.2).

The toxins

These are all extremely potent. In the case of type A toxin death in man occurs on ingestion of between 0.1 and 1.0 μg.

Foods affected

A wide range of foods providing anaerobic and non-limiting growth conditions in respect of their pH, water activity (a_w) and other factors have caused botulism (Table 2.3).

Factors allowing growth and toxin production

Although many growth limiting conditions have been identified in specific circumstances (ICMSF, 1996), it is generally considered that neither growth nor toxin production occur below pH 4.6 under otherwise optimal growth condi-

Table 2.2 *C. botulinum – the differentiating characteristics of metabolic Groups I and II*

	Group I	Group II
Toxins produced	A, B, F	B, E, F
Proteolysis	yes	no
Lipolysis	yes	yes
Fermentation of glucose	yes	yes
Fermentation of mannose	no	yes
Minimum growth temperature	10–12 °C	3.3 °C
Inhibited by NaCl (%)	10	5

Source: ICMSF, 1996.

Table 2.3 *Examples of foods involved in outbreaks of botulism in a number of countries and the dates of occurrence*

	France	UK	USA	Germany	Sweden	Japan
Meats	86% of cases associated with meats: ham (1985, 1986, 1987, 1988, 1989, 1990); pate (1985, 1987, 1989)	Duck paste (1922); rabbit and pigeon broth (1932); jugged hare (1934); minced meat pie (1935)	Vacuum packed luncheon meats (1965) (unconfirmed incident)	More than 75% of cases associated with meat		
Fish	Pickled fish from Mauritius (1955); canned salmon from US (1978)		Vacuum packed smoked ciscoes (1960); vacuum packed smoked white fish chubs (1963)	Vacuum packed smoked trout (1970)	Vacuum packed smoked salmon (1991)	99% of cases associated with fish
Fruit and vegetables	Vegetable conserve (1987, 1990)	Vegetarian nut brawn (1935)	60% of cases associated with vegetables, mostly home bottled Modified atmosphere packaged shredded cabbage (1987)			
Other		Macaroni cheese (1947), kosher airline meal (1987), hazel nut yoghourt (1989)				

Source: ACMSF, 1992.

tions. pH values above pH 4.6 can allow growth and toxin production. Critically, while many strains belonging to Group I are limited to growth above 10 °C, strains belonging to Group II can grow and produce toxin at temperatures as low as 3.3 °C, *i.e.* in foods stored under refrigeration. Close to the permissive limits for growth, toxin is produced extremely slowly allowing food in long storage the potential to become dangerous. Thus Group II strains represent a risk in minimally processed vacuum and modified atmosphere packed chilled foods.

Water activity reduced below 0.99 progressively restrains growth rates, and the limiting value is around 0.94 a_w. Most strains are sensitive to sodium chloride, and its limiting concentration is 10% w/v.

Combinations of values of pH, a_w and temperature can be manipulated to prevent spore germination, growth and toxin production. Some preservatives, particularly sodium nitrite (long used in meat curing), control growth and toxin production but their effects are concentration dependent.

Destruction of cells, spores and toxins

Vegetative cells are destroyed by pasteurisation temperatures. Toxins can be inactivated at 75–80 °C but spores are more heat resistant. Spores of Group II strains can effectively be destroyed in the temperature range 80–100 °C, the D-values being short enough to permit processing in that range, while achieving substantial reduction in numbers. Group I spores can only be effectively eliminated by temperatures above 100 °C (Table 2.4).

Thus, where a food can be stored under chilled conditions below 10 °C which prevents outgrowth of Group I spores and cell multiplication, processing sufficient to destroy the more heat sensitive Group II *C. botulinum* spores is considered adequate. The accepted process of food core temperature 90 °C for 10 minutes, or an equivalent process, provides at least a 6D process with a margin of safety. But where a food is intended for long term ambient storage, *e.g.* canned or bottled food, and is 'low-acid' (*i.e.* at or above pH 4.5) the long accepted 'botulinum cook' is applied. This is a heat process of 121 °C/3 minutes or equivalent, and is deemed to be a 12D process, being the minimal heat treatment which must be experienced by the whole food, including of course the deepest, central parts of the product. This heat processing will not necessarily protect the food from spoilage by other, more heat resistant spore producing organisms.

2.2.2 Meat Pastes

The following information is from *The Practical Grocer* (1906), a compendium of information for the grocery trade (as it then was). It describes preparation stages which, although pre-dating the botulism outbreak by 16 years, may have been similar to those used at that time. It is quoted by permission of Robert Hale Ltd.

"Meat pastes such as ham and tongue are prepared by mixing the particular meat basis with about one third of its weight of bacon or fat pork, the mixture being minced and rubbed or ground to a pasty consistency and then pressed

Table 2.4 *C. botulinum: effect of heat on spores: D and z-values*

(a) General guidelines

Temp. (°C)	Group I, proteolytic non-psychrotrophic strains		Group II, non-proteolytic psychrotrophic strains	
	D-value (min.)	*z (°C)*	*D-value (min.)*	*z (°C)*
80			21.5	9
100	25	various	<0.1	various – strain dependent
120	<0.5	various		

Source: ACMSF, 1992.

(b) Specific examples of strain variability, and variability in different heating systems

Conditions	Strain	Temp. (°C)	Group I, proteolytic, type A toxin producer: effect of heat on spores — D-value (min.)	z (°C)	Group I, proteolytic, type B toxin producer: effect of heat on spores — D-value (min.)	z (°C)	Group II, non-proteolytic, type B toxin producer: effect of heat on spores — D-value (min.)	z (°C)	Group II, proteolytic, type E toxin producer: effect of heat on spores — D-value (min.)	z (°C)
M/15 phosphate buffer pH 7.1	1	79.4							1.3	(not given)
M/15 phosphate buffer pH 7.0	2129B	82.2					32.3	9.7		
M/15 phosphate buffer pH 7.0	CBW25	82.2					1.49	8.3		
M/15 phosphate buffer pH 7.0	ATCC-17844	82.2					4.17	16.5		
M/15 phosphate buffer pH 7.0	213B	120*			0.13*–0.15	11*				
M/15 phosphate buffer pH 7.0	213B	120			0.24	9.6				
Canned corn	62A	121	0.41	12.7						
Mackerel in oil	62A	121	0.15	10.6						
Mackerel in water	62A	121.1	0.089	8.3						
Pea puree	62A	121.1	0.06	8.5						
Distilled water	62A	121.1								
M/15 phosphate buffer pH 7.0	62A	121.1**	0.051**	9.1**						

Sources: ICMSF, 1996.
D value is 'the time required at a given temperature to reduce the number of viable cells or spores of a given micro-organism to 10% of the initial number – usually quoted in minutes' (ACMSF, 1992).
z value is the change in temperature (°C) required for a ten-fold change in the D value (ACMSF, 1992).
Notes: the data labelled * and ** are the bases of the calculated figures shown in Table 2.6.

through a sieve. Salt and various seasonings are added as required, and the paste is then potted. It may either be preservatised with borax, or covered with a layer of fat to exclude the air and keep the article in good condition; or it may be sealed air tight and 'processed'."

Note: At the time of publication no specific limit had been laid down or recommended for boric acid in potted goods.

2.3   EXERCISES

Exercise 1. Identify from the text given in Section 2.1 the exact steps taken by the investigating doctors and the type of evidence they generated from the time of the first case of illness to the proof of its cause. It may be helpful to produce a flow diagram to illustrate how it was possible to confirm that the wild duck paste was the source of the botulism.

Exercise 2. The outbreak of food poisoning at Loch Maree was investigated using the systems in place in Scotland in 1922. What is the system in your country/region/district for investigating food poisoning outbreaks today? How are the overall statistics accumulated?

Exercise 3. From this outbreak it is clear that the incriminated wild duck (meat) paste contained both viable cells of '*Bacillus botulinus*' (*i.e. Clostridium botulinum*) and its toxin. At the public enquiry it turned out the paste was 4 months old which was well within the expected life of the product. Prepare a sequence of questions exploring how this microbial contamination may have occurred and pointing to possible explanations for why the microbial growth and toxin production occurred.

Exercise 4. Although it may have been appropriate in 1922 to comment on the processing techniques used for the wild duck paste that "nothing further could have been done", identify what you think would be required for the safe processing and handling of such a product today.

Study objectives

This case study will allow you to link together the ecological factors which allow growth and toxin production of *C. botulinum* in a food system and how, through using appropriate heating parameters, the organism and its toxin can be destroyed.

You will also be able to evaluate management of the meat paste processing method used in 1922 and consider where we would now consider it lacking in control.

Looking at the case study as a food poisoning outbreak and thinking about its investigation will enable you identify what is needed for the process, what method is currently used in your country, and where difficulties and weaknesses lie.

Your objectives are therefore to find the lessons of a tragic outbreak which occurred in 1922 and relate them to the present.

Table 2.5 *Definitions relating to food poisoning, foodborne illness, and gastrointestinal infection*

Term	Description	Source
Case	Clinical case: an ill person	1
Epidemiology	The study of factors affecting the health and disease in populations and the application of this study to the control and prevention of disease	1
Food poisoning	A notifiable illness normally characterised by acute diarrhoea and/or vomiting caused by the consumption of food contaminated with micro-organisms and/or their toxins	1
Foodborne illness	Illness resulting from the consumption of food contaminated by pathogenic micro-organisms and/or their toxins	1
Gastro-intestinal infection or gastro-enteritis	Infection of the stomach and intestine usually resulting in diarrhoea and vomiting. This may be due to the colonisation of the intestinal tract by pathogenic organisms	1
Laboratory report	Data routinely collected nationally of laboratory identifications of certain pathogenic organisms. Includes all those known to cause foodborne infections	1
Outbreak	Two or more cases linked to a common source	1
Sporadic case	A single case of disease apparently unrelated to other cases	1
Surveillance, epidemiological	The continual watchfulness over the distribution and trends in incidence of disease in a population through the systematic collection, consolidation and evaluation of morbidity and mortality reports and other relevant data. It includes dissemination of information to all those who need to know so that appropriate action can be taken	2

Sources: 1. Richmond, M., 1990; 2. Richmond, M., 1991.

2.4 COMMENTARY

A number of definitions concerning illnesses associated with foods have arisen over the years and can sometimes seem to overlap with one another.

"*Food poisoning* – any disease of an infectious or toxic nature caused or thought to be caused by the consumption of food or water" (ACMSF, 1992).

The definition of food poisoning given above is simple and clear, a valuable description for general purposes. However, in surveillance where statistics are gathered more precise definitions are required and you are referred to Table 2.5 for some of those.

Exercise 1. Identify from the text given in Section 2.1 the exact steps taken by the investigating doctors and the type of evidence they generated from the time of the first case of illness to the proof of its cause. It may be helpful to produce a flow diagram to illustrate how it was possible to confirm that the wild duck paste was the source of the botulism.

In deducing what had caused the deaths, the doctors followed a very logical sequence of thought shown as the numbered points in Figure 2.1, and the microbiologist followed the sequence shown in Figure 2.2.

The work of the doctors and the microbiologist are summarised below:

- formed hypothesis;
- established foods eaten, and their sources;
- established foods in common among those ill;
- identified the suspected food;
- took samples;
- arranged for careful, thorough bacteriological analysis which was designed both to test the hypothesis that botulism was the cause, and to eliminate other organisms also known to cause food poisoning;
- confirmation that the cause was botulism was achieved by testing the toxicity of foods, and of the cultures isolated from the food, using mouse bioassay, and that the symptoms experienced by the mice on both occasions were the same. These were confirmed as botulism through using the anti-toxin which protected the mice against the symptoms.

This follows the classical steps identified by Koch in his 'postulates'. The same type of logic is used today.

> **Exercise 2.** The outbreak of food poisoning at Loch Maree was investigated using the systems in place in Scotland in 1922.
>
> What is the system in your country/region/district for investigating food poisoning outbreaks today? and how are the overall statistics accumulated?

Authorised officers (in England and Wales the doctors, and the environmental health officers (EHOs) who enforce food law) work together. Where there is more than one person affected with similar symptoms, and where they have been to a common event, it is often possible to infer that infected food may be responsible. However, just as it was lucky that a small amount of the food was left over at Loch Maree, it is valuable today to have food samples for analysis. So that the burden of analysis is manageable it is important the EHOs can deduce foods common to those suffering the illness. Where proof of causation becomes difficult is when the symptoms of food poisoning arise many days after the consumption of the food, and also where people do not know they have shared a common event – such as when restaurant food or a food ingredient such as milk powder are implicated. Ideally the causative organisms can be isolated from the sick individuals and be shown to be the same as those in the suspect food, but this is often difficult to achieve. Where proof is achieved the outbreak and cases are recorded in the national food poisoning statistics; where proof is not achieved the cases are recorded as gastroenteritis, or in the case of botulism-like symptoms would be recorded as 'botulism – unproven'.

In England and Wales the sequence of events by which food poisoning is investigated and recorded are shown in Figure 2.3. This Figure also shows how at three links (at least) in the investigating/reporting chain data are lost, implying that the official statistics will never include all cases but will, at best, show trends.

> **Exercise 3.** From this outbreak it is clear that the incriminated wild duck (meat) paste contained both viable cells of '*Bacillus botulinus*' (*i.e. Clostridium botulinum*) and its toxin. At the public enquiry it turned out the paste was 4 months old which was well within the expected life of the product.
>
> Prepare a sequence of questions exploring how this microbial contamination may have occurred and pointing to possible explanations for why the microbial growth and toxin production occurred.

It is clear from the comments made at the public enquiry that there was no evidence to show from where the organisms had come, nor how it was possible that the toxin had accumulated in the jar of wild duck paste. Never the less it is clear that the organism had been present in the jar and had grown and produced toxin.

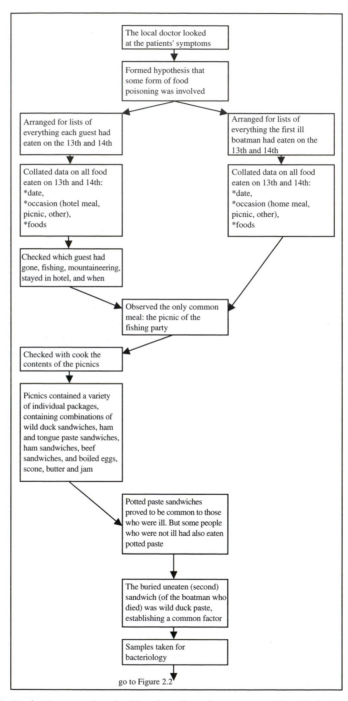

Figure 2.1 *Loch Maree outbreak of botulism: how the cause was identified – Part 1*

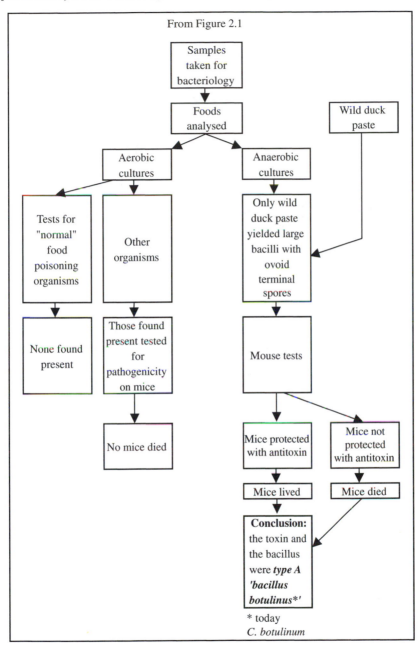

Figure 2.2 *Loch Maree outbreak of botulism: how the cause was identified – Part 2*

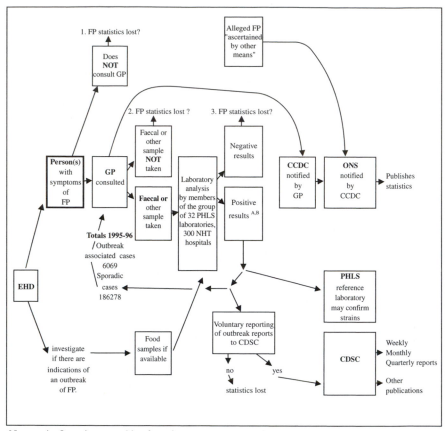

Notes: A : Organisms capable of causing gastro-enteritis (GE)
        B : Identical organisms capable of causing FP found in food and faeces.

Key:
EHD   Environmental Health Department
FP   Food poisoning
GP   General practitioner - "doctor"
CCDC   Consultant in Communicable Disease Control
ONS   Office of National statistics
PHLS   Public Health Laboratory
CDSC   National Health Trust Hospital
NHT   Communicable Disease Surveillance Centre
  ➤ = information flow.

For definitions relating to food poisoning, food-borne disease and gastro-enteritis see Table 2.5

Figure 2.3 *How food poisoning is reported in England and Wales, and where statistics may
               be lost*
               (Adapted from Colley, 1999, with permission)

Looking at it systemmatically:

Sources of the C. botulinum:

- What were the ingredients of the duck paste?
- Could any the ingredients of the duck paste have been the source of the organisms?

It is well known that meats may contain spores of *C. botulinum*, as may spices (ICMSF, 1998). Although when found in meats the concentrations of *C. botulinum* are normally very low, wild ducks may eat a wide range of materials and forage in muds, and their gut flora (and their skin and feathers) could contain *C. botulinum* at higher concentrations. Wild duck is a game bird, and would probably have been 'hung' after shooting to allow it to ripen. In this post-mortem process the gut flora increases and proteolytic enzymes from the gut, which include those from the gut flora, help to tenderise the flesh. This process affects the taste and such ripened meat is less in favour today than it was in 1922.

Spices too may contain microbial spores, including those of *C. botulinum* which survive for very long periods in the dried product but may germinate in the moist conditions of the food. Maybe the spices were the source of the *C. botulinum*.

- Could the *C. botulinum* have entered the paste from any extraneous source – vermin faeces for example – as it was being made? Was all equipment clean?
- Would any of the entrails of the duck – the liver for example – have been used? Would the intestines have been used to make stock? If the gut itself was not used could gut organisms have contaminated the meats used?

Processing

Mincing of the meat should have mixed it well, permitting it to be compacted together, but at the same time the process would have moved organisms from contaminated surfaces into the middle part of the raw paste mixture.

Heat treatment

Cooking was apparently a three stage process – Stage 1, cooking in bulk; Stage 2, dispensing into open containers and sterilising in retorts; then Stage 3, dispensing into glass jars and cooking at boiling point. No temperatures were given in the public enquiry but 'cooking in bulk' suggests temperatures rising up to 100 °C; 'retorting' could have provided temperatures as high as 121 °C; 'cooking at boiling point' suggests the filled bottles were immersed in boiling water pans – product temperatures then rising towards 100 °C the longer they remained in the pans.

- What temperatures were reached and for how long? What was the highest temperature and where was this reached? Were all the pots heated equally? Was heating even, or could inconsistent texture result in uneven heating? Was stirring used at any stage?

Organism numbers

The numbers of organisms present prior to cooking would affect the efficiency of microbial reduction.

- How many spores were initially present?
- Could abnormally high numbers have been present in the raw ingredients?

Organism distribution

- Would patchy distribution of spores in the paste affect the safety margin of the cooking/heating process? Would final temperatures in all parts of the paste have been adequate to destroy all spores present?

Organism D-values

- What were the D-values of the spores at the temperatures experienced?
- What were the D-values for the *C. botulinum* type A in this particular duck paste?
- Was there a high fat content which would tend to make the spores more difficult to destroy?
- If temperatures in the centre of the pots did not exceed 100 °C was this particular strain unusually heat resistant and able to survive? What type of D-values do strains of *C. botulinum* type A have?
- What minimum time of heating would have been needed for a 12D process to be achieved throughout the containers?

Organism z-values

- If the temperature achieved in the duck paste had been a little lower than usual, even if the heating time was the same or longer, the z-values of the contained spores (of *C. botulinum* and of spoilage organisms) would significantly influence which populations survived. Use Table 2.6 to think about the implications of this.

Process management

- Could cross-contamination from raw or undercooked materials have contaminated cooked materials re-introducing viable spores of *C. botulinum*? Could this have happened perhaps at the pot-filling stage?
- How was the process managed? Would the Stage 3 boiling have destroyed spores of *C. botulinum* type A? When were the final pots sealed – at or after Stage 3 cooking?
- How were the staff managed to avoid cross contamination of cooked product?
- What was the actual size of the jars and would the core of the food have been exposed to lethal heat?
- Were the bottles of the same glass as previously used or was it thicker, affecting heat penetration?
- If underheating occurred why was it that there were no signs of spoilage caused by other surviving organisms, and whose growth would have warned the

Table 2.6 *C. botulinum spores: 1D, 12D and z-values for two different strains heated in M/15 phosphate buffer, pH 7.0*

Temp. (°C)	Group I, proteolytic, type A toxin producer: Strain 62A Effect of heat on spores			Group I, proteolytic, type A toxin producer: Strain 213B Effect of heat on spores		
	1D-value (min.)	12D-value (min.)	z (°C)	1D-value (min.)	12D-value (min.)	z (°C)
95		468	9.1	24	288	11
100	11	132	9.1	8.25	99	11
105	3	36	9.1	3	36	11
120*				0.13*	1.56	11*
121.1**	0.051**	0.612	9.1**			

* and **: Data from Table 2.4. Other data values in the table are interpolated graphically from these.
Italic values show that 105 °C D values are the same for strains of different z values.

consumers of danger? Why did the ghillie who died not eat one of his sandwiches? Did it taste odd to him?

- Why was it that only a single pot, of all the pots produced by the factory, contained *C. botulinum* in high numbers, and according to the bacteriologist, in pure culture?
- At Stage 2 of the cooking process was the retort over full, packed with too many containers of paste? Could it have been that one pot – the one which in the end killed eight people – just did not receive enough heat to destroy spores of *C. botulinum* contained within that mass?
- Could one pot have been contaminated after Stage 3 processing – perhaps when the lids were tightened as the pots cooled – and the organisms grew on the top of the pot, tolerating the less than 100% anaerobic conditions, and the toxin produced permeated through the bulk of the product?
- At the factory how were the glass bottles and their lids stored? Were the bottles clean? What was the bottle cleaning and checking process? Is it possible that a single jar was dirty – contained rat or cockroach faeces and that faeces infected a single jar of product into which paste was filled? Would Stage 3 cooking destroy *C. botulinum* type A spores?

Organism growth

- Why could *C. botulinum* spores germinate and the vegetative cells grow in the meat paste? What were the values for the pH, a_w, and sodium chloride concentration? What are the limits for these parameters for growth and toxin production? Was the recipe the same as used on previous occasions? What was the temperature of storage – and was it above 10 °C so allowing *C. botulinum* type A to grow?
- Was a preservative normally used in the paste (for example boracic acid – which is not permitted today)? Were the ingredients properly mixed to achieve a homogeneous mixture, or was it such that uneven heating or uneven preservative concentrations might have occurred?

Storage

- Isolation of the spores from scraps at the bottom and sides of the pot suggests the organisms occurred throughout the pot. Was the paste of a runny consistency at any time in its storage allowing these motile organisms to migrate through the paste? Was storage temperature in the larder in the hotel, in August, so high that the paste, or the fat, or the gelatin became liquid, the organisms grew quickly and toxin was produced? Or did very slow continuous growth occur over the whole four months (120 days) between production and consumption?

What is your hypothesis for how *C. botulinum* was present with such lethal effect in a jar of duck paste?

> **Exercise 4.** Although it may have been appropriate in 1922 to comment on the processing techniques used for the wild duck paste that "nothing further could have been done", identify what you think would be required for the safe processing and handling of such a product today.

Producing safe product today requires management of the total food production system anticipating potential problems and microbial hazards and manages the process accordingly. Questions such as those arising from Exercise 3 would be raised prior to establishing a process, so that the process itself is designed to manage the anticipated microbial hazards.

At each step in production care has to be taken. For processed paste

- hygiene in the handling of all ingredients and knowledge of their normal microbial loads before processing;
- the process flows need to be designed so that cross contamination from raw materials into processed (or semi-processed) product is not possible;
- processing in a manner in which it is categorically known that the minimum heating received in every container of product is adequate to kill the spores of *C. botulinum*, to the 12D level, and desirably using more severe processing in order to also destroy spores of spoilage organisms as well;
- filling all pots of product with the same volume and using the same type of container so that the predicted heating exposure is actually experienced;
- care and hygiene in the cooling process so that contaminating organisms are not drawn into the head space through the gap between the pot and the lid, and avoidance of manual handling of wet pots; all handling equipment to be clean;
- appropriate control of storage and transportation conditions to avoid damage to lids, or exposure to dirt and dust; control of temperature of storage;
- storage and use in a period within the shelf life.

These are some key practices in the safe production of potted meats capable of supporting the growth of *C. botulinum* (ICMSF, 1998). Today there are many codes of practice which relate to the safe processing of low acid canned and bottled foods (DoH, 1994).

2.5 SUMMARY

Eight people died in 1922 due to eating potted wild duck paste infected with *C. botulinum* type A and its toxin. A logical approach applied at the time led to bacteriological confirmation that the cause of the deaths was botulism. This same logical process is used today.

C. botulinum type A has heat resistant spores, is proteolytic, and not psychrotrophic. The intrinsic conditions in the product, together with storage at permissive temperatures, and adequate passage of time allowed growth and toxin production. The failures in the processing and handling of the meat paste allowed survival and/or entry of the spores into the product.

Today it would not be acceptable to suggest that the accident could not have been avoided, that everything that could be done was done. It would be expected that a pro-active management system would be in place, and in the light of such an accident it would be re-examined to determine why it had apparently failed or whether the manufacturer could indeed claim that 'due diligence' had been applied in production.

Notes

1. Esty, R.J. and Meyer, K.F., 1922. The heat resistance of the spores of *B. botulinus* and allied anaerobes. *Journal of Infectious Disease, Chicago*, **31**, 650–663.

The work of Esty and Meyer on the heat resistance of the spores of *C. botulinum* established the standard for heat processing medium and low acid foods. In their experiments, in which more than 100 strains were tested, large numbers of spores, produced under the most favourable conditions, were heated to complete destruction. The heat treatment to reduce the probability of one spore remaining viable to one chance in 10^{12} was found to be 2.5 minutes at 121 °C. However, that is not enough to destroy all spoilage organisms, which can be more difficult to destroy – sometimes requiring prolonged times at 121 °C to destroy them ($D_{121°C} = 5$ minutes or more).

2. Savage, W.G., 1923. Canned foods in relation to health. *The Lancet*, **i**, 527–529.

In this paper Savage says that the Loch Maree case is the only outbreak of botulism which has ever occurred in England [does he mean the UK?], and goes on to say:

"The possibility of its introduction from canned fruit or vegetables demands very careful and detailed consideration. The toxins and the bacilli are easily killed by heat, the spores exceptionally resistant. While most succomb to reasonable temperatures (45 minutes at 105 °C) more resistant strains are known which require 6 hours at 100 °C, or 36 minutes at 110 °C to destroy them. The ingestion of fruit containing undeveloped spores probably does no harm as they do not develop in the animal intestine.* If, however, the conditions in the tin are such that they can multiply they produce toxins and the whole tin is excessively dangerous. Fortunately the other products of growth are unpleasant and definite signs of spoilage are present. In exceptional cases these have not been noted.

In the Loch Maree outbreak the wild duck paste, from its nature would tend to mask such changes, although probably they would have been noticeable to an expert . . .".

---

* But in infant botulism where the child's gut flora is not fully developed *C. botulinum* can grow and cause infant botulism.

2.6 REFERENCES

ACMSF, 1992. Report on vacuum packaging and associated processes, HMSO, London.

Colley, S., 1999. An evaluation of the role of Environmental Health Departments in the prevention of sporadic cases of infectious intestinal diseases. MSc thesis, South Bank University, London.

DoH (Department of Health, UK), 1994. Guidelines for the Safe Production of Heat Preserved Foods. Available through The Stationery Office, UK.

Gaze, J., 1992. The importance and control of *C. botulinum* in processed foods. *British Food Journal*, **94** (1), 8–15.

ICMSF, 1996. *Micro-organisms in Foods*, Volume 5. *Characteristics of Microbial Pathogens*, Blackie Academic and Professional, London.

ICMSF, 1998. *Micro-organisms in Foods*, Volume 6. *Microbial Ecology of Food Commodities*, Blackie Academic and Professional, London.

Leighton, G., 1923. *Botulism and Food Preservation (The Loch Maree Tragedy)*, Collins, London, UK.

Richmond, M., Chairman of the committee on the microbiological safety of food. Report, *The microbiological safety of food, Part I*. Crown copyright. First published 1990, HMSO, London.

Richmond, M., Chairman of the committee on the microbiological safety of food: Report, *The microbiological safety of food, Part II*. Crown copyright. First published 1991, HMSO, London.

Simmons, W.H. (Ed.), 1906. *The Practical Grocer*, The Gresham Publishing Company, London.

CHAPTER 3

Zoonotic Disease

Key issues
- Milk: a primary source of pathogens
- Pasteurised milk and salted telemea cheese
- *Mycobacterium bovis*
- Zoonotic diseases; tuberculosis
- The effect of change on control efficiency

Challenge
This case study addresses the age-old problem of the transfer of disease from animals to man – in this instance *via* milk and milk products. The case study demands that you think about the elements needed in a control system – from where organisms originate, what controls microbial growth in a food system, what can be manipulated to control it, the use of heat for the destruction of organisms, and when we are confident in the safety of a food at local level. But the exercises also demand that you consider these things in order to evaluate the broader regional context in which food control systems are implemented.

3.1 THE CASE STUDY: MILK SAFETY PROBLEM IN ROMANIA

The prevention of dissemination of bovine tuberculosis through drinking milk and eating cheese.

3.1.1 Milk and Its Problems pre- and post-1990

In 1990 Romania – a former Eastern Bloc country – threw off the yoke of the old communist system and started the process of western style capitalism. One immediate effect of this process was that the co-operative farming system, which generally operated with large land holdings, fell apart and the peasant communities reclaimed their historic land holdings. The new government upheld this rapid change and enacted legislation which allowed workers on the former co-operatives and other rural dwellers to become landowners and, if they wished, to farm. Many people were able to claim land, including those people who were salaried employees of the former co-operatives.

Turnock (1997) reports a survey of 500 villages, conducted in 1993–94, which showed that most new land holdings were very small – between 0.5 and 10 hectares, the maximum allowed. The distribution of land found among those new owners was

26% owned less than 1 hectare,
38% between 1 and 3 hectares,
19% between 3 and 5 hectares, and
15% between 5 and 10 hectares.

Yet pre-1990 most farms (usually co-operatives) were much larger, most frequently between 500 and 2000 hectares, and were organisations producing and selling primarily for the Romanian internal market.

Since 1990 the change in the ownership of rural land has also meant that the way farms operate has changed. While some people have grouped together and formed larger co-operating units, sharing machinery and so on, other people have chosen to farm their small pieces of land as individuals for self sufficiency and selling only the surpluses.

This means that dairying has changed too. Before 1990 milk was produced by large co-operatives and sold to the internal Romanian national market. In 1990 the legacy of the pre-1990 era was the existence of much ageing ill-maintained equipment, and a need for much better milking and milk processing equipment. By the year 2000 beside some large farm businesses there were many small farmers producing milk to whatever quality and quantity they could manage.

An effect of the re-distribution of the land to the peasant owners was also that the structures for supply of fodder (and so on) to the dairy cows and the purchasing arrangements for the milk also collapsed, and new systems have had to be developed.

According to Turnock (1997) at the time he was writing milk often failed to meet EU standards, and "milk from private farms rarely reaches the market since peasants prefer to sell cream and cheese which attracts higher prices". He went on to say also that, in 1994 at least, there was a shortage of milk for the capital city, Bucharest, which needed 250 000 litres of milk daily but only received 200 000 litres. But volume was not the only problem – reliability of quality and safety were problems too.

Dairy cows can be infected with 'zoonotic' diseases – that is diseases which not only compromise the health of the animal, but which may also affect the health of people in close contact with the animals, or infect people through the consumption of infected meat, milk or dairy products. The statistics in Table 3.1 show that in Romania deaths from tuberculosis (TB) have steadily risen since the early 1980s. Deaths due to respiratory TB are by far and away the most predominant, but deaths from other forms of TB have shown the same rising pattern. Among those 'other forms' is tuberculosis of the intestine, the form most likely to be contracted through the consumption of infected milk or meat.

The number of cases of TB per annum is not shown in Table 3.1 but, for example, in 1978 while the total number of deaths was 926 the total number of reported cases of TB was fourteen times higher at 13 101. Of these, 2015 cases were 'other forms' of which 74 cases died.

Table 3.1 *Romania – deaths due to tuberculosis*

Year	Respiratory system	Other forms*	Total: all forms
1961	5547	820	6367
1962	5775	719	6494
1963	4861	540	5401
1964	dna		
1965	4208	359	4567
1966	3930	309	4239
1967	4043	306	4349
1968	3895	322	4217
1969	3812	307	4119
1970	3424	317	3741
1971	3244	243	3487
1972	2698	229	2927
1973	dna		
1974	1501	122	1623
1975	1311	107	1418
1976	dna		
1977	dna		
1978	852	74	926
1979	dna		
1980	748	82	830
1981	767	69	836
1982	699	112	811
1983	790	59	849
1984	798	45	843
1985	895	52	947
1986	967	56	1023
1987	1025	61	1086
1988	1129	56	1185
1989	1227	58	1285
1990	1534	68	1602
1991	1628	67	1695
1992	1879	82	1761
1993	2239	73	2312
1994	2313	85	2398
1995	2495	65	2560
1996	dna		
1997	dna		
1998	dna		
1999	dna		
2000	dna		

Note: Romanian population 1961 approx. 18 567 000 based on census 21/2/1956.
Key: dna = data not available.
*'Other forms' are TB of the meninges and central nervous system, of the intestines and peritoneum and mesenteric glands, of bones and joints, and other systems.

Data collated from WHO Health Statistics Annuals 1960–1996. Published by WHO, Geneva 1963–1998 respectively.

In Romania, pre-1990, a system existed for the management of milk which originated from cows possibly infected with TB. This milk was kept separate from other milk and was clearly identified as 'red-label' and it was pasteurised and then used to make a traditional white pickled cheese known as telemea, which was believed to be an appropriate, safe product to make from the suspect milk, and for which there was an internal market. One method of production of telemea cheese is shown in Table 3.2.

In 2000 the large processing dairies in Romania required liquid milk and milk for cheese and yoghourt making. They needed to know the quality of the milk and to buy to a minimum standard with regard to the solids-not-fat (SNF), fat content and microbiological quality.

Prior to 1990 it had not been traditional to sell milk against these standards within the internal market when the transaction between buyer and seller was based on volume. In 2000 many small producers also wished to continue to sell by volume and were tempted to water the milk to achieve the volume of sales

Table 3.2 *The definition of telemea cheese and how it is made*

Definition	Telemea cheese – Romanian brine pickled cheese made in the Danube area. Made from sheep's, buffalo's or cow's milk. Mainly cubical in shape, or 10.5 cm square × 8–9 cm high, weight 1 kg
Rind	None, colour porcelain white but may be yellowish if made from cow's milk. Covered with seeds of the black cumin (*Nigella satura*)
Curd	White, or yellowish
Texture	Soft, almost spreadable
Flavour and aroma	Pleasant, slightly salty and acid when made from sheep's milk, piquant
Milk	Cow's milk (3.5% fat), sheep's milk (7.5% fat), or buffalo's milk (7% fat)
Heat treatment	Long hold, low temperature, 60–68 °C for 20–30 min.
Starter	0.5 l per 100 l milk (includes mesophilic streptococci and *Lactobacillus casei*). Temperature 28–31 °C in summer, 3 °C higher in winter
Additives	50 ml of 40% calcium chloride solution per 100 l milk, and 20 g sodium nitrate
Rennetting	25–30 ml of rennet extract per 100 l milk. Temperature 30 °C summer, 34 °C winter. Setting time 1 hour
Ladling	Coagulum ladled into cloths in layers. Cut the curd crosswise with a long knife. Take up the corners of the cloth together and press the curds for 2 or more hours with weight of 1.5 kg for each 1 kg of curd
Cutting	Cut the curd into blocks 10–11 cm square and 10 cm thick. Cool the cheese blocks in cold water
Salting	Immerse in salt bath (18–22%) salt at 15–16 °C for 15–16 h
Pickling	Layer the curds in casks, barrels, steel or plastic cans. Fill the container with acid whey (1.6–2.0% acid) with 4–5% salt
Ripening	Cheese ripe in one month. Store at 5–10 °C

Reproduced by permission from Scott, R., *Cheesemaking Practice*, 2nd Edition, Applied Science Publishers, London, 1981.

desired, but the bigger commercial purchasers wished to buy by quality and thus may not have been able to find sufficient small producers able to comply. Equally the most attractive way to make income from milk for the small farmers was to 'add value' and convert the milk surplus to their requirements into dairy products and to sell it in the farmers' markets.

The outcome of all this change over that decade was that the dairy industry as a whole had a number of problems, among which were the need to raise the standards of operation of the small producers, to help the industry to buy and sell on the basis of composition, to raise the hygienic quality of the milk, and to ensure that potentially TB-infected milk did not enter the market. When the milk industry as a whole was short of milk, when there was a limit to the size of the market for telemea cheese, and when the new much fragmented structure of the dairy industry had effectively produced shortages of specialists and veterinary officers capable of inspecting and testing the cows for TB and monitoring the situation, this was a difficult situation.

3.2   BACKGROUND INFORMATION ON THE EXPERIENCE AND MANAGEMENT OF TB IN MILKING COWS AND IN MILK IN THE UK: 1930 TO PRESENT DAY

3.2.1   Tuberculosis and the Tuberculosis Organism

Tuberculosis is an infection in man and animals caused by members of the *Mycobacterium tuberculosis* complex of organisms. Respiratory tuberculosis is most frequently caused by *Mycobacterium tuberculosis*. The bovine strain of the organism, *Mycobacterium bovis* (in some texts known as *Mycobacterium tuberculosis* var *bovis*), primarily enters the human body *via* the mouth in milk or meat and causes intestinal tuberculosis. In this condition the intestine can become blocked, thus sometimes requiring major abdominal operations, and death can occur through the gradual development of debility and malnutrition. The organisms also cause a form of meningitis and can affect the joints, the bones and the genito-urinary system. Generalised tuberculosis is observable in the human corpse, or in slaughtered animals because in its growth in the body nodules ('tubercles') are formed. These infected areas lose their normal function and become structurally different, and when cut through are often described as 'caseous, cheese-like'. These are looked for when meat is inspected in the slaughterhouse. The disease was once very widespread and a major cause of human death, but when antibiotic therapy became a developed science the once intractable disease was shown to be treatable. Sometimes the disease lies dormant in apparently healthy people only to become evident when the health of that person is compromised through malnutrition, illness, ageing or other form of debility.

The tuberculosis organism was first isolated in 1882 by Robert Koch. The organism is rod shaped and is difficult to stain. It is described in this context as 'acid fast', having large amounts of lipid in its cell wall, and requires special staining techniques to make it visible under a light microscope. These rods, approximately $0.2–0.6\,\mu$m wide $\times\ 1.0–10\,\mu$m long, are straight, slightly curved or sometimes branched, and can form filaments that are readily fragmented. They are aerobic organisms, non-motile and non-sporing. Infected animals can shed them in their faeces and thus they may occur in soil, in water, and in the environment in which infected animals, cows for example, are reared. In laboratory culture they grow very slowly.

The reservoirs of the mycobacteria which cause tuberculosis are animal hosts such as human beings, primates, cattle and other ruminants including pigs, as well as horses, cats, dogs and badgers. Although each strain primarily causes tuberculosis in its host there is some cross-infectivity between hosts.

Survival in milk and cheese

Mycobacterium tuberculosis is the name given to the organism which causes tuberculosis in human beings. *Mycobacterium bovis* primarily infects cattle, but other animals may act as its reservoir and it is also infective to man. It is transmitted through the consumption of infected milk and meat, and since it is resistant to acid could possibly remain viable in cheese made from raw milk. D'Aoust (1989) pointed out in a review of unpasteurised dairy products that many pathogens can survive in refrigerated cheeses well beyond 60 days (8–9 weeks) of storage dying out only slowly, and it has been noticed also that *M. bovis* is sometimes (Bachman and Spahr, 1995) detectable beyond 90 days. Data from Weinlich *et al.* (1967) indicate that *M. bovis* survives and is detectable in telemea cheese made from ewe's milk after storage for at least 40 days. The organism is sensitive to the presence of salt, and practices in Romania assume that it dies out in telemea cheese if the cheese is stored at least four months prior to consumption, although there is little in the literature to either support or contradict that view. It certainly has been shown to survive in butter stored at 4–6 °C for at least 91 days (Sinha, 1994). In comparison Abdalla *et al.* (1993) have shown that the survival of pathogens such as *Listeria monocytogenes*, *Staphylococcus aureus* and *Salmonella typhimurium* in white pickled cheese (4% salt added and stored in 4% brine) was favoured by production methods which did not use starter cultures. The time periods these organisms survived varied according to circumstances but in all cases were longer than 30 days.

Detection of animal and human TB infection

When the tuberculosis organism infects a host an immune response is initiated. The circulating antibodies can be detected in a skin test, 'the tuberculin test', in which an extract of the organisms – tuberculin – is introduced into the skin. The interpretation of the test is that where a skin reaction of equal to or greater than a certain size is observed the person or animal is deemed either to currently be infected with tuberculosis organisms, or once to have been exposed. The test does not differentiate between the two conditions. The observation of the reaction to tuberculin tests allows decisions to be made – for example treatment or immunisation can be offered to people, or where tuberculosis in cattle eradication programmes are being implemented reacting cattle can be culled. It is now recognised that the tuberculin test is not infallible and that some cattle, which are in fact infected, may be missed. Other testing programmes which are believed to be more reliable are being evaluated, as for example is occurring in Spain (Gonzalez Llamazares *et al.*, 1999).

3.2.2   Transmission of TB and Other Infections by Milk

In Britain, in the 1920s and 1930s enough public health information had been gathered to conclusively indicate that a substantial portion of the tuberculosis suffered by the populace, but particularly by children, was picked up through the consumption of infected raw milk. Recent knowledge underlines the infectivity of the TB organism and has shown that if the milk of only one infected cow is mixed with that of a hundred other cows from the same herd that level of infection is sufficient to cause TB in infants if the milk is consumed raw (Johnson *et al.*, 1990b). The concentration of organisms in the milk of infected cows could of course vary, but Sinha (1994) suggests that an infected udder might excrete tubercle bacilli at between 5×10^3 and 5×10^6 cfu ml^{-1}.

It has also long been clearly recognised that milk is a vector for other infections:

"Moderate estimates [in 1934] place the number of deaths in Great Britain attributable to infection with tuberculosis of bovine origin at over 2500 a year. This is by no means the only danger. If it were, one might look forward to a time when by Herculean efforts tuberculosis would be stamped out by drastic slaughtering of infected cattle and rigorous veterinary control. But there are other dangers from the milk yielded by the herds of selected and carefully nurtured cows that supply the special grades of TT (tuberculin tested) and certified milk in this country. Of the milks of this type examined in one laboratory no less than 73.5% were found to contain the organism of contagious abortion, which is regarded as the causative organism of undulant fever in man, and 63.5% carried organisms of the type known as 'haemolytic streptococcus', which cause serious epidemics of infective sore throat. If this is the potential danger in raw milks of the best types one may imagine what risks are run in drinking raw milk from farms where conditions are far from good.

A great deal has been done in recent years to improve the quality of milk as produced on the farm. Conditions are for the most part very different from what they were fifty or even twenty years ago. The necessity for washing the hands and the udder before milking, for sterilising vessels and protecting the milk from dust and dirt, is fairly well known, but no amount of care, important though it is, can give the consumer absolute confidence that there is no risk of a chance infection leading to illness and perhaps death. Milk borne infections are nothing like as common as they were in the days when all milk was heavily contaminated with dung and other filth but unfortunately they are by no means a thing of the past . . .

There is one simple measure that could be taken. It should be made a punishable offence to sell milk containing live tubercle bacilli, or any other organisms known to cause disease in man."

[from *An Englishman's Food*, by J.C. Drummond and A. Wilbraham, published by Jonathon Cape. Used by permission of The Random House Group Ltd.]

Possibly the immediate answer to prevent the transmission of TB through milk could have lain in the compulsory pasteurisation of milk – a process which can be defined as "any heat treatment of milk which secures the certain destruction of TB without markedly affecting the flavour or the cream line" (Davis, 1963).

The science of this process had been worked out by successive scientists, starting with Theobald Smith in 1898, who determined that at 60 °C the tuberculosis organisms was inactivated, and later in 1927 by North and Park who published figures for different time/temperature regimes which would render infected milk non-infective. Other scientists studied how the mild heating required to kill the TB organism affected the nutritional qualities of the milk. Broad consensus was reached that the organism could effectively be eliminated from milk while retaining optimal nutritional quality and not affecting the cream line which was at one time for the public an important indicator of quality. Agreement appears to have been reached world wide that the low temperature holding method (LTH) at 63.5 °C for 30 minutes, or the high temperature short time method (HTST) 71.7 °C for 15 seconds, can provide an acceptably low level of risk from the known range of pathogens affecting milk. Published D-values and z-values for *M. bovis*, although differing from each other (Table 3.3), indicate that such processes give a very considerable safety margin with respect to destruction of that organism and, of course, other organisms of equal or lesser heat resistance. Pasteurisation at 71.7 °C/15 seconds could give as much as an 8–10D process with respect to *Mycobacteria* although the impact of heating under either LTH or HTST regimes is dependent on the strain of organism present, its numbers and the fat content of the milk.

Of course if the milk originates from a herd of affected cows where the average excretion rate of the TB organism is high, for example above 10^5 cfu ml^{-1}, then it is possible that, even after pasteurisation, milk could contain living TB organisms at low level. This organism may be able to grow in milk held between 30–38 °C (Johnson *et al.*, 1990b), although Sinha (1994) states categorically that it does not.

However in the UK in the 1930s and 1940s pasteurisation of milk was a matter of massive controversy. Davis (1963) says, "Probably no subject outside religion and politics has been the cause of more prolonged and bitter controversies than the proposal to compulsorily pasteurise milk". The controversy started at the beginning of the century (Newman, 1901) and raged for decades (Savage, 1929) ranging from whether *M. bovis* was actually transmissible to human beings, to anxiety about loss of nutritional value and change in taste of milk, to the concerns of small farmers that the costs of the necessary equipment would put them out of business.

Faced with such controversy successive UK Governments adopted in legislation an incremental approach with the objectives of protecting the public from tuberculosis and from other milk borne infections transmitted through milk while progressively also raising the standards of hygiene in milk production.

Table 3.3 *D-values for pathogens in milk*

Pathogen	$D_{55}\,°C$ (min.)	$D_{60}\,°C$ (min.)	$D_{63.5}\,°C$ (min.)	$D_{65}\,°C$ (min.)	$D_{65.5}\,°C$ (min.)	(s)	$D_{95}\,°C$ (min.)	z °C	Ref.
Mycobacterium bovis			4						1
Mycobacterium tuberculosis		1.75			0.2–0.3	12–18		4.4–5.5	2
Mycobacterium tuberculosis bovis				0.28					3
Brucella species					0.1–0.2	6–12		4.4–5.5	2
Bacillus cereus							19.1		4
Campylobacter in skim milk	0.74–1.00								4
Salmonella other than Salmonella senftenberg 775W					0.02–0.25	1.2–15		4.4–5.5	2
Salmonella senftenberg 775W					0.8–1.0	48–60		4.4–6.6	2
Staphylococcus aureus					0.2–2.0	12–120		4.4–6.6	2
Streptococcus pyogenes					0.2–2.0	12–120		4.4–6.6	2

Sources:

1. Grant, Ball and Rowe. 1996.
2. ICMSF, 1980.
3. Johnson, Nelson and Johnson, 1990b.
4. ICMSF, 1996.

The Government sought through the Milk (Special Designation Order), 1922, to protect the public from tuberculosis infection by allowing 'choice', the so-called option of the purchase of premium milk grades, which were the most expensive and produced only from licensed dairies. This 'choice', however, only protected those affluent enough to buy it, and the epidemic of TB continued. By controlling the source of the milk – for example only allowing sales of raw milk from TT (tuberculin tested) cows or cows which appeared, through veterinary inspection, to be free from TB was another cautious step. Gradually the Government changed the designations of milk permitted while encouraging the sale of pasteurised milk, and they implemented agricultural programmes of regular tuberculin testing of the cows and culling of the positive reactors to ensure progressive upgrading in the numbers of attested herds.

At the same time they put pressure on the dairy industry to improve hygienic standards in milk production and, through legislation, achievement of certain minimum microbiological standards were required for dairies to be able to retain the licence to produce the superior grades of milk. But these requirements for microbiological standards, measurable through plate count techniques, were shown to be unworkable at the time and were rescinded in 1936. Instead dye reduction tests were introduced to give the possibility of more immediate answers to the microbiological quality of the milk as produced or as supplied to the processing dairies and payment was linked to microbiological quality. The dye reduction tests provided, in periods shorter than 4.5 hours, indication of microbiological quality, which meant that milk could be accepted for processing, or rejected, the same day.

Only after the Second World War, in the mid-1940s, were the processing parameters for the milk designation 'pasteurised milk' defined in law. By the 1960s it was possible to claim that the majority of herds of milking cows in England and Wales were tuberculosis free and also by that time the volume of pasteurised milk sold had increased to form by far the greater part of drinking milk sold. But at no time in England and Wales, even to the present day, has the pasteurisation of milk been compulsory.

Even with careful monitoring of the temperature of exposure, pasteurisation of milk was not a reliable process until a processor could be sure that all the milk reached the critical pasteurising temperature, and all of it was held for the right period of time. It is reported, for example, that 7–10% of 'pasteurised' milk sold in London in 1946 still contained live tubercle bacilli (Dormandy, 1999) showing that in spite of undergoing a heating regime the whole process was implemented in such a manner that a significant proportion of retail milk remained infective.

Batch heat processing, commonly used in the 1940s, did not really facilitate the needed fine control, whereas continuous flow methods, in which thin films of milk are heated using heat exchange plates and which were developing, did. Today, in the UK, most pasteurised milk is produced by processing through plate heat exchangers. Provided that the milk is pumped at a known rate and is raised to a known temperature it is possible to ensure that the milk is held at the required temperature for long enough to destroy the target organism, *M. bovis*. The invention of the flow diversion valve in the early 1950s meant, provided the

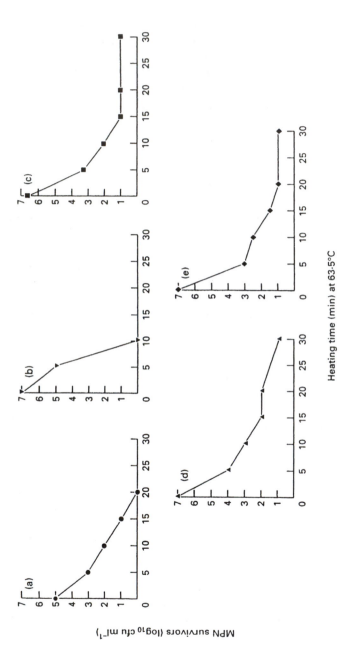

Figure 3.1 *Thermal death curves for several Mycobacterium spp. in milk at 63.5 °C: (a) M. bovis T/94/163C; (b) M.fortuitum NCTC 10394; (c) M. avium NCTC 8552; (d) M. intracellulare NCTC 10425; (e) M. kamsasii NCTC 10268 (Source: Grant, I.R. et al., 1996. Thermal inactivation of several Mycobacterium spp. in milk by pasteurisation. Letters in Applied Microbiology, **22**, 253–256. Reprinted by permisison of Blackwell Science Ltd.)*

valve was correctly set, that under-heated milk would be automatically diverted back to the raw milk side of the process for further heat treatment. But even this device can be manually over-ridden with ill-effect as was demonstrated when a large outbreak of salmonellosis (due to *Salmonella typhimurium* phage type 10) was traced back to that act and which resulted in the transfer of contaminated raw milk into vats of pasteurised cheese milk (D'Aoust, J.-Y., *et al.*, 1985).

Properly controlled pasteurisation, as defined in Table 3.4, destroys the more heat resistant pathogens *Mycobacterium bovis* and *Brucella abortus* as well as other significant though more heat sensitive organisms such as *Salmonella*, *Escherichia coli* and *Campylobacter* strains in the milk. Thus pasteurised milk, even if from TB infected cows, should not be capable of causing TB. Recent work (Grant *et al.*, 1996) has confirmed that the death kinetics of *M. bovis* during the low temperature holding method (LTH) of heating milk, namely 63.5 °C for 30 minutes, are linear and that the holder method of pasteurisation should render milk safe even if the initial population of *Mycobacterium bovis* in the milk were as high as 10^7 cfu ml^{-1}. This is a worst-case scenario and such milk would only exceptionally enter the food chain. But the same work also shows that the death kinetics of other strains of *Mycobacteria* – some also capable of causing tuberculosis tend to show a 'tail' when heated and indicate that if similar high numbers of those strains were present in milk the LTH method of pasteurisation could leave some survivors present (Figure 3.1).

Later work by the same authors (Grant *et al.*, 1999) shows heating at 72 °C for in excess of 25 s (more rigorous than the HTST method of pasteurisation) to be necessary for the inactivation of populations of 10^6 cfu ml^{-1} of *Mycobacterium paratuberculosis* – a strain potentially present in milk and which causes Johne's disease in cattle and may cause Crohn's disease in humans. Thus current legal regimes for milk may not eliminate some strains of mycobacteria in milk.

Furthermore, although it has been illegal in Scotland since 1983, in England and Wales in the year 2000 pasteurised milk originating from non-reactors in a herd that contained reactors could be sold as drinking milk, while the only raw milk which could be sold had to have originated from attested herds. Compliance with these requirements is monitored in England and Wales by government inspectors from the government Ministry (currently known as DEFRA – Department of the Environment, Food and Rural Affairs).

The phosphatase test – which allows verification of the pasteurisation process – became available by 1938, and by 1949 a sound reliable basis for the test in cows' milk was established (Aschaffenberg and Mullen, 1949) (Figure 3.2).

The test, a colour reaction, evaluates whether the enzyme alkaline phosphatase (whose pattern of thermal destruction is very similar to that of the tuberculosis organism) has been inactivated by the heat treatment the milk has received. If the milk passes the phosphatase test as now required in law (Anon, 1995) it is deemed to have been satisfactorily pasteurised. Historically the development of the flow diversion valve, together with the use of the phosphatase test meant that processors could reliably pasteurise milk, and check that they had done so, while the market was steadily growing as consumers were accepting it more widely.

The phosphatase test involves the incubation of milk with disodium p-nitrophenol phosphate under alkaline conditions. If active alkaline phosphatase enzyme is present a yellow coloration develops which is due to the formation of p-nitrophenol. The enzyme is destroyed by heating and the parameters of its destruction are very similar to the destruction of the organism *Mycobacterium tuberculosis* in milk. The degree of destruction of the enzyme is assessed by comparison of the tested milk with control milk against standard colours in a comparator disc. The test requires the incubation of the test and control milk for 2 hours at 37 °C, and is accepted as satisfactory – *i.e.* the milk is deemed satisfactorily pasteurised if 10 μg or less p-nitrophenol ml^{-1} is detected.

The test is not satisfactory for tainted or soured milk. It is also only appropriate to use on freshly pasteurised milk since in their growth some psychroptrophic Gram negative organisms, which could contaminate and grow in chilled pasteurised milk, can produce phosphatase enzymes.

The original test was described by Aschaffenberg, R. and Mullen, J.E.C. (1949). Full details can be found in Kirk, R.S and Sawyer, R. (1991).

Figure 3.2 *The phosphatase test*

3.2.3   Milk and the Incidence of TB in Cattle in the UK in the 1990s

Inspectors from DEFRA keep a close eye on the health of dairy cattle. From the 1930s when approximately 40% of cows were infected with *Mycobacterium bovis*, the numbers of cows infected have steadily dropped and rest in the year 2000 at below 0.01% of the national herd. Disease in people in the UK caused by *M. bovis* has dropped to less than 1% of all TB cases, and it is largely found in the elderly who may have contracted it many years ago before pasteurisation of milk was widespread (DEFRA website 1). However, there is ongoing concern that a small percentage of dairy cattle are infected with TB and that the badger population acts as a natural reservoir for the disease (Krebs, 1997; Hancox, 1998) and that the cycle of infection still remains.

Most milk sold in the UK is heat treated, and only a very small proportion of milk is consumed raw. Since 1990 it is been a legal requirement that raw milk sold to the public must bear a health warning on the packaging. This must clearly state that children, pregnant women, elderly people and those who are currently unwell or have chronic illness should not consume raw cows' milk. All herds producing milk either for raw consumption or for processing are tested annually for TB (Table 3.4).

Table 3.4 *Numbers of animals tuberculin tested, and the number of positive reac-*
tors slaughtered (England and Wales) 1995–1998

Year	Total cattle tested	Total herds tested	Cattle compulsorily slaughtered
1995	2 287 722	30 677	3451
1996	2 311 636	33 185	3881
1997	2 222 595	30 276	3760
1998	2 502 956	32 573	6083

Source: MAFF website No. 2. Crown copyright. Reproduced with permission.

3.3 EXERCISES AND STUDY OBJECTIVES

Having read Section 3.2 you should be able to appreciate the nature of what has been done in the England and Wales to control milk-borne TB. In Romania before1990 TB occurred to some extent in the milking herds, and one aspect of control was, for example, that milk from TB infected herds could be used, but only after pasteurisation with subsequent use in the salted telemea cheese process.

Exercise 1. Could obligatory pasteurisation of all milk in any country be described as an adequate strategy for the control of human infection from zoonoses, particularly tuberculosis, originating from milk?

Exercise 2. Why would the risk of the transmission of tuberculosis through milk products, particularly milk and telemea cheese, be of greater concern in a period of great change such as Romania has experienced in the decade post-1990 than in the period pre-1990? How does this provide a general lesson for the control of food safety?

Study objectives

Prepare to address the two problems set by re-reading the case study (Section 3.1) and reading the material provided in Section 3.2.

Determine whether the information supplied is sufficient to address **Exercise 1** or whether additional information needs to be found.

Address **Exercise 1** and prepare a summary of the key elements which point towards an answer.

Evaluate your analysis of **Exercise 1**, and after consideration produce a list of key elements to argue **Exercise 2**.

3.4 COMMENTARY

In thinking about how a zoonotic infection can be controlled, in this case *Mycobacterium bovis* which is spread through milk (and meat), it is useful to try to think of the whole circumstances in which the milk is produced.

Exercise 1. Could obligatory pasteurisation of all milk in any country be described as an adequate strategy for the control of human infection from zoonoses, particularly tuberculosis, originating from milk?

First ask yourself some questions like these:

- What is the micro-flora of raw milk likely to comprise? Which types of organisms, and at what cell concentrations?
- How are zoonotic organisms associated with cows (and other milking animals) transmitted to human beings?
- Since milk is the product in question is the oral route the only significant infection route?
- Should other routes of infection be considered in the control of the zoonoses (milking, touching animals, farm environment, air inside farm buildings housing animals, dust, insects, other animals infected from the primary sources . . .)?
- What is needed to control infection *via* these routes?
- Do I understand what pasteurisation is?
- What are the heating parameters universally recognised as minima for the pasteurisation heat treatment of milk? Are these equally appropriate to all types of milk?
- What are the limitations of these heat treatments with regard to the destruction of *Mycobacteria*?
- What about how pasteurisation is implemented – what ensures that treated milk and non-treated milk do not mix?
- Does pasteurisation render milk safe from all zoonotic infections?
- Is pasteurised milk suitable for making milk products such as cheese?
- Does cheesemaking control zoonoses?
- If obligatory pasteurisation of all milk were decided upon by a country what would be necessary to ensure it was implemented?

You may well think of other questions – areas of knowledge which you feel you need to explore in thinking about these problems. But after you have drawn up a list of questions some answers to them may be evident on re-reading the earlier parts of this case study – Sections 3.1 and 3.2. You may also need to go to other sources of information. But think carefully and in a structured way.

In answering **Exercise 1** read the text again – particularly Section 3.2 – and draw up for yourself a diagram such as that shown in Figure 3.3, which summarises how control of the transmission of TB through milk is implemented in England and Wales today.

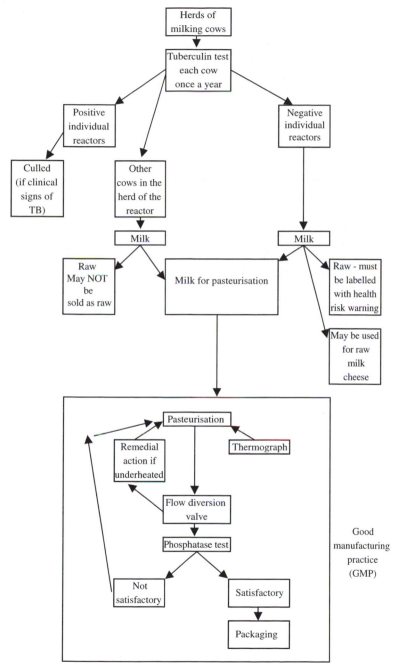

Figure 3.3 *Safeguards against transmission of TB organisms into milk in England and Wales*

Notice that the success of the pasteurisation process depends on monitoring the process (recording the processing temperature), diverting milk back to the raw side if the correct temperature is not reached, and verification that the phosphatase enzyme is destroyed. But confidence in those processes depends on them being implemented properly – the thermograph being calibrated, the flow diversion valve being maintained in correct working order, not being manually over-ridden for any reason, the phosphatase test being carried out correctly. Notice also that Figure 3.3 does not show the LTH method of pasteurisation – equally valid in law. Batch processing is more difficult to control so you will need to think about that. The milk has to be heated to the correct temperature and all of it held for 30 minutes at that temperature. None of it must be contaminated by un- or under-heat treated milk. Small capacity pasteurisers capable of the HTST method have been designed but are a capital investment which has to be afforded. The milk from either method of pasteurisation must subsequently be handled hygienically, filled into clean containers and chilled before distribution – all under a system of good manufacturing practice (GMP).

Also in answering **Exercise 1** you need to think about the types of target organisms and whether the heating process is actually adequately designed to destroy them too. Are the levels of target organisms in the unheated milk always low enough to be completely removed from the product? Additionally you need to be aware of new knowledge – such as indicated in Grant (1999) where attention is drawn to pathogenic *Mycobacteria* of greater heat resistance than *M. bovis*, and you need to consider whether the heating parameters for the target organism are in fact appropriate for all zoonotic organisms potentially present. Does, for example, pasteurisation destroy *Brucella abortus*, if present?

Finally, while obligatory pasteurisation of milk as a strategy for the control of human infection from zoonoses originating from it would, in theory, remove many zoonotic infections before use or consumption, in practice pasteurisation is not enough and needs to be supported both by the improvement in the quality of the raw milk and by protective strategies which prevent post pasteurisation contamination. Better raw milk quality could be achieved by the progressive eradication of the zoonotic organisms from the animals which would reduce direct transmission of infection from animals to man, together with improvement in farm and dairy hygiene. Processing poor microbiological quality milk has safety implications for dairying workers who may become infected from raw milk, for the processability of the milk, and challenges the ability of the time/temperature regimes defined for pasteurisation to render the treated milk safe.

Thus obligatory pasteurisation would be a very powerful step in the control of human infection from zoonoses originating in milk – but would need to be applied within the context indicated above.

Exercise 2. Why would the risk of the transmission of tuberculosis through milk products, particularly milk and telemea cheese be of greater concern in a period of great change such as Romania has experienced in the decade post-1990 than in the period pre-1990? How does this provide a general lesson for the control of food safety?

The types of questions which you may need to ask are listed below. Asking questions helps you to analyse a problem and indicates which facts need to be ascertained to develop answers. Remember: you are not trying to be an expert in either Romania or dairy farming but using the period of change Romania experienced as a model to illustrate and analyse the needs of a continuing microbiological control problem.

In respect of Romania pre-1990 ask yourself

- What systems of management were in place on the co-operative farms to manage and reduce the incidence of tuberculosis in dairy cows? Or, what would have been needed for that purpose where the main system of farming was in large units such as the co-operatives?
- Was there supportive legislation? What should the key elements of such legislation have contained?
- How would such legislation be enforced?
- What incentives were there or could there have been to encourage the farmers to comply?
- How was milk, suspected to contain *M. bovis*, managed?
- Why was this milk used at all?

And thinking about Romania post-1990 ask yourself

- What effects would the change from the system of farming on large co-operative units to numerous smaller farming units have, particularly when many of them were very small indeed?
- Would the systems in place pre-1990 for the management of tuberculosis in cows be equally operable in these new conditions?
- Where larger numbers of smaller farming units exist what would the effect be on the workload of the available experts – such as vets, dairying experts and so on?
- Would small farmers be able to provide the same quality of barns, milking equipment *etc.* as the large co-operatives? Would they be able to pay for the services of the available experts? Would they feel the experts' attentions necessary?
- Would changes in legislation be necessary to continue to manage the reduction of tuberculosis in dairy cattle?
- What about the processing of the milk – in what essential ways would the

production, collection and processing of milk differ post-1990 from pre-1990?
- Is the production of telemea an assured barrier against the transmission of tuberculosis from milk?
- Would there be constraints on the widespread production of telemea cheese?

Approaching an understanding

A useful first step could be to put together diagrams from the evidence supplied in this chapter which indicate milk production in Romania pre-1990 and post-1990.

Look at Figure 3.4 relating to the pre-1990 period and which is an example of how you might approach this. The diagram is simple and indicates the scheme to detect whether the cows were infected with TB – in other words a test such as the tuberculin test and/or veterinary inspection indicated the health status of the cows. Evidently milk from suspected TB positive cows was separated from other milk and clearly labelled 'red label'. Furthermore the evidence supplied in Section 3.1 tells you red label milk was pasteurised prior to making telemea cheese and further heat treated within the cheese-making process. So milk from tuberculous cows was diverted into making telemea cheese.

What happened to that cheese?

A number of questions arise relating to the control of its production:

- How much of it was made?
- Did the farming co-operatives produce it themselves or was it made by dairies, distant from the point of milk production?

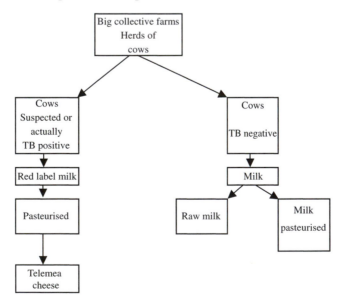

Figure 3.4 *Milk management system pre-1990 in Romania*

- Where was it sold and was there a surplus or shortage of it nationally?
- Was telemea also made from 'non red label milk' and/or from unpasteurised milk and were the three sorts differentiated in any way by labelling and/or by price?

If the cheese was made by the process shown in Table 3.2 would that process in your opinion have guaranteed safety from transmission of viable cells of *M. bovis*?

Now consider the situation post-1990 (Figure 3.5). Farming structure changed after 1990 and the few large farms were replaced by many small ones. The question asked within the Figure 3.5 ('can the milk be pasteurised?') relates to the many small farmers. Some would have been able, and others would not have been able to invest in milk processing equipment. All would have to comply with any state scheme regarding eradication of tuberculosis. Yet tuberculosis eradication depends on the ability to implement and enforce the scheme, on testing cattle regularly to see whether they are infected, on culling the animals which appear to be positive, compensating farmers for the culled animals. All that has to run smoothly. It is often easier to implement systems where organisations are large, and there are hierarchies of people performing specialised roles. Large organisations can afford to invest in equipment – they have the volume of product which justifies this. So the old, large co-operatives would have been able to milk large numbers of cows, employ a veterinary officer, manage a tuberculin testing scheme, have a processing dairy.

Post-1990 it is likely that much of that organisation became fragmented – small farmers would then have had to become capable of understanding all those roles, while the state had to enforce the law. The job of the veterinary officers to test and track TB infected cows became harder because of the fragmented nature of the farming and dairying industry, with so many small farming units having only few (*e.g.* 1–5) animals. Another control problem may have been that the peasant farmers took to drinking their own raw milk. They may not have been able to pasteurise it, not having the equipment. In order to pasteurise it they may have had to invest in pasteurisation equipment or transport it to a distant processing dairy. Equally, now outside the large management structure of a huge co-operative farm they may not even have had the knowledge about the risks associated with raw milk. Another problem of control is that they may have tried to sell their milk to the processing dairy who would only accept it if it was up to defined quality standards. If a farmer could not sell to the large processors, is a small farmer going to throw his milk away? He may strive to improve his quality and safety, or may choose not to sell to that dairy. Equally he may choose instead to make products, maybe telemea cheese, which he hopes to sell independently in the open market. Is the telemea process adequate to protect against the tuberculosis organism if present? But what if everyone is making telemea cheese on the farm? Could farmers turn to other products – like non-pickled white cheeses, yoghourt, cream, or cream cheese, and what would be the fate of *M. bovis* if present? If this were so what could the implications for the health of the people as a whole be?

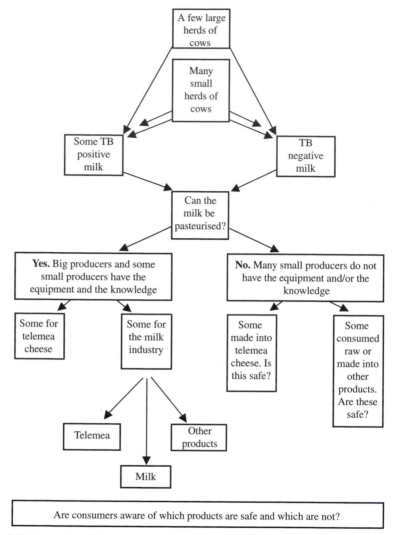

Figure 3.5 *Romania: milk management problems post-1990*

Pre-1990 milk from tuberculous cows was subject to the double safeguards of pasteurisation followed by the making of the telemea cheese, and the maturation of the cheese in brine. But in the decade post-1990 because of the new structure of farming and dairying it could be envisaged that some telemea cheese was made without the double heating regimes and salt concentration safeguards in place. The safety of the telemea cheese depends on the microbiological quality of the milk, on the actual temperature regime to which the milk is subjected, on the hygiene of production, and, if TB organisms are present, whether the telemea cheese is made and handled in such a way that they die out. So the question is whether the salt and acid systems in the cheese are restrictive enough to be effective for that purpose. What effect do the storage temperature and duration have on *M. bovis*? Is every small farmer producing the cheese aware of these requirements?

It is also worth thinking about why in pre-1990 Romania the use of milk from tuberculous cows was permitted only if it was pasteurised separately from other milk, and made into telemea cheese – which involves heating, and storage in brine (up to 20% w/v). In contrast, in England and Wales when a herd is known to contain reactors the law permits milk from non-reactors to be used for drinking milk, provided it is pasteurised and provided that the white cell count is no greater than a certain threshold. Do you think both systems, in Romania, and in England and Wales, are equally protective?

Thus in answering **Exercise 2** you should consider the advantages and disadvantages of the big co-operatives, selling to the country's own fixed internal market. But you should also consider the requirements for the management of the safety of milk under the new farming system. While it is of course possible to implement good control in a small farm selling its milk to a processing dairy, the overall problem is that there were thousands of small concerns all of which needed to be run well. Britain still has many small farms as well as large estates and it took Britain between 50 and 60 years from first identifying the problem of bovine TB (at the turn of the 1900s) to the 1960s to be able to claim that most cows were free from TB. Of course it need not take that long for it took the USA only 10 years to achieve a similar outcome. The key is organisation and willpower as well as the ability to enforce whatever law is put in place, and a willingness to make it worthwhile for farmers to ensure their animals are tested, and to compensate them well financially when it is necessary that an animal is culled. Legislation and its enforcement may also need to be strengthened to ensure the safety and hygiene of farm products.

From the information supplied it is not possible to rank the risk pre- and post-1990 but you may be able to form an opinion and you will see that the fragmentation of dairying imposed new requirements. When you think about the two situations you will fill in details both from facts you ascertain and from anticipating possible scenarios. In both these conditions you should then be able to anticipate what causes risk to increase or decrease through better or weaker (inappropriate) controls. However, under circumstances of change the old systems (good or bad) are replaced by new ones. Nothing is certain. Until it is shown to be otherwise it is always a good policy to assume that change affects control systems and brings with it risk of new problems.

3.5 SUMMARY

Pasteurisation of milk is a heat treatment which, if properly implemented, reduces the microbial population in the milk, and destroys the majority of the contained heat sensitive pathogens.

- Raw milk is a primary source of milk spoilage organisms, pathogens, and lactic fermentative organisms;
- It can be a source of zoonotic organisms *e.g. M. bovis*, *B. abortus* and others;

- Good farming and dairy practice reduce the numbers of organisms contaminating or growing in the milk;
- In pasteurisation the extent of reduction in the numbers of each species present in the milk depends on the D-values of those organisms in that milk type at the temperature used;
- Change in the health of the cows, or in farming practice, in milk quality or in processing practice, if not managed consciously and properly, could lead to loss of control in a food safety system.

Telemea cheese – control of *M. bovis* depends on its absence from milk or the destruction of low numbers in the milk by pasteurisation; an added safeguard is a salt concentration in the cheese brine high enough to prevent the growth of *M. bovis*, and to cause slow death of any viable cells over an adequate period of storage time.

- Unobserved change which allows slippage outside the controlling values of these parameters may lead to increase in risk of transmission of zoonotic organisms such as *M. bovis* through the cheese.

Control of a zoonotic disease requires application of knowledge of the disease organism in an integrated approach combining farming practice and food processing practice with economic policy, financial resources, subsidies and compensation payments. Governments have to provide a lead by enacting appropriate legislation which aims to raise standards at a pace which can be implemented and enforced. The organisational structure of the industry has resource implications for financing the numbers of inspectors, vets and other experts. Scientific knowledge has to be up to date to ensure processing parameters are targeted and adequate. Technical provision has to be appropriate too so that equipment functions properly and processed product can be handled hygienically. Staff have to be trained to implement what is required under good manufacturing practice.

Control of milk-borne tuberculosis in Romania before 1990 depended on a nationwide scheme implemented through the large co-operative farms. Political change post-1990 completely altered that system. This massive structural change meant that the whole system of control had to be re-assessed, and the weaknesses in control rectified. This illustrates on a large scale what happens in any food control system. Change of any sort – in processing method, in volumes, in packaging, in target customer, in shelf life and so on may undermine the effectiveness of a once secure system causing it to become insecure, and risk from microbiological hazards to increase.

Control of milk-borne TB in England and Wales. The safety of milk depends on several factors:

- The reduction of TB in the herds through obligatory slaughter of infected cows;
- Awareness and monitoring of potential sources of re-infection of the animals from vermin, wild life or other sources;

- Testing the cows for their exposure to tuberculosis by giving the cows the tuberculin test. This test detects 'reactors' – that is cows which have antibody to TB in their blood – and which can mean they are actively suffering from TB even if they show no clinical symptoms;
- Prevention of the sale of raw milk from herds found to be, or suspected to be, positive for tuberculosis, *i.e.* 'reactors';
- Processing from milking stage onwards should be managed on the principles of good manufacturing practice;
- Pasteurisation of the milk by heating regimes defined in law for that milk designation. In England and Wales the permitted regimes are: 71.7 °C/15 seconds or equivalent processes (such as 62.8 °C/30 minutes);
- Incorporation of a flow diversion valve linked to a continuing recording thermometer (a thermograph) at the end of the holding section of the pasteuriser to divert any underheated milk back to the raw side;
- Verification of pasteurisation by the use of the phosphatase test to confirm satisfactory pasteurisation;
- Process management to ensure that cross contamination of pasteurised milk by raw milk does not occur;
- Regular inspection and enforcement of the law at the farm and the processing premises;
- Distribution under temperature controlled and hygienic conditions.

Change in any of the processing parameters, plant layout, or product protection systems can lead to loss of control with the outcome of increased risk of zoonotic disease transmission to the consumer, as well as loss of product quality.

3.6   REFERENCES

Abdalla, O.M., Davidson, P.M. and Christen, G.L., 1993. Survival of selected pathogenic bacteria in white pickled cheese made with lactic acid bacteria or antimicrobials. *Journal of Food Protection*, **56** (11), 972–976.

Anon, 1995. The Dairy Products (Hygiene) Regulations 1995. Statutory Instrument 1995, 1086, HMSO, London.

Aschaffenberg, R. and Mullen, J.E.C., 1949. A rapid and simple phosphatase test for milk. *Journal of Dairy Research*, **16**, 58–67.

Bachmann, H.P. and Spahr, U., 1995. The fate of potentially pathogenic bacteria in Swiss hard and semi-hard cheese made from raw milk. *Journal of Dairy Science*, **78**, 476–483.

Bryan, F.L., 1983. Epidemiology of milk-borne diseases. *Journal of Food Protection*, **46**, 637–649.

D'Aoust, J.-Y., Warburton, D.W. and Sewell, A.M., 1985. *Salmonella typhimurium* PT10 from cheddar cheese implicated in a major Canadian food borne outbreak. *Journal of Food Protection*, **48**, 1062–1066.

D'Aoust, J.-Y., 1989. Manufacture of dairy products from unpasteurised milk: a safety assessment. *Journal of Food Protection*, **52**, 906–914.

Davis, J.G., 1963. *Dictionary of Dairying*, 2nd Edition, p. 787. Leonard Hill, London.

Davis, J.G., 1975. *Cheese*. Volume IV. *Annotated Bibliography with Subject Index*. Churchill Livingstone, Edinburgh, London and New York.

Dormandy, T.L., 1999. *The White Death. A History of Tuberculosis*. The Hambledon

Press, London and Rio Grande.

Drummond, J.C. and Wilbraham, A., 1939. *The Englishman's Food*, pp. 554–555. Jonathan Cape, London.

Gonzalez Llamazares, O.R., Gutierrez Martin, C.B., Aranez Martin, A., Liebana Criado, E., Dominguez Rodriguez, L. and Rodriguez Ferri, E.F., 1999. Comparison of different methods of diagnosis of bovine tuberculosis from tuberculin- or interferon-reacting cattle in Spain. *Journal of Applied Bacteriology*, **87**, 465–471.

Grant, I.R., Ball, H.J. and Rowe, M.T., 1996. Thermal inactivation of several *Mycobacterium* spp. in milk by pasteurization. *Letters in Applied Microbiology*, **22**, 253–256.

Grant., I.R., Ball, H.J. and Rowe, M.T., 1999. Effect of higher pasteurization temperatures, and longer holding times at 72°C on the inactivation of *Mycobacterium paratuberculosis* in milk. *Letters in Applied Microbiology*, **28**, 461–465.

Hancox, M., 1998. Of Bourne, badgers and bovine TB. *Letters in Applied Microbiology*, **27**, 383.

Harrington, R. and Karlson, A.G., 1965. Destruction of various kinds of mycobacteria in milk by pasteurization. *Applied Microbiology*, **13**, 494–495.

ICMSF, 1980. *Microbial Ecology of Foods*, Volume 1. *Factors Affecting Life and Death of Micro-organisms*. Academic Press, London.

ICMSF, 1996. *Micro-organisms in Foods*, Volume 5. *Characteristics of Microbial Pathogens*. Blackie Academic and Professional, London.

Johnson, E.A., Nelson, J.H. and Johnson, M. 1990a. Microbiological safety of cheese made from heat-treated milk, Part 1. Executive summary, introduction and history. *Journal of Food Protection*, **53** (5), 441–452.

Johnson, E.A., Nelson, J.H. and Johnson, M., 1990b. Microbiological safety of cheese made from heat-treated milk, Part II. Microbiology. *Journal of Food Protection*, **53** (5), 519–540.

Johnson, E.A., Nelson, J.H. and Johnson, M., 1990c. Microbiological safety of cheese made from heat-treated milk, Part III. Technology, discussion, recommendations, bibliography. *Journal of Food Protection*, **53** (7), 610–623.

Kirk, R.S. and Sawyer, R., 1991. *Pearson's Composition and Analysis of Foods*. 9th Edition. Reprinted 1997. Longmans, London.

Krebs, J. 1997. Report: *Bovine Tuberculosis in Cattle and Badgers*, 1997, HMSO, London.

MAFF* website no. 1, 1999. maff.gov.uk/animal/tb/public/sheeta2.htm – dated 7 April 1999.

MAFF* website no. 2, 1999 maff.gov.uk/animalh/tb/statsi.htm (as at 24 October 1999.

Newman, G., 1901. Tuberculous milk and meat, and preventive measures against consumption. Reprinted 1994, in *British Food Journal*, **96** (5/6), 10–14.

North, W.E and Park, W.H., 1927. Standards for milk pasteurization. *American Journal of Hygiene*, **7**, 147–173.

Northolt, M.D., 1984. Growth and inactivation of pathogenic micro-organisms during manufacture and storage of fermented dairy products. A review. *Netherlands Milk and Dairy Journal*, **38**, 135–150.

Savage, W.G., 1929. The prevention of human tuberculosis of bovine origin. Macmillan, London.

Scott, R., 1981. *Cheesemaking Practice*. Applied Science Publishers, London.

Sharpe, J.C.M., 1987. Infections associated with milk and dairy products in Europe and North America 1980–86. *Bulletin of the World Health Organisation*, **65**, 397–406.

*MAFF = Ministry of Agriculture, Fisheries and Food [in the UK] is now renamed Department of Environment, Food and Rural Affairs (DEFRA).

Sinha, R.N., 1994. *Mycobacterium bovis*. In *The Significance of Pathogenic Micro-organisms in Raw Milk*, pp. 141–166. International Dairy Federation, Brussels.

Smith, T., 1898. *Journal of Experimental Medicine*, **3**, 451.

Turnock, D., 1997. *The East European Economy in Context. Communism and Transition*, pp. 263, 290–291. Routledge, London and New York.

Weinlich, M., Deac, C. and Draganescu, V., 1967. 'Viability of *Mycobacterium bovis* in telemea cheese from ewe's milk'. *Igiena*, **16** (3), 163.

WHO Health Statistics Annuals 1960–1996. WHO, Geneva, 1963–1998 respectively.

Acknowledgement

The paper by Weinlich *et al.*, 1967 was translated from the Romanian for me by Dr Carmen Tudorica, University of Plymouth.

CHAPTER 4

Should Pasteurisation of Drinking Milk Be Obligatory?

Key issues
- Drinking milk
- *Campylobacter jejuni*
- Developing strategies for control
- The need for a holistic approach to control

Challenge

Three outbreaks of *Campylobacter jejuni* infection contracted through milk are presented to you. While some of the issues raised in Chapter 3 are also relevant here the main challenge to you is to consider the role pasteurisation has in managing drinking milk safety and whether that role is unique.

4.1 THE CASE STUDIES: *CAMPYLOBACTER JEJUNI* OUTBREAKS ASSOCIATED WITH RAW AND PASTEURISED MILKS

4.1.1 The Jogging Rally

Stalder *et al.* (1983) reported an outbreak of *Campylobacter* enteritis involving over 500 participants in a jogging rally, in which 800 runners took part in Switzerland in the early 1980s. Of these 800 people over 500 became ill 3 to 4 days afterwards with enteritis illness. An immediate investigation was launched which involved interviewing both the ill and the well and recording through the questionnaire what food they had eaten prior to the event. This identified that 659 participants had consumed a raw milk drink, and of these 510 were ill giving an attack rate of 77.4%. The questionnaire showed that the mean duration of the illness was 4.4 days, symptoms commonly being diarrhoea, abdominal cramps, fever, headache and, sometimes, vomiting. Stool cultures from 22 participants were tested and *C. jejuni* was the only organism isolated. 14 isolates were serotyped and of these 13 were identified as type 2, and one remained untypable. Investigations into the source of the milk showed that it originated from three farms. Of the 44 cows tested seven carried *C. jejuni*, and of these one carried the incriminated *C. jejuni* type 2, a strain which was rare in that region of Switzerland. It seemed that it was due to this one cow whose milk was presumed to be faecally contaminated that so many people had suffered illness.

4.1.2 The Boys' School

An outbreak of milk-borne enteritis in a residential school in the UK was described by Wilson *et al.*, 1983, and contained the following key information:

In the south of England, one day in March 1982, a number of boys attending a boarding school reported illness with symptoms of diarrhoea, abdominal pain or vomiting. The outbreak affected a total of 189 children, with 29 reporting on the first day, 67 on the following day which proved to the peak of the outbreak, and others becoming ill over the rest of the week.

There were 20 staff and 782 boys in the school, all of whom were asked to complete a questionnaire. 99% (775) of the boys completed it, and it showed that 518 boys had one or more symptoms, 57% suffering diarrhoea, 30% pain without diarrhoea and 13% symptoms other than pain or diarrhoea. The incidence of diarrhoea correlated with drinking milk ($p < 0.05$), although it also demonstrated that the older boys who drank greater quantities were less likely to suffer symptoms that the younger ones.

Forty faecal samples were subjected to microbiological examination and *C. jejuni* was isolated from 88% of them, and found to be of the same serotype [Lior scheme 7; modified Penner scheme (type 13)]. The milk the boys had consumed had come from the school's farm, and it had been pasteurised. But investigation showed that, while the equipment was functioning properly, some milk was not pasteurised because it was sold raw to a large milk processor. It seemed as if somehow raw milk may have been provided to the children because *C. jejuni* of the same strain was also isolated from the rectal swabs and milk of one cow.

4.1.3 The Day Nursery

Summary
A point source outbreak of *Campylobacter jejuni* affected 11 children in a day nursery. Milk consumed by the children was known to have been pecked by magpies on occasions. Illness was significantly associated with the consumption of milk on a single morning. Examination of milk from a bottle pecked after the outbreak yielded campylobacters. The level of contamination was approximately six cells of *C. jejuni* per 500 ml of milk.

The following extract, from Riordan, T., Humphrey, T.J. and Fowles, A., 1993. A point source of campylobacter infection related to bird pecked milk. *Epidemiology and Infection*, **110**, 261–265, is quoted with permission of Cambridge University Press.

"At the end of May 1991, a telephone report was received of an outbreak of diarrhoeal illness at a day nursery. The initial report was of 15 children being affected over a period of 4 weeks. Stool samples had been submitted from two of the more severely affected children and had yielded *Campylobacter* sp. An investigation was instigated.

The nursery had a total of 69 children on the register although the maximum attendance per session was 42. The children were divided into four units by age group. Each unit functioned independently having its own main room, toilets and washing facilities. Paper hand towels were provided. A rigorous environmental cleaning programme was in operation for all areas including the toilets.

The following meals were served each day: mid-morning milk and fruit, cooked lunch, afternoon milk with sandwiches.

Each day 20 pints of pasteurised milk in glass, one-pint bottles with silver caps were delivered before 8.00 a.m. when staff arrived. The nursery was surrounded by mature trees and magpies were known to be prevalent in the area. A system of covering the bottles to prevent pecking was in operation but apparently the milkman sometimes forgot to use it. Questioning of the cook revealed that, on occasions, pecked milk was given to children to drink without further treatment. Milk was distributed to the four units in jugs, the allocation being about 1.5 to 2 pints per unit for the older age groups' . . .

Not all children attending the nursery drank bottled cows' milk. A number of babies drank expressed breast milk, other children drank soya milk or skimmed milk. [In those children who were ill] illness was associated with consumption of bottled cows' milk for those who were present on the morning of 17 May 1991 (putative time of exposure) . . .

Over the three week period when the outbreak was investigated three bottles of milk not for consumption were deliberately left uncovered each day. The milk in two bottles where the tops were pecked was examined for the presence and numbers of campylobacters. *C. jejuni*, at the level of six cells per 500 ml of milk, was isolated from one of the two pecked milk bottles examined.

[The outbreak] highlights the need for public education since simple measures to prevent consumption of pecked milk would eliminate this source [bird pecking] infection."
[end of quotation from Riordan *et al.*, 1993]

4.2 BACKGROUND

4.2.1 *Campylobacter jejuni* and the thermophilic campylobacters

These Gram negative, microaerophilic, curved rods have come to be recognised as a group of organisms which occur very widely, and which may be isolated from water and water supplies, from environmental sources and from food animals and poultry. Their taxonomy, which has continued to change from the 1960s when their systematic study began, has been shown to be complex, and to include a number of groups currently including *Campylobacter*, *Arcobacter*,

Helicobacter and others (On, 2001). Among the *Campylobacter* are some which are catalase positive and of those the thermophilic strains – ones in which the optimum growth temperature for the group is close to 42 °C and which cause acute enteritis in otherwise healthy adults and children – are the strains of main concern in considering food safety.

This latter group of organisms, which includes *C. jejuni* and *C. coli*, normally occur in the gastrointestinal tract of many types of animals and particularly in both poultry (Corry, 2001) and wild birds (Hudson *et al.*, 1990, 1991), but can sometimes cause mastitis in milking cows and be shed in the milk (Orr *et al.*, 1995).

Human beings who

- work in close proximity with live animals and birds, for example on farms,
- work in slaughter houses,
- handle raw meats in processing plants and butchers shops,
- work in catering and handle meats
- consume foods, particularly red meats, poultry meats, eggs and milk, which are infected with the organisms

may become affected by campylobacters and suffer acute gastroenteritis.

Campylobacters are the most frequent cause of infectious gastro-intestinal disease in the England and Wales and predominate over *Salmonella* (Table 4.1)

Table 4.1 *Campylobacter and Salmonella infections in England and Wales, 1986–2000*

Year	*Campylobacter sp. Laboratory reports, faecal isolates*	*Salmonella in humans. Faecal and unknown reports excluding S. typhi and S. paratyphi (faecal and unknown reports)*
1986	24 809	16 976
1987	27 310	20 532
1988	28 761	27 478
1989	32 526	29 998
1990	34 552	30 112
1991	32 636	27 693
1992	38 552	31 355
1993	39 422	30 650
1994	44 414	30 411
1995	43 876	29 314
1996	43 337	28 983
1997	50 177	32 596
1998	58 059	23 728
1999	54 987	17 532
2000[a]	53 858[a]	14 844[a]

[a]Provisional data.

Source: PHLS website at www.phls.co.uk (updated to 16 March 2001). Reproduced with permission of the PHLS Communicable Disease Surveillance centre. © PHLS.

and this fact has been associated with the widespread consumption of fresh and
frozen poultry meat, although water and other foods have also been implicated
(see Table 4.2).

Table 4.2 *Food vehicles, settings and nature of evidence associated with Campylobacter outbreaks in England and Wales, 1992–1994[a,b]*

Vehicle	Setting	Evidence
Water	College	Microbiological/cohort
Water	College	Microbiological
Water	Community	Microbiological
Water	Adventure camp	Cohort
Water	Function	Descriptive
Water	College	Descriptive
Unpasteurised milk	Music festival	Case control
Unpasteurised milk	Music festival	Descriptive
Unpasteurised milk	Farm visit	Cohort
Pasteurised milk	Community	Case control
Poultry	Function	Cohort
Poultry	Restaurant	Descriptive
Shellfish/poultry	College function	Descriptive
Pate vol-au-vent	Restaurant	Cohort
Meat products	Restaurant	Cohort

[a] Reported to Communicable Disease Surveillance Centre (CDSC), Central Public
Health Laboratory, Colindale, London.
[b] Excluding five outbreaks in which the vehicle of infection was not identified.
Adapted with permission from Frost, J.A., 2001. Current epidemiological issues in
human campylobacteriosis. *Journal of Applied Microbiology*, **90**, 85S–90S. Blackwell
Science Ltd.

Robinson (1981) estimated the infective dose in 180 ml of milk consumed to be
as low as 500 cells, yet the case studies instanced in Section 4.1 indicate that it
may be lower than that (see Table 5.7).

Campylobacter organisms are considered not to grow in foods but they do
survive in cooled and frozen foods and waters, while being sensitive to drying
conditions. Laboratory isolation has to take account of their microaerophilic
and thermophilic characteristics, as well as the possibility that the cells may be
injured and need resuscitation and, finally, must inhibit other populations of
cells present in the samples (On, 1996). The combination of their survivability
and the low infective dose contributes to the reasons for the high numbers of
infections associated with them. Yet the organism is very heat sensitive and is
readily destroyed by mild heating processes, such as would be experienced in the
early stages of cooking foods. Thus cooked foods handled hygienically should
not be sources of campylobacter infection (Table 4.3).

Table 4.3 *Campylobacters, thermophilic: characteristics, growth ranges and D-values*

Growth ranges:

	Temperature °C	pH	Water activity (a_w)	NaCl (%)	Atmosphere
Range	32–45	4.9 to ca. 9.0	>0.987	0<1.5	
Optimum	42–43	6.5–7.5	0.997	0.5	5% oxygen and 10% carbon dioxide

D-values (examples of):

Substrate	Temperature °C	D (minutes)
0.1 M phosphate	50	0.88–1.63
Skim milk	50	1.3–4.5
Skim milk	55	0.74–1.00
Cooked chicken	55	2.12–2.25
Cooked chicken	57	0.79–0.98

Source: ICMSF, 1996.

4.2.2 Milk and Outbreaks of Infectious Intestinal Disease

In England and Wales outbreaks of food poisoning associated with the consumption of milk and dairy products regularly occur as exemplified by the twenty outbreaks reported in the years 1992 to 1996 (Djuretic *et al.*, 1997) (Table 4.4). Djuretic *et al.* (1997) also state that "the number of general outbreaks of food poisoning in which milk and dairy products have been implicated as a source of infection has fallen over the last few decades". Observation of Table 4.4 indicates that cheese, milk and ice cream were implicated as vehicles, and that there were a variety of causative organisms. The following can also be seen:

	Unpasteurised milk	Pasteurised milk
General outbreaks	10	6
People affected	218	265

(From Table 4.4)

4.3 EXERCISE

Exercise. After considering the three outbreaks of *Campylobacter jejuni* infection described in Sections 4.1.1, 4.1.2, and 4.1.3 evaluate why these incidents arose, and form an opinion on whether pasteurisation of drinking milk should be obligatory to protect public health.

Cross refer to Case Study 3.

Table 4.4 *General outbreaks of infectious intestinal disease associated with milk or dairy products in England and Wales 1992–1996*

Year/outbreak number	Month	Pathogen	Setting	Number of people affected	Suspected vehicle	Evidence
1992						
1	June	Campylobacter	Outdoor festival	72	Unpasteurised milk	Statistical
2	June	Salmonella livingstone	Psychogeriatric hospital	10	Cheese	Descriptive
3	August	Campylobacter	Doorstep delivered farm milk	110	Pasteurised milk*	Statistical
4	September	S. enteritidis PT4	Hotel	25	Icecream	Statistical
5	November	S. enteritidis PT4	School	44	Pasteurised milk*	Statistical
1993						
6	February	S. enteritidis PT4	Hotel	7	Icecream	Statistical
7	May	E. coli O157	Doorstep delivered farm milk	7	Unpasteurised milk	Microbiological
8	May	S. typhimurium DT103	Doorstep delivered farm milk	13	Unpasteurised milk	Microbiological and statistical
9	June	Campylobacter	Outdoor festival	22	Unpasteurised milk	Descriptive
1994						
10	January	S. typhimurium DT12	Doorstep delivered farm milk	11	Unpasteurised milk	Microbiological
11	March	Campylobacter	Farm visit	23	Unpasteurised milk	Statistical
12	June	S. typhimurium DT104	Doorstep delivered farm milk	4	Unpasteurised milk	Microbiological and statistical
13	July	S. typhimurium DT104	Doorstep delivered farm milk	26	Pasteurised milk**	Microbiological

1995						
14	May	Campylobacter	RAF base	35	Unpasteurised milk	Statistical
15	September	Cryptosporidium parvum	School	67	Pasteurised milk**	Statistical
16	November	S. typhimurium DT104	Nursery school	26	Unpasteurised milk	Microbiological
1996						
17	January	S. typhimurium DT104	Doorstep delivered farm milk	5	Unpasteurised milk	Microbiological
18	April	E. coli O157	Doorstep delivered farm milk	6	Pasteurised milk**	Microbiological
19	December	E. coli O157	Community	12	Pasteurised milk**	Descriptive
20	November/ December	S. goldcoast	Farm produced cheese	75	Cheese	Microbiological and statistical

*Post-pasteurisation contamination.
**Failure in pasteurisation.

Source: Djuretic, T., Wall, P.G. and Nicholls, G., 1997. General outbreaks of infectious intestinal disease associated with milk and dairy products in England and Wales: 1992 to 1996. *Communicable Diseases Report*, 7, Review No. 3, R41–R45. Reproduced with permission of the PHLS Communicable Disease Surveillance Centre. © PHLS.

4.4   COMMENTARY

> **Exercise.** After considering the three outbreaks of *Campylobacter jejuni*
> infection described in Sections 4.1.1, 4.1.2 and 4.1.3 evaluate why these
> incidents arose and form an opinion on whether the pasteurisation of
> drinking milk should be obligatory to protect public health.

4.4.1   The Three *Campylobacter jejuni* Outbreaks

Campylobacter jejuni is an organism which can be associated with cows, and
therefore with milk. But it is a very heat sensitive organism (see Table 4.3) and
even populations in excess of 10^6 ml^{-1} would be destroyed by the HTST process.
Thus pasteurisation is a means of destroying this zoonotic organism (Anon,
2000). But its infective dose is low, and so cross-contamination from a source of
the organism to foods, including pasteurised milk, which will be subject to no
further heat processing is a problem that milk processors and other food businesses need be aware of.

Case study 4.1.1 was associated with the consumption of infected raw milk –
the organisms originating from the milk of an infected cow. Campylobacter
infections in animals can often occur without signs of disease in the animal. You
need to consider why raw milk was considered safe to drink, and, if it was
habitual practice in Switzerland at the time, to wonder whether this outbreak
had characteristics which made it remarkable and was therefore published as an
article. Were campylobacter outbreaks rare? Were they commonplace but normally small, so was this an unusually large outbreak? Consider how the organisms are detected and when methodology for detection was developed and how
this would affect the reporting of campylobacter outbreaks.

Case study 4.1.2 was associated with pasteurised milk in which *C. jejuni* would
have been absent, destroyed by the heat treatment – but in this outbreak the
design of the processing plant was poor for it was possible to bypass the
pasteuriser. Additionally it also seems that the management of the process was
poor too, because the pipe layout and flow directions should have been known to
the management and such poor design been rectified. Furthermore the training
of the staff running the plant together with appropriate management systems
should have ensured that raw milk was pasteurised, and not confused with that
other raw milk which was supplied to the external dairy, and that only pasteurised milk was supplied to the school. Probably raw milk was supplied to the
school but how did such a muddle occur?

Case study 4.1.3 demonstrates that new knowledge needs to be incorporated
into current ways of managing food safety risk. Here the organisms entered
pasteurised milk well after the pasteurisation stage – indeed just before consumption. Pasteurisation had removed the heat sensitive pathogens from the raw milk.
But for the milk to be safe to consume it needed to be protected from contamina-

tion post-pasteurisation, through filling into sterile containers, and through the use of well designed packaging robust enough to resist damage and leakage of contamination into the container. But here birds were attracted by the shiny metal tops of the bottles in the period between delivery at the day nursery and the time they were taken indoors when the staff arrived.

Not only did contamination arise through birds pecking the tops, the organisms evidently survived in the milk. Did you consider survival and whether the organism can grow in foods? When milk as described in Section 4.1.3 was allowed to be pecked for experimental purposes the organism was detected at a level of 6 cells per 500 ml. Do you think this very low cell concentration could be infective? Remember too that the children were very young, and vulnerable. In Case 4.1.2 the younger children were the more likely to display symptoms. Yet in Case study 4.1.3 the staff used the milk not considering that the pecked tops had any significance to the health of the children.

Different types of tops might have prevented this risk. But in the 1990s, in the UK, shiny metal aluminium tops on glass bottles were still widely used – indeed different colours coded for different grades of milk – full cream, semi-skimmed and so on. Yet at the same time it was already known that birds carried campylobacter, and that some types of birds are attracted to shiny objects.

You have to ask yourself whether such information was considered to be too abstruse to disseminate to the public, whether the risk was considered at all, whether the costs of changing the capping method were considered too costly (or unnecessary) and whether ordinary people had knowledge of the risk but ignored it.

Thus although pasteurisation has a role to play in the management of the risk of milk-borne organisms – in these examples *Campylobacter jejuni* – it is the whole chain, from the quality of the raw milk right to the point of consumption, which has to be managed effectively.

4.4.2 Should Pasteurisation of Drinking Milk be Obligatory to Protect Public Health?

In Chapter 3 you were asked to form an opinion about whether obligatory pasteurisation of milk would be an adequate strategy to protect people from milk-borne zoonotic disease, particularly *Mycobacterium tuberculosis*. You may have formed the view that pasteurisation alone is not enough and needs to be supported by other strands to the overall policy of the control of a serious disease. You can also see in this chapter from Table 4.4 that pasteurised milk, in spite of its pasteurisation, can still be involved in food-borne zoonotic disease and from Table 4.4 that the numbers of people affected can be significant (see Section 4.2.2).

Thus once again it is to be emphasised that pasteurisation is a method of destroying heat sensitive pathogens occurring in the raw milk but is only effective in a well managed overall system. But safe drinking milk is only achieved when all the links in the production chain are sound: good quality raw milk, well managed storage and transportation which minimises microbial growth in the raw milk, well designed and well maintained pasteurisation plant which avoids the mixing of raw and pasteurised milk, correct pasteurisation processes, appropriate management post pasteurisation so that milk of filled into near sterile leak proof containers from near sterile fillers, and stored and distributed cool, all implemented by well trained and managed staff.

The overall strategy to ensure drinking milk is safe is one which sees the production chain as an integrated whole and the product as vulnerable to contamination at all points. Pasteurisation is only an extremely valuable tool within that process – a point at which the microbiological quality and safety of milk improves. That gain in quality can be lost by poorly managed subsequent processes.

But pasteurisation has also proved to be a controversial process since the beginning of the 1900s to today (see Section 3.2.2, and Neaves, 2000). Today it is fully recognised that raw milk can contain a range of pathogens including *Mycobacterium tuberculosis*, *M. paratuberculosis*, verocytotoxic *E. coli*, *Campylobacter jejuni* and *Salmonella*. The illnesses these organisms cause can be serious, and may cause death. Set against that risk is the wish of some people to taste and enjoy the flavour of untreated milk with its full complement of nutrients and health giving properties.

In some countries the retail sale of raw milk is prohibited, and heat treatment before sale obligatory. But in England and Wales mandatory heat treatment of all drinking milk has never been implemented and so those people who wish to enjoy the unique taste of unprocessed milk can do so by purchasing directly from the point of production. The volume of raw milk sold is very small relative to the pasteurised milk market and food law still imposes on the producer the obligation to sell safe products.

In France most drinking milk sold is treated by the UHT process and pasteurised milk and raw milks represent only about 3.0% of the total market. However, the new process of tangential microfiltration has been developed (Gesan-Guiziou *et al.*, 2000), assessed as being an appropriate tool to reduce the microbial counts in milk (Anon, 1998), and is finding application not only in France but also, for example, in China (Wang Min *et al.*, 2000).

In France the current commercial focus is a minority market claiming (Anon, 1996) to provide an alternative to pasteurisation, producing a milk which tastes comparable to raw milk with a shelf life of 14 days under refrigeration, twice that of pasteurised milk. The manufacturers of the process claim a microbial reduction of 99.99%.

This process depends on the filtration of milk using technology similar to ultra filtration in which milk is forced under pressure through a ceramic membrane whose pore size ($1.0\,\mu$m) is small enough to prevent the passage of microorganisms, yet large enough to allow through the milk proteins. Yet the microfiltration stage is followed by recombination of the microfiltered skimmed milk with heat treated cream, and then by packaging (Figure 4.1). These post filtration stages could re-introduce organisms so the hygiene of the post microfiltration processes is just as important as the hygiene of the post-pasteurisation of milk processes. The technology is not a high volume technique, one plant producing

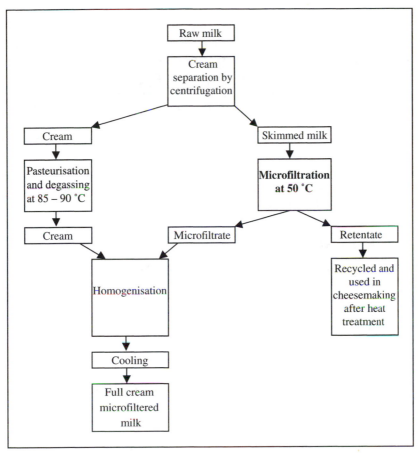

Figure 4.1 *The principles of production of microfiltered milk*
(Based on data in Anon, 1996)

2000 l per day, and cannot at the moment replace the high volume capabilities of pasteurisation by plate heat exchangers where throughputs in tens of thousands of litres per day are possible. Furthermore the microfiltered milk sells in France at a higher price than UHT milk and so because of both the small capacity of the plant and the higher price the producers are focusing on a specialised sector of the overall drinking milk market.

Nevertheless the process demonstrates that the application of pasteurisation to whole milk can be avoided, and that what is important is the management of the whole production-to-consumer chain using appropriate tools – animal management, farm hygiene, hygienic handling throughout production and distribution, appropriate processing to reduce microbial load including the pathogens within that load, together with good management and staff training to correctly implement the actions needed to achieve safe milk, and in these integrated ways to protect the public from milk-borne zoonotic diseases.

4.5 SUMMARY

Pasteurisation of milk offers a method of destroying heat sensitive pathogens in milk. It has had, and does have, a very important role in the control of zoonotic disease such as milk-borne tuberculosis and other zoonotic infections caused by *Campylobacter jejuni, E. coli, Salmonella* and other organisms. But it is only effective when a holistic view of the food chain is taken and all parts of the food chain are well managed including the avoidance of post-pasteurisation contamination. It is only essential for what it can do, but other technologies which remove organisms and additionally protect the unique flavour and nutritional qualities of milk are being developed. But until they reach large scale commercial viability pasteurisation or more severe heat treatments will remain extremely valuable public health barriers.

4.6 REFERENCES

Anon, 1996. Microfiltration à la Co-operative Laiterie de Villefranche. *Cahier des Industries Alimentaires*, **38**, 19–22.

Anon, 1998. Knock out microbes in milk. *Food Review*, **25** (8), 33.

Anon, 2000. Zoonoses Report 1998. Produced by the Ministry of Agriculture, Fisheries and Food, London, UK.

Corry, J.E.L. and Atabay, H.I, 2001. Poultry as a source of *Campylobacter* and related organisms. *Journal of Applied Microbiology*, **90**, 96S–114S.

Djuretic, T., Wall, P.G. and Nicholls, G., 1997. General outbreaks of infectious intestinal disease associated with milk and dairy products in England and Wales: 1992 to 1996. *Communicable Disease Report*, **7**, Review No. 3, R41–R45.

Frost, J.A., 2001. Current epidemiological issues in human campylobacteriosis. *Journal of Applied Microbiology*, **90**, 85S–95S.

Gesan-Guiziou, G., Daufin, G. and Boyavel, E., 2000. Critical stability conditions in skimmed milk crossflow microfiltration: impact on operating modes. *Lait*, **80** (1), 129–138.

Hudson, S.J., Sobo, A.O., Russell, K. and Lightfoot, N.F., 1990. Jackdaws as a potential source of milkborne *Campylobacter jejuni* infection. *Lancet*, **335**, 1160.

Hudson, S.J., Lightfoot, N.F., Coulson J.C. *et al.*, 1991. Jackdaws and magpies as vectors of milkborne human campylobacter infection. *Epidemiology and Infection*, **107**, 363–372.

ICMSF, 1996. *Micro-organisms in Foods*, Volume 5. *Characteristics of Microbial Pathogens*. Blackie Academic and Professional, London.

Neaves, P., 2000. Unpasteurised milk: do the risks outweigh the benefits? *Food Science and Technology Today*, **14** (1), 38–40.

On, S.L.W., 1996. Identification methods for campylobacters, helicobacters and related organisms. *Clinical Microbiology Reviews*, **9**, 405–422.

On, S.L.W., 2001. Taxonomy of *Campylobacter, Arcobacter, Helicobacter* and related bacteria: current status, future prospects and immediate concerns. *Journal of Applied Microbiology*, **90**, 1S–15S.

Orr, K.E., Lightfoot, N.F., Sisson, P.R., Harkis, B.A., Tweddle, J.L., Boyd, P., Carroll, A., Jackson, C.J., Wareling, D.R.A. and Freeman, R., 1995. Direct milk excretion of *Campylobacter jejuni* in a dairy cow causing cases of human enteritis. *Epidemiology and Infection*, **114**, 15–24.

Riordan, T., Humphrey, T.J. and Fowles, A., 1993. A point source of campylobacter infection related to bird pecked milk. *Epidemiology and Infection*, **110**, 261–265.

Robinson, D.A., 1981. Infective dose of *Campylobacter jejuni* in milk. *British Medical Journal*, **282**, 1584.

Stalder, H., Isler, R., Stutz, W., Salfinger, M., Lauwers, S. and Vischer, W., 1983. *Outbreak of Campylobacter enteritis involving over 500 participants in a jogging rally*. In *Camplyobacter II. Proceedings of the Second International Workshop on Campylobacter Infections*, 6–9 September 1983. Edited by A.D. Pearson, M.B. Skirrow, B. Rowe, J.R. Davies and D.M. Jones. Published by Public Health Laboratory Service, London, UK.

Wang Min, Yu Jing-hua, Zhang Li-li and Dou Jun, 2000. The preliminary experiments of microfiltration of milk for the removal of micro-organisms in raw milk. *China Dairy Industry*, **28** (2), 13–14 [quoted from *Food Science and Technology Abstracts*, **32** (2000) No. 12 Abstract 2000-Pe1924].

Wilson, P.G., Davies, J.R., Hoskins, T.W., Lander, K.P., Lior, H., Jones, D.M. and Pearson, A.D., 1983. *Epidemiology of an outbreak of milk-borne enteritis in a residential school*. In *Campylobacter II. Proceedings of the Second International Workshop on Campylobacter Infections*, 6–9 September 1983. Edited by A.D. Pearson, M.B. Skirrow, B. Rowe, J.R. Davies and D.M. Jones. Published by Public Health Laboratory Service, London, UK.

B. Techniques for Control

Surveillance and Microbiological Analyses

Key issues
- Meat: grilled kebabs
- Meat: a primary source of pathogens
- *Salmonella* and other organisms
- Microbiological surveillance as an aid to control
- Microbiological analysis as a tool
- Hygiene

Challenge

Meats, like milk, carry zoonotic organisms. Three case studies challenge you to think about the microbiology of grilled meat. They demand consideration of the significance of the organisms in the raw and cooked meat. They also ask you to think about how microbiological tests produce results, what the results mean, and when surveillance is of value.

5.1   THE CASE STUDIES: DONER KEBABS, UK, 1992–1995

5.1.1   An Outbreak of *Salmonella mikawasima* Associated with Doner Kebabs

The following extract is quoted with permission from Synnott, M., Morse, D.L., Maguire, H., Majid, F., Plummer, M., Leicester, M., Threllfall, E.J. and Cowden, J., 1993. An outbreak of *Salmonella mikawasima* associated with doner kebabs. *Epidemiology and Infection*, **111**, 473–481. Cambridge University Press.

"During October 1992 an increase in the number of isolates of *Salmonella mikawasima*, a rare serotype, was noted including a cluster of nine cases in the South West Thames region.

All nine primary cases, six males and three females, suffered diarrhoea and abdominal pain. Other symptoms reported were fever (8), anorexia (6), headaches (5) and vomiting (4). Duration of symptoms ranged from 5 to 21 days. All cases were seen by a GP, but none was hospitalised and there were no deaths. The dates of onset of illness in cases were between 17 September and 26

September. [. . .] Eight cases reported eating doner kebab from takeaway 'A' and one had bought and eaten chicken kebab there prior to the onset of illness. The date of the kebab meal was known in seven cases and the interval from consumption of kebab to onset of illness ranged from 20 to 72 h with a mean of 39 h.

"Outcome

The epidemiological investigation of [the] cluster of cases [. . .] pointed strongly to an association between illness and eating a kebab meal at takeaway 'A'.

Firstly, a statistical association between illness and eating at takeaway 'A' was demonstrated in the case control study using household controls ($P = 0.003$) and neighbourhood controls ($P = 0.025$).

Secondly, a comparison of cases with household controls with respect to exposure to food items at takeaway 'A' showed cases were more likely to have eaten kebabs [. . .]

Thirdly, using knowledge about the pattern of trade in an average week at takeaway 'A', a comparison of consumption of kebabs by cases with the probability expected also supported the hypothesis that a kebab meal was the incriminated food item ($P = 0.0000076$).

Finally, the epidemiological association was supported by laboratory analyses which showed the cases to have identical plasmid profiles.

"Discussion

An association between consumption of kebab (a meat containing product) and salmonella gastroenteritis was biologically plausible. The method of handling and cooking kebab meat at takeaway 'A' would have favoured the multiplication of any bacterial pathogens. Finally, the illness could not be explained by a history of consumption of any other food item. While there was a statistical association between illness and consumption of beef outside the home, in the case/household control study, only three cases had eaten beef and the same association was not found in the case/neighbourhood control study.

The source of *Salmonella mikawasima* in this outbreak could not be determined. It is possible that raw lamb delivered to the manufacturer was contaminated. Isolation of salmonella from 5.3% of samples of imported packs of boneless frozen mutton has been reported, but not from samples of raw chilled carcase lamb or mutton taken at the wholesale market.[1] In a 1981–85 study in the Manchester area salmonellae were isolated more frequently: from 18% of raw lamb samples.[2] In a study of the incidence of salmonella in abattoirs, butchers' shops and home-produced meat and their association with human infections in England and Wales, two strains of *S. mikawasima* were isolated from drain swabs from abattoirs dealing with cattle, sheep and pigs.[3] However there were no reports of isolations of *S. mikawasima* from human infections in the area of the abattoirs.

It is possible that *S. mikawasima* was introduced into the raw meat during the handling at the kebab manufacturing unit. Microbiological investigations were negative; however all samples were taken 2–3 weeks after onset of illness in cases.

Another possibility is that contamination was introduced into the kebab meat during handling at takeaway 'A'. Deficiences conducive to multiplication and transmision of bacteria were identified in the handling and cooking practices at takeaway 'A'.

S. mikawasima is a rare serotype of salmonella. Between 1985 and August 1991 only 55 human isolations had been indentified by the PHLS LEP* in England and Wales,[4] and it has not been recovered from domestic animals or poultry in Great Britain.[5] In the United States only 21 human isolates were reported to the Centers for Disease Control and Prevention between 1980–91. During the same period only 3 of 16,646 non-human isolates represented this serotype: one from a turtle, one from an unspecified reptile and from an unspecified 'other' source.[6] In Canada *S. mikawasima* has been isolated from aquarium snails imported from Florida.[7]

This outbreak came to attention because the serotype was unusual. It is possible that many cases of food poisoning may be related to the consumption of kebab meals, but the link is not easily recognized with common organisms. Assessments of hygiene risks in restaurants and takeaway shops have shown unsatisfactory results in 23% (36 out of 154) of shops selling kebabs.[8,9] For kebabs, poor temperature control and risk of contamination (*via* hand contact or cross contamination) were the most frequent problems encountered. While the temperature for reheating the kebabs (66 °C) was generally satisfactory, some doner kebabs were heated at much lower temperatures.[8] Thus further studies are needed to assess whether there is a more widespread risk of illness in association with consumption of doner kebabs.

"References

1. Hobbs, B.C, Wilson, J.G. Contamination of wholesale meat supplies with salmonella and heat-resistant *Clostridium welchii. Monthly Bulletin of the Ministry of Health Laboratory Service* (1959); **18**, 198–206.
2. Barrell, R.A.E. Isolations of salmonellas from humans and foods in the Manchester area: 1981–1985. *Epidemiology and Infection* (1987); **98**, 277–284.
3. Report of a working party of the Public Health Laboratory Service on salmonellae in abattoirs, butchers' shops and home-produced meat, and their relation to human infection. *Journal of Hygiene* (1964); **62**, 283–302.
4. PHLS Communicable Disease Surveillance Centre. *Salmonella mikawasima. CDR* (1992); **43**, 193.
5. Written correspondence with MAFF (Ministry of Agriculture, Fisheries and Food).
6. Written correspondence with CDC (Center for Disease Control, USA).
7. Bartlett, K.H., Trust, T.J., and H. Lior. Isolation of bacteriophage 14-lysogenized salmonella from the fresh water aquarium snail *Ampullaria. Applied and Environmental Microbiology* (1978); **35**, 202–203.
8. Tebbutt, G.M. Assessment of hygiene risks in premises selling takeaway foods. *Environmental Health* (1991); **99**, 97–100.

*PHLS LEP = Public Health Laboratory Service, Laboratory of Enteric Pathogens.

9. Tebbutt, G.M. Development of standardised inspections in restaurants using visual assessments and microbiological sampling to quantify risks. *Epidemiology and Infection* (1991); **107**, 393–404."
[End of extract, adapted from Synnott, M. *et al.*, 1993]

5.1.2   A Small Local Survey in London, January to May 1994

(Acknowledgement is made to the London Borough Environmental Health Department who kindly let me use their work.)

In October 1992 the UK Public Health Laboratory Service (PHLS) reported that an increase in suspected cases of one strain of Salmonella – *S. mikawasima* (not a very common strain) – was associated with reports of eight persons with gastro-intestinal illness, and it was noted that they had eaten doner kebabs (see Section 5.1.1, Synott, 1993). It was this report, together with the observations by diligent Environmental Health Officers of some of the cooking and handling practices in their London Borough that stimulated a small observational and sampling programme of doner kebabs sold out of small fast food shops in its area, which was carried out in January to May 1994.

The surveillance programme was modest and within the budget of the Borough, visiting 20 doner kebab outlets, without prior notification, taking samples and arranging for the microbiological analyses to be undertaken in the professional laboratories they normally used. The EHO had the right to inspect and take samples, but of course had to declare his role, which may have made staff at the premises change their practices, be a little more careful of how things were done. Samples of raw, and ready to eat kebab meat were taken, as well as samples of chilli sauce and salad fillings. The temperature of every sample was taken at the moment of sampling and recorded. In the survey samples were always taken between 12.30 and 2.00 p.m., and were packed into sterile plastic bags, and retained on ice until they reached the laboratory later the same day.

The microbiological methods of analysis used

The testing procedures were carried out in accordance with relevant parts of BS 5763, (BSI, 1987–1993).

Quantitative counts per gram of material were carried out testing for the aerobic plate count, coliforms, *Escherichia coli*, *Staphylococcus aureus* and *Clostridium perfringens*, while presence/absence tests for *Salmonella* in 25 g, for *Campylobacter* in 10 g and for *Listeria* per 25 g were undertaken.

The results generated

Over the period January to May 1994 samples of uncooked and cooked doner meat were taken from the twenty premises visited. The ranges of results are shown in Table 5.1.

Additionally the temperatures of 15 out of 19 cooked samples taken were recorded and these are shown in Table 5.2.

Table 5.1 *Ranges of microbiological results for uncooked and cooked kebab meat. London Borough Survey (refer to Section 5.1.2)*

Microbiological test (18 samples)	Uncooked doner kebab meat (16 samples)	Cooked doner kebab meat
TVC at 30 °C (cfu g^{-1})	<30–3.0×10^7	<30–4.6×10^6
Coliforms (cfu g^{-1})	<30–1.5×10^7	<30
E. coli (cfu g^{-1})	<30–9.5×10^3	<30
S. aureus (cfu g^{-1})	<30–3.0×10^5	<30
C. perfringens (cfu g^{-1})	<30–5.0×10^2	<30
Salmonella in 25 g	not detected	not detected
Listeria in 25 g	one sample positive *L. innocua* found	not detected
Campylobacter in 10 g	not detected	not detected

Reproduced with permission, courtesy of a London borough.

Table 5.2 *Cooked doner samples (15): temperature at time of sampling and TVC of sample*

Sample number	Temperature of sample (°C)	TVC: Log_{10} cfu g^{-1}
1	40	4.86
2	44	2.61
3	45	6.66
4	47	2.6
5	55	1.47
6	56	2.95
7	57	3.04
8	58	1.47
9	60	1.47
10	60	1.47
11	60	1.47
12	60	2.75
13	60	2.77
14	61	1.47
15	62	1.47

Reproduced with permission, courtesy of a London Borough.

Interpretation of results

The analytical microbiologist felt it would be of help to the EHOs to interpret the results found, reporting on each sample. The whole set of results obtained, which were originally recorded separately on individual sample record sheets, are collated, sample by sample, in Table 5.3 and are shown as recorded.

Table 5.3 *The London Borough's kebabs survey: summary results sheet*

Date	Food	Lab. code	TVC	Colif.	E. coli	SA	S	C.p.	L	Camp.	Analyst's interpretation of quality
940111	cooked	25456	73000	<30	<30	<30	ab	<30	ab	ab	Satisfactory
940118	cooked	25581	410	<30	<30	<30	ab	<30	ab	ab	Satisfactory
940125	cooked	25636	1100	<30	<30	<30	ab	<30	ab	ab	Satisfactory
940131	cooked	25781	560	<30	<30	<30	ab	<30	ab	ab	Satisfactory
940201	cooked	25806	4600000	<30	<30	<30	ab	<30	ab	ab	The micro quality was satisfactory. The TVC at 30 °C was too high
940208	cooked	25920	400	<30	<30	<30	ab	<30	ab	ab	Satisfactory
940111	cooked	25978	<30	<30	<30	<30	<30*	NI*	ab	ab	Satisfactory *<30 cfu g^{-1}; *NI = not isolated in 25 g
940215	cooked	25983	<30	<30	<30	<30		<30	ab	ab	Satisfactory
940222	cooked	26064	NI*	<30	<30	<30	ab	<30	ab	ab	Satisfactory *NI = not isolated in 25 g
940307	cooked	26391	900	<30	<30	<30	ab	<30	ab	ab	Satisfactory
940308	cooked	26431	<30	<30	<30	<30	ab	<30	ab	ab	Satisfactory
940322	cooked	26606	NI*	<30	<30	<30	ab	<30	ab	ab	Satisfactory *NI = not isolated in 25 g
940412	cooked	26832	<30	<30	<30	<30	ab	<30	ab	ab	Satisfactory
940419	cooked	26836	<30	<30	<30	<30	ab	<30	ab	ab	Satisfactory
940419	cooked	26840	590	<30	<30	<30	ab	<30	ab	ab	Satisfactory
940503	cooked	27255	<30	<30	<30	<30	ab	<30	ab	ab	Satisfactory

Sample	Type	TVC	Colif.	E. coli	S	SA	L	Camp.	Result
940111	raw	410	<30	<30	ab	<30	ab	ab	Satisfactory
940111	raw	3200000	<30	<30	ab	<30	ab	ab	Satisfactory
940118	raw	160000	<30	<30	ab	<30	ab	ab	Satisfactory
940118	raw	850	<30	<30	ab	<30	ab	ab	Satisfactory
940125	raw	610000	<30	<30	ab	<30	ab	ab	Satisfactory
940201	raw	4600000	<30	<30	ab	<30	ab	ab	Satisfactory
940228	raw	2500	150	<30	ab	<30	ab	ab	Satisfactory
940214	raw	300	<30	<30	ab	<30	ab	ab	Satisfactory
940215	raw	30000000	1300	<30	ab	<30	ab	ab	Not satisfactory. The TVC at 30 °C and coliform level too high
940222	raw	<30	<30	<30	ab	<30	ab	ab	Satisfactory
940228	raw	290000	1100	20	ab	<30	ab	ab	Not satisfactory – coliforms and *S. aureus* isolated
940307	raw	3700000	9500	<30	ab	500	ab	ab	Not satisfactory
940308	raw	1900000	<30	<30	ab	<30	ab	ab	Satisfactory
940323	raw	15000000	18000	<30	ab	<30	ab	ab	
940411	raw	590	<30	<30	ab	<30	L.I.	ab	Not satisfactory – *Listeria innocua* isolated
940412	raw	180	<30	<30	ab	<30	ab	ab	Satisfactory
940419	raw	120000	75	<30	ab	<30	ab	ab	
940419	raw	730000	<30	<30	ab	<30	ab	ab	
940503	raw	30000000	<30	310000	ab	<30	ab	ab	

Key: TVC = 'total viable count' cfu g^{-1}. Colif. = coliforms cfu g^{-1}. E. coli = *Escherichia coli* cfu g^{-1}. S = *Salmonella* – presence/absence in 25 g. C.p. = *Clostridium perfringens* cfu g^{-1}. SA = *Staph. aureus* cfu g^{-1}. Camp. = *Campylobacter* presence/absence in 10 g. L = *Listeria* presence/absence in 25 g. L.I. = *Listeria innocua*.

Reproduced with permission, courtesy of a London borough.

5.1.3   A National Survey of the Microbiological Quality of Doner Kebabs, 1995

(Adapted from LACOTS/PHLS Study of the microbiological quality of doner kebab meat, LAC 13/96/3. With permission.)

A national survey was conducted under the auspices of LACOTS (the (UK) Local Authority Co-ordinating Committee for Trading Standards) and the PHLS during October and November 1995, in order to establish the microbiological quality of ready-to-eat doner kebab meat samples from takeaways, restaurants, cafes and mobile vendors, at the point of sale, and to assess its fitness for human consumption and whether it was of the quality demanded by the purchaser. Only kebab meat which did not come into contact with other constituents was sampled. The sampling method was in accordance with that detailed in the Food Safety Act 1990, Code of Practice No 7: Sampling for Analysis and Examination, with reference to Part III, Samples for Examination.

Outcome

The study of ready-to-eat kebab meat sampled in restaurants and other retail premises found 1652/2538 samples were satisfactory on all criteria applied. The remaining 886 samples were less than satisfactory, not meeting at least one criterion. 307/2538 samples (12%) were unsatisfactory and 5/2538 samples (0.2%) were of unacceptable microbiological quality.

Quality criteria

Analyses for the aerobic plate count, coliforms and presumptive *E. coli*, *Staph. aureus*, *C. perfringens* and *Salmonella* were undertaken.The quality of the kebabs was additionally judged on the basis of the criteria shown in Table 5.4.

The 886 samples which were less than satisfactory, and the categories into which they fell, are shown in Table 5.5.

Discussion

Quoting the authors, the study said:

"This study has shown that most kebabs from retail take-away premises in England, Wales or Scotland are safe to eat, although the fact that 12% were classed as unsatisfactory using PHLS Guidelines for ready to eat foods [see Table 5.4] suggests that improvements could be made in the hygiene of these products. Only 0.2% of the samples were potentially hazardous, and the main reason for unsatisfactory results was a high aerobic plate count which is not itself a health risk but does indicate an overall lack of hygiene. The three samples with high counts of *C. perfringens* had levels which could cause food poisoning. All three were sliced meat chicken kebabs from take away premises. This, together with a considerably higher level of unsatisfactory samples in chicken kebabs, suggests that these products present a greater risk than lamb kebabs. This study

Table 5.4 *Microbiological guidelines for some ready-to-eat foods sampled at the point of sale (1996)*

	Aerobic count	E. coli	Staph. aureus	Listeria monocytogenes	Listeria (other species)
Satisfactory	$<10^3$ g^{-1}	<20 g^{-1}	<20 g^{-1}	Not detected in 25 g	Not detected in 25 g
Borderline/limit of acceptability	10^3 to $<10^4$ g^{-1}	20 to $<10^2$ g^{-1}	20 to $<10^2$ g^{-1}	Present in 25 g to <200 g^{-1}	Present in 25 g to <200 g^{-1}
Unsatisfactory	$>10^4$ g^{-1}	10^2 to $<10^4$ g^{-1}	10^2 to $<10^4$ g^{-1}	200 to $<10^3$ g^{-1}	200 to $<10^4$ g^{-1}
Unacceptable/ potentially hazardous		$>10^4$ g^{-1}	$>10^4$ g^{-1}	$>10^3$ g^{-1}	$>10^4$ g^{-1}

For 1992 guidelines see Table 11.9; for 2000 guidelines see Table 11.10 and Table 11.11.
Source: Gilbert, R.J., de Louvois, J., Donovan, T., Hooper, W.L., Nichols, G., Peel, N.R., Ribeiro, C.D. and Roberts, D., 1996. Microbiological guidelines for some ready-to-eat foods sampled at the point of sale. *PHLS Microbiology Digest*, **13**, 41–43. Reproduced with permission of the PHLS Communicable Disease Surveillance Centre. © PHLS.

Table 5.5 *Results of the LACOTS/PHLS cooked kebab meat samples survey and their microbiological categories based on PHLS guidelines*

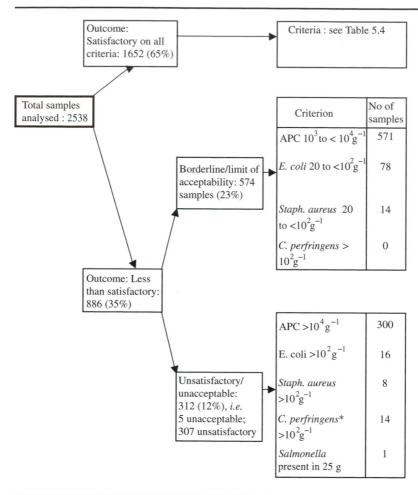

Criterion	No of samples
APC 10^3 to $< 10^4 g^{-1}$	571
E. coli 20 to $<10^2 g^{-1}$	78
Staph. aureus 20 to $<10^2 g^{-1}$	14
C. perfringens $> 10^2 g^{-1}$	0

APC $>10^4 g^{-1}$	300
E. coli $>10^2 g^{-1}$	16
Staph. aureus $>10^2 g^{-1}$	8
*C. perfringens** $>10^2 g^{-1}$	14
Salmonella present in 25 g	1

Outcome: Satisfactory on all criteria: 1652 (65%)

Criteria : see Table 5.4

Total samples analysed : 2538

Borderline/limit of acceptability: 574 samples (23%)

Outcome: Less than satisfactory: 886 (35%)

Unsatisfactory/ unacceptable: 312 (12%), *i.e.* 5 unacceptable; 307 unsatisfactory

Notes: 1. Criteria applied – see Table 5.4.
         2. A sample can fail on more than one criterion.
* Of which three samples had levels that could cause food poisoning.
Source: table collated from LACOTS/PHLS 1996.

shows that the greatest food poisoning risk to the public from doner kebabs is from *C. perfringens*. Kebabs are usually manufactured in a production unit and stored frozen. They are put onto the spit while still frozen, thawed for a while, then cooked on the outside. At the end of the day the kebab is returned to the freezer or to the fridge until the following day. These cooking practices allow variations in the temperature control of the product, and these could be one of the reasons for differences between the results between different premises. The doner kebab is a large quantity of meat which will heat up and cool down rather slowly. As better results were obtained from premises where the whole kebab was returned to the freezer between opening hours, it is likely that this practice allows the large amount of meat to cool relatively quickly. *C. perfringens* is the commonest form of food poisoning associated with cooked meat products.[1] The cooking process drives off oxygen and this favours the growth of anaerobic bacteria, particularly spore bearing species. Cross contamination of the cooked meats can occur from raw meats or from contaminated equipment. The salmonella isolated from a single sample could have derived from the original meat, been cross contaminated from uncooked food, work surfaces or equipment, or from a food handler. Cross contamination in these types of premises remains a potential source of sporadic infections with *Salmonella* and *Campylobacter* [. . .]

High bacterial counts were associated with kebabs which had been on the spit for more than 12 hours and may reflect growth of micro-organisms in the doner kebab, or on its surface due to post cooking contamination. Although the practices used to prepare and cook kebabs are probably adequate, there has been some concern that these practices may not be adequate to prevent occasional outbreaks of food poisoning.[2]

"References

1. Roberts, D., Hooper, W., Greenwood, M. *Practical Food Microbiology*, PHLS London, 1995.
2. Synnott, M., Morse, D.L., Maguire, H., Majid, F., Plummer, M., Leicester, M., Threllfell, E.J. and Cowden, J. An outbreak of *Salmonella mikawasima* associated with doner kebabs, *Epidemiology and Infection* (1993), **111**, 473–481."

[End of extracted summary]

5.2 BACKGROUND

The doner kebab is traditional in Greece and Turkey, and in London it is in the Greek and Turkish communities where the doner kebab shops tend to predominate. But some owners of the traditional British fish and chip shop have diversified and they sell doner kebabs as well. All these shops tend to be small, and tend to be run by owner managers, and they supply their fast food to the communities who live and work in the area.

The doner

A 'doner' is a formed block of raw minced meat – often lamb, but it can also be beef or chicken – which is gradually cooked as it is turned on a vertical spit in front of a lighted chargrill. The custom is that the block turns slowly on the grill and customers buy small portions of the cooked outside meat, sliced off as they wait. That meat is then filled into a pitta bread and it may be served like that or with added salads and sauces.

The doner, when fresh, is normally a cylindrical mass of chilled or frozen meat, centrally drilled, which at nearly one metre high is placed over the central spindle of the grill. The new doner is between 50 and 75 cm in diameter and weighs anything from 25 to 40 kg, and the grill is curved and forms an arc of about 120° around it. The vertical heating elements are sectioned in such a way that any or all of its areas may be alight at one time. It is part of the experience of enjoying a doner kebab to watch the shop owner with his enormous round ended double bladed knife, all 70 cm of it, delicately slice the hot meat off the turning spit, a process which leaves the ever smaller doner spirally ribbed like an imitation of a barber's pole. The shavings of meat fall into the curved scoop he holds in his other hand at the base of the rotating block of meat. If you are a lucky customer he will shoot his pile of hot meat directly into an open pitta bread, but the temptation is there to stuff the meat in with judicious use of the free thumb, while cradling the pitta in the fingers before serving it to you.

The pattern of business may be intermittent. It may be such that in preparation for the rush of customers at peak times, at mid-day for example, the kebab shop staff may shave the doner ahead of need and carve off piles of meat shavings and keep them warm in a dish semi-immersed in hot water, a bain-marie. And when the day is slack, and the lunch time rush is past, fuel may be saved by turning off the grill, stopping the block rotating, and allowing it to stand there warm and ready to go when the next customers come in hungry off the street.

At the end of the long working day, perhaps as late as two in the morning, fourteen or more hours after starting business for the day, it is not uncommon for a much reduced doner still to be on the spit. What the manager does with this still useable meat is his daily problem – to leave it on the spit, to remove it from the spit and refrigerate it, freeze it or to throw it away? That problem solved the day may then be over, but some businesses may have to make new doners to use for the next day.

The doners are made from raw meat using a recipe such as minced lamb (from boned shoulder and breast meat), salt, herbs, and spices. The whole mass of the meat is churned together in a large mixer ensuring the spices and the salt are evenly distributed. Sometimes milk is added to make it easier to work, and when thoroughly mixed the whole mass is turned into a mould and the meat bound together by compression, a process which works better if the meat is cold. This block is either moulded around a central steel spit on site or, if it is being made in a factory, ready prepared with its central hole and then further chilled or frozen and kept until it is to be distributed to the retail outlet.

5.2.1 Microbiology of Raw Meats

Meat is subject to much microbial contamination in its processing from abattoir to shop. Moist, high a_w, nutritious, of pH reducing after slaughter to an eventual 5.4–5.5, it provides an environment in which many of its contaminating flora can grow, and which would, if unchecked cause the meat to become unacceptable. Even if strategies such as refrigeration or vacuum or modified atmosphere packaging are adopted to prevent spoilage and prolong shelf life, those very strategies, while preventing one sort of quality loss, can themselves permit microbial spoilage unique to that storage condition.

Organisms which are associated with the living animal, its faeces and its environment, together with organisms associated with the slaughter process and environment are those which contaminate the meat (Table 5.6).

The aerobic viable counts found on carcases vary according to the hygiene of the handling practices, and the animal species. Meat animals are warm blooded having body temperatures similar to that of human beings (which is 37 °C). After slaughter the carcases must be chilled to remove the body heat and achieve deep temperatures of 7 °C or below within 24 h of slaughter, and thus discourage microbial growth. Reported counts on chilled carcases range between 10^2 and 10^5 cfu cm^{-2}, and when all slaughter and carcase dressing operations have been carried out it is considered advisable that initial counts on carcases should be less than 10^4 cfu cm^{-2} (ICMSF, 1998). Among this flora, which has the ability to spoil the meat, pathogens are also likely to be present (Table 5.6). Surveys evaluate the presence of selected pathogens on meat and indicate their prevalence (FSIS, 1994; McNamara, 1995). Baseline data are sought in order to establish the current levels of contamination of carcase meat and to work to reduce them.

Because meat is stored under chill conditions to inhibit the aerobic mesophilic population, it is the growth of psychrotrophic organisms such as pseudomonads which then becomes significant. But under vacuum packaging of meat the Gram negative aerobic flora (*Pseudomonas, Acinetobacter-Moraxella*) is suppressed and the organisms more tolerant of those conditions such as *Lactobacillus* and *Leuconostoc* tend to become the spoilage flora. Similarly in modified atmosphere packaging, where the concentration of carbon dioxide is raised to 20–25% and the oxygen concentration is also raised above that of air to encourage the meat to keep its red colour, only organisms such as *Brochothrix thermosphacta*, lactic acid bacteria and Enterobacteriaceae tolerant of both the high carbon dioxide

Table 5.6 *Organisms which occur in raw meats*

Alcaligenes spp.	*Enterobacter* spp.
Alteromonas spp.	*Escherichia coli*
Bacillus cereus	*Escherichia coli* O157:H7
Bacillus circulans	*Flavobacterium* spp.
Bacillus licheniformis	*Hafnia* spp.
Bacillus subtilis	*Lactobacillus* spp.
Brochothrix thermosphactum	*Leuconostoc* spp.
Campylobacter jejuni	*Listeria monocytogenes*
Campylobacter coli	*Moraxella-Acintobacter* spp.
Clostridium botulinum	Salmonella
Clostridium perfringens	*Staphylococcus aureus*
Clostridium sporogenes	*Streptococcus faecalis*
Coliforms	*Yersinia enterocolitica*
Corynebacterium-Microbacterium spp.	Yeasts and moulds

Source: Al-Sheddy *et al.*, 1995.

Table 5.7 *Infectivity/toxicity of organisms associated with food-borne disease*

Organism	Infectivity/toxigenicity
Bacillus cereus	Symptoms arise after ingestion of food containing large numbers of toxigenic bacteria ($>10^5\,\text{g}^{-1}$), or preformed toxin
Bacillus subtilis-licheniformis group	Symptoms arise after eating food containing large numbers of bacteria $>10^5\,\text{cfu}\,\text{g}^{-1}$
Campylobacter jejuni, C. coli, C. lari	As few as 100 organisms can cause illness if consumed with milk or other foods that may neutralise gastric acidity
Clostridium botulinum toxins (A, B, E, F)	The toxin is potentially lethal at very low doses. For example 0.1–$1.0\,\mu\text{g}$ Type A toxin causes death in man
Clostridium perfringens (formerly *C. welchii*)	Usually $>10^6$ micro-organisms are required to cause illness. Enterotoxin types A and C are produced in the intestine when the organism sporulates there
Cryptosporidium species	Virulence of oocysts not known under environmental conditions. Infectivity appears to be high: infective dose can be as low as one oocyst
Entero-invasive *Escherichia coli* (EIEC)	Infectivity relatively low, but can vary with gastric function and acidity of gastric contents
Enteropathogenic *Escherichia coli* (EPEC)	Relatively high numbers of bacteria required to cause illness
Enterotoxigenic *Escherichia coli* (ETEC)	Not highly pathogenic, usually $>10^6$ bacteria required to cause illness
Verocytotoxin-producing *Escherichia coli* (VTEC), otherwise called enterohaemorrhagic *Escherichia coli* (EHEC)	Relatively high. Small numbers of bacteria can cause illness because toxin is elaborated as the population increases in the gut
Giardia lamblia	As few as 25–100 cysts may cause illness
Hepatitis A virus	Infectivity relatively high
Listeria monocytogenes	Not highly pathogenic for healthy adults outside high risk groups. Minimum infective dose >100 cells
Salmonella species excluding *S. typhi* and *S. paratyphi.*	Normally relatively large numbers of bacteria are required to cause illness in healthy adults, but vulnerable groups (babies, immuno-suppressed) can be infected by lower numbers. Infection can occur from very low doses, *e.g.* <1000 organisms, particularly in foods that protect salmonellae from the acidity of the stomach, *e.g.* salami, chocolate
Salmonella typhi., Salmonella paratyphi. types A–C	Infectivity variable. 10^5–10^9 bacteria may be required to cause illness, depending on strain and host susceptibility. Infants and immuno-compromised people may be more susceptible

Table 5.7 (*cont.*)

Organism	Infectivity/toxigenicity
Shigella sonnei and other shigellas (*e.g.* *S. flexneri, S. boydii, S. dysenteriae*)	Small numbers of bacteria (10–100) have caused illness in volunteer groups
Staphylococcus aureus enterotoxin	Food poisoning can occur in the absence of live cells of *Staphylococcus aureus*; toxin may have been produced and the organisms die out. Sufficient toxin to cause illness may be produced if bacterial numbers reach 10^5–10^6. 0.1–1.0 μg enterotoxin per kg of the patient's body weight can cause illness
Vibrio cholerae serotype 01 and non-01 strains	10^6 organisms have caused illness in volunteers, if given with alkali to neutralise stomach acidity
Vibrio parahaemolyticus	Infectivity relatively low – at least 10^5–10^7 organisms of a virulent strain may be required to cause illness
Viruses – small round structured viruses (SRSVs) including Norwalk agent, rotavirus, adenoviruses, coronaviruses and astroviruses	Infectivity relatively high

Source: Department of Health, UK, 1994. Crown copyright. Reproduced with permission.

concentration and the chilled storage conditions will be able to grow.

Signs of deterioration are likely to appear once the spoilage flora reaches 10^7 cfu cm^{-2}, leading to unacceptability. During the growth of the spoilage flora, potentially pathogens present can increase, although their rate of growth depends on their tolerance of the storage conditions. But their successful growth in meat also seems to depend on whether the background spoilage flora permits or antagonises their growth. There is now some evidence to suggest that measures taken in abattoirs, such as spraying the carcases with organic acid solutions to reduce the surface flora and which are viewed as decontamination measures, may be counterproductive (Jay, 1997). That spoilage flora may then be too sparse to antagonise, inhibit or kill any pathogens which may subsequently contaminate the carcase. Where the infective dose of pathogens is also low, maybe as few as 10–100 cells (Table 5.7), rather than being outgrown and suppressed the survival and growth of the pathogens on the raw meat may contribute to the risk the meat represents when prepared for consumption.

Recognition of microbial interference as a potential aid in risk reduction from the pathogens among the microflora of raw meat challenges the conventional view that the lower the overall microbial counts the better. Rather it poses the view that the composition as well as the numbers of organisms present is an important consideration.

Because of the pathogenic microbial flora meat is liable to carry, meat has predominated among the foods associated with outbreaks of food poisoning (see examples in Chapters 2, 11, 17 and 20), but since it is widely customary to cook meat those outbreaks either arise as a result of microbial survival in the meat or from post-cooking contamination. Many of the pathogens contaminating meat are heat sensitive and easily destroyed in cooking, but the sporing organisms are more resistant (Table 5.9). Both to ensure that chilled raw meats keep for longer periods, and to reduce the risk of gastro-intestinal disease associated with meat, minimum acceptable microbiological standards for raw meat products have been developed which recognise that the subsequent experience of the meat product prior to cooking and consumption will tend to allow increase in microbial numbers. These standards have advisory rather than statutory status (Table 5.8).

Table 5.8 *Microbiological criteria for raw meat*

Product examples	Joints, mince, diced meats, offal, burgers, sausages, bacon, marinaded products, cured products
Storage	Frozen or chilled
Use	To be cooked

Pathogens: Salmonella, Campylobacter, VTEC and parasites may be present; *Y. enterocolitica* may also be important in pork and *S. aureus* in bacons. Monitoring the incidence of bacterial pathogens may be useful for trend analysis

Organism	GMP	Maximum
Bacterial pathogens:	Criteria for absence not generally applicable	

Indicators and spoilage organisms:
APC can be used to indicate quality. Counts are generally higher in minced meats than on whole cuts. *E. coli* may be useful as an indicator of hygienic slaughterhouse practices. Coliform/Enterobacteriaceae may be useful for trend analysis

Organism	GMP $cfu\,g^1\,or\,ml^{-1}$	Maximum $cfu\,g^1\,or\,ml^{-1}$
APC	$<10^5$	10^7
E. coli	$<10^2$	10^4
Yeasts (sausages and marinaded products)	$<10^4$	10^6

Notes: GMP (good manufacturing practice) values are those expected immediately following production of food under GM (good manufacturing) conditions.
Source: PFMG, 1997. Reproduced with the permission of the publisher, the Institute of Food Science and Technology, UK.

5.2.2 The Control Framework for the Sale and Consumption of Cooked Kebab Meat

In the UK meat products are controlled under a framework of legislation and its implementation and monitoring. These are outlined in Figure 12.4.

5.3 EXERCISES

Exercise 1. Thinking about the scale and context of the London Borough survey (Section 5.1.2) of the microbiological quality of kebabs, what do you think was, or could have been achieved by it?

Exercise 2. Evaluating the London Borough data (Tables 5.1–5.3) and the methods by which they were achieved, what do the results mean, and what determines your confidence in them?

Exercise 3. Since many fast foods are grilled products (*e.g* burgers, chicken pieces . . .) what lessons from three kebabs Case studies (Sections 5.1.1, 5.1.2 and 5.1.3) can be applied to the control of other meat based fast foods?

Exercise 4. The surveillance exercise reported by LACOTS/PHLS (Section 5.1.3) was clearly a costly, time consuming exercise to undertake, involving co-ordination of 44 Food Liaison Groups, engaging the time of hard pressed environmental health officers requiring them to take samples, depleting their budgets for analysis of other food samples, and taking time away from the analytical laboratories. In the light of the evident costs and effort involved in surveillance exercises such as that reported by LACOTS/PHLS (1996) what definable contribution to public health does such surveillance make?

Table 5.9 *D values for some pathogens in meats (D-values in minutes)*

Pathogen	°C											
	50	51.6	54.4	56	57.2	60	62.8	64.3	70	80	100	115
Campylobacter (thermophilic) (ground beef) (lamb cubes)	5.9–6.3 5.9–13			0.62–0.96 0.96–1.26		0.21–0.26						
E. coli O157:H7 (ground beef)			39.8			0.75		0.2				
Listeria monocytogenes (ground beef) (beef) (beef steak homogenate)						3.12 3.8 6.27			0.14 0.14–0.2			
Salmonella (ground beef)		62			4.2		0.7					
C. perfringens (spores) (*water or **phosphate buffer)										120*	6*	0.6**

Source: ICMSF, 1996.

5.4 COMMENTARY

There are a number of issues which arise in considering these case studies concerning the microbiological quality of kebabs which link to other case studies considered in this book.

In the *Salmonella mikawasima*/kebab case study (Section 5.1.1) it is worth noting that the typing techniques facilitated determination of the probable source of infection causing the gastroenteritis in the eight people affected. It was the fact that salmonellas could be typed and that the same, unusual type of salmonella was common to those cases which led to epidemiological investigation. Serotyping and plasmid profiling of the salmonella strains isolated from the infected persons permitted exact identification of the organisms and the conclusion that the cases were probably linked to the same source of organisms. But no proof was achieved. In the case control study careful questioning of cases and controls led to the probability that the kebab meat purchased from one kebab takeaway ('A') was the common factor. Other approaches can also be used which may indicate (Chapter 1) or recognise (Chapter 3) that a common source of pathogen exists.

Exercise 1. Thinking about the scale and context of the London Borough survey (Section 5.1.2) of the microbiological quality of kebabs, what do you think was, or could have been achieved by it?

Raise a number of questions such as:

- What were the aims of the survey?
- How/why would I set up such a survey?
- How was the survey conducted?
- How were the premises chosen?
- How long between sample taking, achievement of results in the laboratory, and reporting of results from the laboratory to the environmental health officer?
- What was the basis of the analyst's judgement 'satisfactory/not satisfactory'?
- What action could the EHO have taken on receiving the data?

In considering the survey itself it seems that the EHO simply aimed to visit and take samples for microbiological analysis from all twenty of the kebab shops in his district.

Examination of the survey results collated in Table 5.3 gives dates of when individual samples were collected but does not connect the data to particular premises. Presumably those records were held separately. The survey of the 20 premises was undertaken over a period from 11 January to 3 May – 16 weeks – allowing the officer to spread the work between other duties, while from his point of view gaining repetitive experience of how the kebab takeaways normally operated.

Depending on the nature of the individual tests the microbiological data on the individual samples would have been available three to seven days after presenting the samples to the laboratory. These data would be retrospective, the kebab meat would have been eaten and the only immediate use to which the officer could put the data would be to form a value judgement on the meat quality as used (the uncooked meat) or as sold (the cooked meat). By comparing the data with quality guidelines, or at least by accepting the opinion of the microbiologist analyst, he would have been able to inform the takeaway staff whether the samples were satisfactory or possibly hazardous, and in so doing perform a role in guidance and education. Normally therefore the feedback should have been factual, providing a copy of the laboratory results for the food business's records. If the samples were doubtful or potentially hazardous the opportunity existed to provide guidance as to how better assurance of safety could be achieved, and to give general hygiene advice, remembering that the result would relate to a block of kebab meat which had already been retailed and eaten.

By collation of all the data gathered the officer would learn more about the microbiology of the products, and also through visiting the various premises gain detailed information on normal practices. It would have been possible, through reading other literature, to put the local information gained into a broader context – perhaps a national context, and establish whether recommendations for the safety management of this type of food existed. Indeed from his own observation and data evaluation his own list of recommendations could have been drawn up to include:

- use of fresh meat of good microbiological quality, and with storage at $<5\,°C$;
- the hygiene of preparation operations including keeping the temperature of the meat during preparation of the kebab cone at as low a temperature as possible, and its storage below $5\,°C$;
- consideration of the size of cone used on the grill, using a size which can be used within one trading day;
- cooking the meat so that the meat which is sliced from the kebab cone is cooked thoroughly;
- cleanliness of the utensils and knives used in slicing and handling the cooked meat;
- ensuring that any storage of cooked meat keeps the temperature of the meat above $63\,°C$, and that minimal quantities of such meat are stored;
- cooking continuously once the cone is on the spit.

You may find the following references of help: Matossian and Kingcott, 1979; Bryan *et al.*, 1980; Todd *et al.*, 1986.

Exercise 2. Evaluating the London Borough data (Tables 5.1 to 5.3) and the methods by which they were achieved what do the results mean, and what determines your confidence in them?

This exercise requires you to look very carefully at the data in Table 5.3, to decide on their validity and in so doing interpret the use to which the data generated for each sample could have been put.

Individual microbiological results for single samples

For each sample eight analyses have been done, six of which are for specific pathogens, and 'TVC' and 'coliforms' to estimate the overall microbial quality. Looking at these data some inferences can be drawn.

'TVC' stands for 'total viable count' – but since the method actually only estimates the aerobic organisms which can grow under the conditions of test it is usually preferable to use the term 'APC' ('aerobic plate count') or ACC ('aerobic colony count'). The TVC is recorded as 'cfu' (colony forming units) per gram or ml of sample calculated from the number of colonies seen on culture plates. The colonies may be derived from single cells, or from pairs or clumps of them. The term 'cfu' is a recognition that this type of microbiological test is only an estimate of the concentration of organisms in the original material (see discussion of this issue in Adams and Moss, 2000).

Each set of results means that those organisms were detected at the levels indicated. The minimum numbers detectable ('the sensitivity of the test') are limited by the conditions of the test – the medium used, the sizes of the inocula, the temperature and duration of incubation, the ability of the analyst. The results found for each sample raise the question: was the food safe to use and/or eat?

Where pathogens were found to be present in the sample, consumption of similar meat would have carried a risk of infection. However, it is as well to remember that the raw kebab meat is grilled before consumption, and the risk from pathogen infected raw meat lies in its undercooking, or cross-contamination from it to other food. Many of the pathogens liable to occur in raw meat (Table 5.6) are heat sensitive, and would be destroyed in the moist heat of many cooking methods, and even spore bearers could be destroyed by the fierce burning heat of grilling. But the extent to which the grilling heat penetrated the meat would determine whether spores or even heat sensitive pathogens survived. In these samples at the levels of sensitivity of the tests applied no pathogens were found in the ready to eat meat.

Were the kebab meat samples of acceptable microbiological quality?

The uncooked kebab meat sample counts reflect the microbiological quality of the raw meat used; the microbiological quality of the other ingredients (milk, spices *etc.* . . .); the freezing time; the microbiological cleanliness of the equipment used; the hygiene standards of the operatives and the practices associated with making, handling, storing and using the kebab; the time taken to prepare it; the temperature at which it was prepared, its moisture content and localised a_w and pH values; the presence or absence of preservatives as allowed by law, the length of time held on a spit during which continuous or intermittent cooking occurred; the temperatures at the surface and penetrating into the depth of the block of meat; the methods of microbiological analysis; the way the sample for analysis was handled; the reliability and appropriateness of the laboratory method; and the accuracy of recording the results.

Take for example 'Uncooked doner – Sample code 26432'. This raw meat is recorded as being 'satisfactory' – having a TVC of 1.9×10^6 cfu g^{-1}. In this population neither coliform organisms nor any pathogens were detected. There were no further tests other than specific tests for pathogens to check the nature of the microbial flora but clearly this large diverse population of organisms was present, and probably growing. Time and warmth would have allowed further growth and would have led, eventually, to a population where spoilage of the raw meat was evident through some form of organoleptic change. Judgement of the maximum upper microbial limits for acceptable quality is to some extent variable – allowing for differences in preference between the development of tolerable flavours and those associated with deterioration. In chilled meats, under aerobic conditions, maximum levels of around 10^7 cfu g^{-1} is a commonly accepted threshold above which quality defects will appear, but in meats stored warmer, or under vacuum packing conditions, or chilled and vacuum packed, different floras will develop and because of the nature of those, different thresholds of acceptability will be be derived (Table 5.4).

For 'Uncooked doner – Sample code 25451' – a TVC of 4.1×10^2 cfu g^{-1} was recorded, very much lower than the TVC for 'Uncooked doner – Sample code 26432' (discussed above). These two results represent a ten thousand fold difference in microbial concentrations and indicate a considerable difference in the microbial status of these two samples.

Look also at 'Uncooked doner – Sample code 26225', where in spite of a detected TVC of 2.9×10^5 cfu g^{-1}, ten fold lower than for Sample code 26432, the sample is deemed 'unsatisfactory', presumably on the basis of the 1100 coliforms g^{-1}. In raw meat coliform organisms could be accompanied by other gut dwelling (enteric) organisms such as *E. coli* and *Salmonella* – although it is probable they would be present in lower numbers than the coliform organisms. Note also that 'Uncooked Sample code 26392' is the only uncooked sample to show the presence of *C. perfringens*, a pathogen whose spores are heat resistant (Table 5.8) and whose survival in the cooked kebab meat would depend on the temperature at the grilling surface and the temperature gradient into the meat.

The method (BS 5763) used in this survey for the detection to 'presumptive *Staphylococcus aureus*' identifies coagulase positive strains, the majority of which are also toxigenic. While *Staph. aureus* is often a human skin-borne organism, it can also be derived from animal skin and be present in raw meat and so its presence does not indicate its origin. The majority of samples showed less than 30 cfu g^{-1} yet four samples are recorded as having between 20 and 3.1×10^5 cfu g^{-1}. These were present at the same time as higher numbers of 'TVC' and were part of that growing population, and probably would not compete well with it, and would be outgrown. Being heat sensitive they would normally be destroyed in cooking, but if they chanced to contaminate cooked meat they might then grow in the absence of competing flora. *Staph. aureus* has more significance in cooked products, in products which are salty (*e.g.* bacon) or products of reduced water activity where it may grow and produce toxin in the absence of effective competitors.

Overall the lower the microbial counts, together with the absence of detected *Salmonella, Listeria* and *Campylobacter* in the raw kebab meat, the lower the risk from the cooked product (Table 5.9) because of the variability in the heat resistances of the pathogens and the variability in the grilling process, and the varying depths of the sliced kebab meat.

In the cooked meat samples the flora is derived from the raw product and its subsequent heating experience and hygiene of handling. In addition what is identified in the samples also depends on the thickness of the slivers of 'cooked' meat and whether any sliver incorporated uncooked meat; the microbial loading of the knife used to cut the meat shavings; the temperature at which the bain-marie was maintained; the temperature of the cooked meat while remaining in the bain-marie; the duration for which the cooked meat remained in the bain-marie; and any cross contamination experienced.

In the case of the cooked doners 'satisfactory' implies 'safe to eat' – that is to say meeting standards of acceptable risk (Table 5.4).

Cooked meat should have very few organisms present – only those which have survived the cooking process. The D-value of vegetative cells present is a matter of a few seconds at temperatures above 70 °C (Table 5.9) but there is no evidence available here to indicate what the surface and below-surface temperatures of the grilling kebab would have been. Remember also that kebab meat is not a solid block in the sense that a side of beef or lamb is. It is formed from mince and its microbial population is distributed throughout – from surface to the deep interior. Thus the combination of the temperature of the meat at and below the heating surface and the thickness of the shavings, which potentially contained semi-cooked and uncooked meat, would influence the microbial counts obtained. Thus it is a cause for comfort that the samples of 'cooked meat' had low TVCs, the majority below 1×10^4 cfu g^{-1} and would on the criteria defined in Table 5.4 have been described either as 'acceptable' or 'borderline acceptability'. But it is a cause for concern that one sample contained over 4 million 'TVC' per gram, 'unsatisfactory'. That sample may have originated from raw meat of less good microbial quality, may not have received effective cooking, or could have experienced microbial growth while retained in a bain-marie at below 60 °C.

The presence and counts of coliforms in cooked meat (and in cooked foods) is indicative either of undercooking or of poor hygienic practice, and the consumer is thus at risk of exposure to an unknown range of organisms deriving from those practices. Thus upper limits for the presence of coliforms in foods at the point of use are set (Table 5.4).

The laboratory records do not record 'chicken', 'lamb', 'beef' *etc.* In retrospective analysis like this it would have been of value. But at the time the survey related to 'doners' and maybe it was felt to be irrelevant. However, different types or strains of organisms are more strongly associated with different types of meat and in a repeat survey or a larger one it would be appropriate to record this.

Confidence in the results

The collated data shown in Table 5.3 appear to show inconsistencies in the recording of the dates on which the samples were drawn. Look for example at 'Cooked doners' samples 25456 and 25978. These are both recorded as being taken on 11 January. Yet the sample numbers are 522 numbers apart. It is conceivable that a laboratory might analyse 522 samples in a day, but looking at the rest of the sequence of numbers and the date codes it seems unlikely that all the other codes are wrong. So perhaps a human error has been made in transcribing the results from laboratory record books to official sheets. Yet a similar anomaly has occurred in respect of uncooked samples 25921 and 26225. Could it be that the results sheets are indeed faithful to the laboratory record? Were these both taken on 28 February or not? Was one frozen (or chilled) and analysed weeks later? Are the codes wrong? What has happened?

Look again at Table 5.3. Remember that this was a survey in which all samples should have been analysed on the same basis. Records for samples 25451, 25456, 25978, 26064, 26225, 26392, 26432, 26829, 26837, 26841, 27254 all record eight microbiological tests as being done. 25983 records six tests.

Record 26225 records *Staph. aureus* as '20 cfu g^{-1}' whereas all the others record the minimum detectable level of *Staph. aureus* as 'less than 30 cfu g^{-1}'. Notably the 20 cfu g^{-1} *Staph. aureus* found in 26225 is commented upon; no comment is made in the others. It is worth here looking again at the *Staph. aureus* methodology and determining for yourself what the limit of sensitivity of the test, as described, is.

Record 25978 records *Salmonella* as 'less than 30 cfu g^{-1} and *C. perfringens* as 'not isolated in 25 g', results, if believed, contrary to the methodology used, and probably in fact a recording error. Record 25983 does not record any tests for *Salmonella* and *Campylobacter* – were these not done, or were they not recorded?

In reality dialogue between the laboratory and the recipient of the data would have occurred and clarified these points. But it is for this type of reason that laboratories themselves should have quality systems in place such as accreditation to a national system to assure the reliability of the data they generate.*

Exercise 3. Since many fast foods are grilled products (*e.g.* burgers, chicken pieces . . .) what lessons from three kebabs case studies (Sections 5.1.1, 5.1.2 and 5.1.3) can be applied to the control of other meat based fast foods?

Knowledge of the nature of the microflora of meat including the pathogens potentially present, their heat resistances and ability to grow in the cooked, or (accidentally) semi-cooked product can be applied to other meat based fast food.

'Fast food' is designed to be cooked and served fast.

But in the implementation of cooking meat products such as burgers, consideration has to be given to whether to cook from frozen, or from chilled raw product. When that decision is taken, then the intensity of heat which can be applied to the burger, and the depth, dimensions and internal temperature of the burger itself influence the necessary time both to cook the meat and to kill the

---

* Such as that implemented in the UK through UKAS (United Kingdom Accreditation Service) at www.ukas.com

microflora. The overriding factor in the achievement of safe 'fast food' meat products is that if sold to the public on the basis of the rapidity of service then the design of the whole cooking process and its implementation have to be very rigorously controlled. This does not allow for variability in any of the factors listed above.

In contrast in the sale of doner kebab meat, variability is built in, and hence the control of the process is more dependent on the implementation of the recommendations outlined in the commentary for Exercise 1.

Exercise 4. The surveillance exercise reported by LACOTS/PHLS (Section 5.1.3) was clearly a costly, time consuming exercise to undertake, involving co-ordination of 44 Food Liaison Groups, engaging the time of hard pressed environmental health officers requiring them to take samples, depleting their budgets for analysis of other food samples, and taking time away from the analytical laboratories. In the light of the evident costs and effort involved in surveillance exercises such as that reported by LACOTS/PHLS (1996) what definable contribution to public health does such surveillance make?

The first case study presented in Section 5.1.1 showed that surveillance identified an existing problem which led towards actions to bring it under control.

In the opinion of Berkelman *et al.* (1994) surveillance is the foundation for control of infectious disease, serving as the basis for responses to new threats such as identifying contaminated food or other products. They see surveillance as the vital tool 'needed to assess the extent of illness and death associated with infectious diseases so that priorities can be assigned to control efforts', and 'is also critical assessing the effectiveness of regulatory and advisory measures to safeguard public health'.

They point out cases of where surveillance, had it been operating effectively, would have alerted the authorities to outbreaks of serious food poisoning permitting implementation of controls such as alerting the public to necessary hygiene practices and/or requiring the withdrawal of suspected food from the food chain. The problems of inadequate surveillance and co-ordination of actions were illustrated in the 1995 Lanarkshire, UK outbreak of *E. coli* O157:H7 poisoning which resulted in 496 cases and 17 deaths directly attributable to the organism (see Chapter 12). In the USA effective surveillance revealed the infection of ground beef by *E. coli* O157:H7 resulting both in four deaths and in the recall of 250 000 contaminated burgers which then terminated the outbreak (Anon, 1993).

Berkelman *et al.* (1994) recommend the USA's Centers for Disease Control strategy (CDC,1994) for improved surveillance:

1. strengthening the national notifiable disease system;
2. establishing sentinel networks (which link groups of health care providers or laboratories to a central data receiving and processing centre);

3. establishing population based centres focussed on epidemiology and prevention of emerging infections, and
4. developing a system for enhanced global surveillance.

The definable contribution which surveillance makes to public health in general,
and food-borne illness control in particular, is that of the development of
awareness of existing and emerging problems, which then facilitates the implementation of control or eradication schemes. These can be immediate responses
to detected outbreaks, or long term strategies to establish base line data, or
directed to the control and eradication of infectious disease. Thus the
LACOTS/PHLS (LACOTS/PHLS, 1996) survey on kebabs (Section 5.1.3) was
an example of surveillance to test whether there was an on-going problem in one
particular sort of food, doner kebabs, and its outcome permitted an appraisal of
the microbiological risks associated with them, together with the development of
recommendations for their safer use in retail food.

Useful internet addresses of infectious disease surveillance centres:

World Health Organisation   http://www.who.int
United Kingdom   http://www.open.gov.uk/cdsc/site_fr4.htm
United States of America   http://www.cdc.gov

5.5 SUMMARY

Grilling of raw meat rapidly exposes the surfaces to sterilising temperatures, but
the effect of the grilling on the microflora within the meat depends on the extent
to which the heat penetrates the depth of the block. Where a doner kebab is
frozen solid ($-20\,°C$) when placed on the spit there will be a narrow layer of meat
near the surface whose temperature lies with the growth temperature zone
(approximately $1-60\,°C$), while the grilling surface itself will rapidly rise to
temperatures well above $100\,°C$. Heat penetration depends on the grill temperature, the time of exposure and the thermal conductivity of the meat.

In grilling, control over the microflora of the product can be better achieved by
standardising the size and thickness of the grilled items (burgers, sausages . . .),
grilling from both sides simultaneously, standardising on the use of either a
frozen or chilled product, and using a very clearly defined grilling time and
temperature regime. Such standardisation of the process would achieve fast-food
products in which the failure rate of adequate destruction of the microbial
population would be very low. But this in turn also depends both on the size, the
composition and heat resistances of the microbial flora in the comminuted meat.

Raw meat is a primary source of pathogenic organisms, and it also carries a
wide range of spoilage organisms. Comminuted meat products such as doner
kebabs, burgers and sausages tend to have microbial counts which are higher
weight for weight than carcase meat.

The production, distribution and use of raw meat and raw meat products take place under a legal framework which sets minimum hygiene standards and practices, and through Codes of Practice provides guidance as to how those standards are to be achieved. In the UK compliance is enforced and monitored by Environmental Health Officers.

Surveillance at national and local levels has an important role to play in the identification of new food borne hazards and in their prioritisation for management. Food sampling and analysis are essential elements in surveillance and standardisation of the manner in which that is conducted is essential, as is the quality assurance system of the laboratory practice.

In the UK national surveillance of doner kebab meat demonstrated that the greatest microbiological risk was from *C. perfringens* surviving the cooking process, but prevention of outbreaks of food poisoning from doner kebab meat depends on the use of high quality raw meat and on its sound hygienic management. An outbreak of *Salmonella mikawasima* food poisoning, associated with doner kebabs illustrated failure in these respects.

5.6   REFERENCES

Adams, M.R. and Moss, M.O., 2000. *Food Microbiology*, 2nd Edition. Royal Society of Chemistry, Cambridge, UK.

Al-Sheddy, I.A., Fung, D.Y.C. and Kastner, C.L., 1995. Microbiology of fresh and restructured lamb meat: a review. *Critical Reviews in Microbiology*, **21** (1), 31–52.

Anon, 1990. Food Safety Act, HMSO, London. Code of Practice no. 7: Sampling for analysis and examination, HMSO, London.

Anon, 1993. Update: Multistate outbreak of *Escherichia coli* O157:H7 infections from hamburgers – Western United States, 1992–1993. Morbidity and Mortality Weekly Report, Centers for Disease Control and Prevention, MMWR, **42** (14) April 16, 258–263.
[accessible at http://vw.cfsan.fda.gov/~mow/O157up.html]

Berkelman, R.L., Bryan, R.T., Osterholm, M.T., LeDuc, J.W. and Hughes, J.M., 1994. Infectious disease surveillance: a crumbling foundation. *Science*, **264**, 368–370.

British Standards Institution. BS5763: Microbiological examination of food and feeding stuffs:

Part 0. General Laboratory Practices, BSI, London, 1986.

Part 1. Colony count technique at 30 °C, BSI, London, 1991.

Part 2. Enumeration of coliforms. Colony count technique, BSI, London, 1991.

Part 3. Enumeration of coliforms. Most probable number technique. BSI, London, 1991.

Part 4. Detection of Salmonella. BSI, London, 1990.

Part 8. Enumeration of presumptive *Escherichia coli*. Most probable number method, BSI, London, 1994.

Part 10. Enumeration of Enterobacteriaceae, BSI, London, 1986.

Part 13. Enumeration of *Escherichia coli*. Colony count technique at 44°C using membranes, BSI, London, 1989.

Part 14. Detection of *Vibrio parahaemolyticus*, BSI, London, 1991.

Part 17. Detection of thermotolerant Campylobacter, BSI, London, 1996.

British Standards Institution. BS EN ISO 11290-1:1997, BS 5763-18:1997. Microbiology of food and animal feeding stuffs. Horizontal method for the detection and enumeration of *Listeria monocytogenes*. Detection method.

British Standards Institution. BS EN ISO 7932:1998. Microbiology. General guidance for the enumeration of *Bacillus cereus*. Colony count tehnique at 30°C.

Bryan, F.L., Stanley, S.R. and Henderson, W.C., 1980. Time/temperature conditions of gyros. *Journal of Food Protection*, **43**, 346–353.

CDC (Centers for Disease Control and Prevention), 1994. *Addressing emerging infectious disease threats: a prevention strategy for the United States.* US Department of Health and Human Services, Washington, DC, USA.

Department of Health, UK, 1994. *Management of outbreaks of food borne disease.* A booklet published by Two Ten Communications, West Yorkshire, UK.

FSIS (Food Safety and Inspection Service, Microbiology Division, US Department of Agriculture), 1994. *Nationwide beef microbiology data collection program: steers and heifers* (October 1992–September 1993), USDA (United States Department of Agriculture), Washington, DC, USA.

Gilbert, R.J., de Louvois, J., Donovan, T., Hooper, W.L., Nichols, G., Peel, N.R., Ribiero, C.D. and Roberts, D., 1996. Microbiological guidelines for some ready-to-eat foods sampled at the point of sale. *PHLS Microbiology Digest*, **13**, 41–43.

ICMSF, 1998. *Micro-organisms in Foods*, Volume 6. *Microbial Ecology of Food Commodities*. Blackie Academic and Professional, London, UK.

Jay, J.M., 1997. Do background organisms play a role in the safety of fresh foods? *Trends in Food Science and Technology*, **8** (12), 421–424.

LACOTS/PHLS, 1996. [Nichols, G., Monsey, H., de Louvois, J.] *Study of the microbiological quality of doner kebab meat,* LAC 13/96/3.

Matossian, R. and Kingcott, E.W., 1979. The doner kebab – a possible food poisoning hazard. *Environmental Health*, **86**, 67–68.

McNamara, A.M., 1995. Establishment of baseline data on the microbiota of meats. *Journal of Food Safety*, **15**, 113–119.

PFMG (Professional Food Microbiology Group), Institute of Food Science and Technology, UK, 1997. Development and use of microbiological criteria for foods. *Food Science and Technology Today*, **11** (3), 137–177.

Synnott, M., Morse, D.L., Maguire, H., Majid, F., Plummer, M., Leicester, M., Threllfall, E.J. and Cowden, J., 1993. An outbreak of *Salmonella mikawasima* associated with doner kebabs. *Epidemiology and Infection*, **111**, 473–481.

Todd, E.C., Szabo, R. and Spiring, F., 1986. Donairs (gyros) – potential hazards and control. *Journal of Food Protection*, **49** (5), 369–377.

Microbial Hazards

Key issues
- Fish – caviar and fish roes
- Microbial hazards – spoilage organisms, and pathogens
- Fish: a primary source of organisms
- Points of safety and quality loss
- Microbial hazards – what are they?
- Controls for microbial hazards
- Hazard analysis

Challenge

Caviar is a delicate, rare, expensive commodity and as such you would expect that great care would be taken at every stage of its production. This case study should lead you to explore which microbial hazards need be controlled to assure its safety and quality. It should also cause you to consider how economic forces influence perceptions of safety and quality, and thus how control systems bend to those forces, whether in an ideal world they should or should not.

6.1 THE CASE STUDY: THE ILLEGAL CAVIAR INDUSTRY IN THE LATE 20TH CENTURY

6.1.1 Sturgeon and Caviar

- Caviar – the pickled roe of sturgeon or other large fish eaten as a delicacy (Oxford Concise Dictionary, Anon 1995b).
- The name 'caviar' unqualified may be applied only to the eggs of the sturgeon prepared by a special process (FDA definition: Anon, 1997).

The eggs of the various sturgeon fish species differ in size, colour, flavour, texture and commercial value. Because of the rarity of the source fish, caviar attracts extremely high prices per kilogram. The three most expensive types are beluga caviar from the sturgeon species *Huso huso*, followed by osietra from *Acipenser gueldenstaedtii* and sevruga from *Acipenser stellatus*. Small portions served in luxury restaurants demand prices which fluctuate according to availability, but examples are $69.50 for 30 g at Petrossian in New York in 1995 (DeSalle and Birstein, 1996); 150 000 lire for 35 g beluga caviar in the Splendido Hotel, Porto

Fino, Italy in November 1998 and €139 for 30 g in the Shelbourne Hotel, Dublin, March 2000 (personal observation). These figures put the 'street value' of caviar in the millions of pounds sterling per tonne.

Sturgeon are large fish which can exceed 4 m in length and live as long as 200 years. Fish of this size and age are rarely caught for they have been overfished for decades to a point where now, at the turn of the new century, they are severely endangered. As the fish stocks are depleted so the rarity of the fish and their product, caviar, increases, and the value goes up again. All sturgeon fish, of which there are at least 20 species, are now protected by international treaty and since 1 April 1998 have been listed in Appendix II of CITES* (Anon, 2000). Under this treaty all international trade in sturgeon fish and their products, including caviar, must comply with its provisions.

According to a Traffic/WWF report (De Meulenaer and Raymakers, 1996) just seven countries provide 99% of the legal caviar trade most of which (90%) originates from the Caspian Sea region. On the consumer side 95% of legally exported caviar goes to just four major importers: the European Union, Japan, Switzerland and the USA.

However, the high value of the eggs together with massive political change seen in the primary producing areas (former USSR and Iran) in the last part of the 20th century have led to an illegal trade in caviar. In 1997 the World Conservation Union (IUCN) held a workshop (Birstein *et al.*, 1997) considering 'Sturgeon stocks and caviar trade', generally evaluating the state of the fish populations and the forces driving these fish to extinction.

One of the papers presented at that workshop (Taylor, 1997) states an opinion of the state of the caviar trade in 1997, and it forms the basis of this case study.

6.1.2   The Historical Development of the Caviar Trade and the Caviar Industry

Source: S. Taylor of Dieckmann and Hansen Caviar GmbH, Hamburg, Germany. (Abstracted with abbreviations with permission from Birstein, V.J., Bauer, A. and Kaiser-Pohlmann, A. (Eds.), 1997. *Sturgeon stocks and caviar trade workshop.* IUCN, Gland, Switzerland and Cambridge, UK, viii + 88 pp.)

"History of the caviar trade and the present situation

By the end of the 19th century, the amount of Russian caviar being exported already exceeded 100 tonnes. Even then, Hamburg was the main trans-shipment port for Russian caviar. From Hamburg, caviar was exported to Paris, New York, Rome, London and Stockholm. Most of the caviar was produced from sturgeon females caught in the northern part of the Caspian Sea, some from the Baku region (capital of Azerbaijan), and an even smaller number captured in the rivers of eastern Siberia. After the Russian Revolution of 1917, the Soviet Union owned all the fishing rights on the Caspian sea. Only in 1953, Persia (Iran) obtained the right to fish in the southern part of the Caspian Sea. The Iranian

---

*CITES Convention on International Trade in Endangered Species of Wild Fauna and Flora.

state-owned fishing industry employed Russian scientists to organise caviar production until the Iranian Revolution in 1979–1980. Even several years after the Revolution, the Soviet Union was still buying caviar from Iran in order to meet its own very high domestic demand.

As the Western countries became more affluent from the 1950s onwards, the demand for caviar grew considerably. During the First and Second World Wars sturgeon fishing had had to be halted and as a result fish stocks increased. The Soviet Union traditionally exported only 10% of its produced caviar. The total amount of caviar produced in the USSR in the 1980s was over 2000 tonnes a year, 140 tonnes of which were of export quality. In Iran, the sturgeon catch increased greatly after 1981. Almost the entire amount of caviar produced (about 300 tonnes) was exported. Although sturgeon fishing and caviar production are still under the control of the Iranian state-owned fishery (SHILAT), caviar started to be smuggled out of Iran in the 1980s; the illegal trade reached its peak (70 tonnes) in 1983. This so called 'bazaar caviar' was produced from the roe of the illegally caught sturgeons. Roe processed at 'home kitchens' in unhygienic conditions reached Western countries through unusual channels in extremely primitive tins with rubber ring seals made from old tyres. In the Western countries, this caviar was sold by Iranian refugees who had fled the revolution. The reputation of Iranian caviar suffered from this poor quality, and hence its price was low. Draconian measures conducted by the Iranian Government over ten years eventually brought smuggling back (down) to its pre-revolution levels of about 2–4 tonnes annually. At about the same time, however, sweeping changes began in the Soviet Union, and by the time the country had collapsed, caviar was starting to be smuggled from there. This trade reached about 100 tonnes in 1993, even though caviar was on the Russian list of strategic export commodities (or, precisely because it was on that list).

By 1994, mainly the Poles, along with citizens from other countries of the former Eastern bloc smuggled caviar out of Russia. They repacked it into their own tins and jars which imitated original packaging. So, the old caviar was sold as the 'new catch'. This, in turn, not only damaged the previously more or less spotless reputation of Russian caviar, but also caused a sharp decline in caviar prices on the world market. The decline in prices caused an increase in worldwide demand for caviar outside the producing countries from approximately 300 tonnes per year (which until then had been a 'normal level') to about 450–500 tonnes. Two international markets have developed: one for so-called 'people's caviar', and another one 'top-class caviar' for expensive hotels and restaurants. The 'people's caviar' was sometimes sold for a third of the price of the 'top-class caviar'. The demand from individual airlines rose to 18 tonnes a year, and supermarkets, especially in France, were buying up to 8 tonnes a year to be sold as a special offer during the Christmas season, mostly in 50 g tins costing DM12 per tin, or DM240 per kg net weight, duty and tax paid. In Germany, the customs duty alone accounted for 30% of the price, and in France for 18% of the price. The quality of the caviar was (and still is) below any acceptable level!

From the early 1950s and until the 1970s, there were about 10 caviar trade companies in the Western countries; Dieckmann and Hansen GmbH is the oldest of them. The caviar trade was strictly regulated. Only experienced companies with the necessary expertise were involved in the business. For decades, the Soviet Union had one or two general importers in the Western countries: two in Germany, one in France, one in Britain and one in Japan. For political reasons, the USA was not supplied directly, but traditionally mostly *via* Germany. Also, there were shipments of caviar from Iran every 5 years to the United States, South American countries, Europe, Japan and the Middle East countries until the 1979/80 revolution. The total amount of caviar exported from Iran was 115 tonnes per year, and the Soviet Union bought 30–70 tonnes annually from Iran. Dieckmann and Hansen GmbH was the only general agent for both the Russian and Iranian caviar. After the revolution, the management of SHILAT was replaced. The despised 'traders' were disposed of or had already emigrated. The Persian businessmen who had emigrated set up the cheap trade in exile, just as Russian emigrants became caviar importers in various countries (primarily the USA) after the collapse of the USSR.

Caviar trade and crime

As a result of developments, particularly in the CIS* the caviar trade gradually became a victim of organised crime. Many former Persian smugglers and Russian and Polish emigrants and caviar dealers organised caviar shipments from the large former Soviet companies in Astrakhan and Guryev (Kazakstan), which were then, however, not paid for. These activities were partly run as barter transactions. For example, a German construction firm in Kazakstan signed a contract to build 40 houses in return for 24 tonnes of caviar. These 24 tonnes were never paid for, *i.e.* the houses were never built. A barter transaction in Astrakhan (road construction in return for caviar) ended up with the caviar being exported to Alaska, where it was received, without the road construction ever being started. Altogether, bartering totally ruined the caviar trade as there was and still is, fraud and deception on all sides. The Iranian Company SHILAT has never traded caviar in barter transactions but only in return for direct payment in hard currency.

In the CIS, caviar is today only supplied after 100% prepayment. Trade is continuing to suffer due to the privatisation of caviar production in the CIS. The vast majority of producers are only interested in earning 'quick bucks' through the export trade. The World Bank's requirement of embracing the market economy has almost inevitably led to privatisation of the caviar industry; it remains unclear whether the bank's loans to caviar producers were used properly or not.

Present quality of caviar

The fact that producers in the CIS have no knowledge of the regulations and standards by which the caviar business is conducted in the importing countries does not help in the international caviar business. More than ever, production

*Commonwealth of Independent States (of the former Soviet Union).

today is not geared to world market requirements, but still follows the old soviet quality and packaging standards. This means that the importers need to repack caviar themselves, which is very expensive, and moreover further damages the already inferior quality of the caviar.

At present, properly packed caviar of export quality is available in small quantities only in Kazakhstan. Russia, with its main caviar production centre in the city of Astrakhan, no longer produces any caviar of acceptable quality. In Azerbaijan, the amount of legally and illegally produced caviar has increased significantly since 1991. However, due to very poor conditions in the production, packaging and transportation stages, almost 80% of the Azerbaijan caviar is fit only for disposal. During the past few years, smuggling on a large scale and the export of caviar from Azerbaijan *via* Turkey and Dubai, as well as Germany and the USA, have developed. These processes are depleting the already extremely scarce sturgeon stocks in the Caspian Sea.

The prices at which caviar is sold by legal and illegal producers in their home markets have risen considerably too. The smugglers, however, have an important advantage over Western buyers of exported goods: in the old Soviet Union, and now Russia, the beluga, osetrova or sevruga caviars are sold for the same price. However, because of the terrible quality of the Russian caviar, especially that coming from Astrakhan (in particular, the osetrova from the Russian sturgeon caught in summer), the smugglers could barely make a profit in 1995 since the importers refused to pay more for osetrova than for sevruga (which is cheaper on the international market) due to its poor quality. The price of osetrova will probably soon reach the same low price as that of sevruga. As the selling price for beluga has already been very high for years, the demand for this caviar has sharply declined in the Western countries, and now the world traders buy only the tiniest quantities of it in Russia and Iran. [. . .]

Preservation of sturgeon caviar

According to the traditional Russian method of caviar processing salt was added to the eggs for caviar preservation. Over the past few decades, a mixture of salt with borax has been used for this purpose in order to reduce the salt content in caviar. Borax is the salt of boric acid, and its use in caviar has been made legal again in Europe since February 1995, although its maximal content should be less than 4 g/kg. The United States and Japan continue to prohibit the addition of borax, allowing only caviar with salt to be imported. Since the Soviet Union produced caviar according to its old GOST standards, it was clear which caviar contained salt and which contained an admixture of borax. Currently, every second shipment contains either too much or too little borax. This creates another difficulty for the importers, who now more than ever must comply with the European laws and regulations governing food stuffs. As the imported goods have to be paid for in advance, importers are now buying 'a pig in a poke'. Any hope of retrieving money paid to the CIS countries is so illusory that no attempt has been made.

European hygiene regulations

The European hygiene regulations relating to caviar production have been in force since July 1993. They stipulate the conditions required in the processing of fish and fish products in order that exporters are eligible to export their products to member countries of the European Community (EC). Despite numerous efforts made by the caviar industry (including efforts at the political level), it has not been possible to enforce these regulations. Even where the factories are in relatively good shape (as in Astrakhan, for instance), health and safety conditions throughout the CIS are still disastrous by European standards. According to the provisional regulations, a state from which a processing factory may export its fish production over a certain period of time must use a health certificate issued by local authorities. This practice has been undermined by the fact that every stamp and every signature in the CIS and such 'transit countries' as Dubai, Turkey and Poland can be bought. It is virtually impossible for importers and the EC authorities to determine which stamps are genuine and which are not. At present, the provisional regulations will apply another two years. After that every factory in a third country must be licensed by the EC authorities before it will be allowed to export its production to the EC countries. It is doubtful that there will be any improvement in the conditions of caviar production in the CIS in these two years."
[end of extract]

6.2 BACKGROUND

Information concerning the production, processing and quality of caviar.

6.2.1 Production and Processing

(Abstracted, and abbreviated with permission from De Meulenaer, T. and Raymakers, C. (1996). *Sturgeons of the Caspian Sea and the international trade in caviar.* TRAFFIC International, Cambridge, UK.)

"Caspian caviar production

The fragility of caviar requires that the oocytes are extracted from the sturgeon and processed with great care immediately after death. The quality of the final product depends on the degree of maturity of the oocytes and on the working conditions which should follow strict sanitation rules. Refrigeration at below 4 °C should be maintained and oxidization prevented: successive reconditionings (the trade term equating roughly to 'repacking') do not always respect these requirements, in which case quality degradation ensues.

Sturgeons are usually still alive when they arrive at the processing plant, or as sometimes in the CIS, factory ship. In Iran the sturgeon are measured, weighed and individually registered with a number relating to the processsing plant and the fish, and this number is marked on the side of each can of Iranian caviar.

Immediately after the killing the oocytes are extracted from the gonads of the female, and then washed and sieved to separate them from surrounding membrane, and re-washed. They are then examined by a specialist who will determine their quality according to hardness, size and colour. If preservation with salt or borax is intended, the oocytes are stirred in brine of 2.8–4.4% salt and borax (a preservative added in a ratio of nine units of salt to one unit of borax) for two to three minutes, until they reach the right firmness. If stirred too long, the caviar becomes too sticky for consumers' taste. Alternatively, a certain proportion undergoes pasteurisation as a means of preservation.

Once preserved, the caviar is packed in tin boxes (in the former USSR all of which were numbered), standard sizes 0.5 or 1.8 kg (but net contents varies), coated on the inside with an anti-oxidant for enhanced keeping qualities. On arrival at destination, most caviar is re-packed in smaller tins or jars (typically for quantities of 30 g, 50 g, 125 g, 250 g, 500 g and 1 kg) by importers, and labelled under their name for retail. Repacking from wholesale containers to retail jars and tins is done by specialists and must be carried out within minutes and according to precise temperature and sanitary specifications. The distribution of caviar in retail packs continues within an uninterrupted 'cold chain' to the retail outlet. In Iranian practice the tins are stored at $-2\,°C$ or $-3\,°C$ which does not freeze the caviar because of its natural oil content and salt additives.

Packaging (in the CIS) has allegedly undergone serious changes since the days of the USSR. The traditional 1.8 kg or 0.5 kg tins are slowly being replaced as Russia and Kazakhstan try to introduce their own retail packaging. Conveyor belts have been installed for packing caviar in glass jars and tins for retail sales. About 50% of Russian caviar is pasteurised, before being vacuum packed in glass jars of 56 g or 113 g capacity, or in 90 g tins. In the long run this new trend could eliminate repacking by importers, making quality control more difficult for them.

Poor quality caviar

Owing to the fragility of fish eggs, it is easy to produce poor quality caviar, which falls below market standards. Typical factors leading to sub-standard caviar are:

- use of oocytes which are either too mature, or not ready to be harvested from immature females;
- poor preparation, for instance, through insufficient washing and sieving, which leaves traces of blood and membrane, or by using dirty salt, or too much salt;
- use of unclean water for washing the oocytes;
- use of poor packaging, which arrives at its destination rusted or with holes;
- inappropriate temperature control, for instance, interruption of the cold chain, where temperatures rise above $4\,°C$;
- repacking of caviar under careless sanitary and temperature conditions on importation;
- the mixing of oocytes from different sturgeon species.

Such conditions occur mostly with caviar produced and transported illegally by smugglers who do not have proper facilities for the trade, and by new 'official producers' who may lack the necessary expertise. Once again, therefore, the chaotic regulation of the caviar industry in parts of the Caspian basin has created a situation which is potentially or actually threatening the viability of sturgeon populations there."
[end of extract]

6.2.2   Pasteurisation of Caviar

This is a mild heat treatment normally received as a final step when the caviar is vacuum packed within its container. According to Hauschild and Hilsheimer (1979) the maximum temperature used is less than 70 °C. But great care has to be taken to ensure that the eggs are not exposed to temperature/time combinations which 'cook' some or all of the eggs causing total quality loss. Some producers ensure that the central temperatures inside the packages do not rise above 63 °C, the duration of exposure of the containers to the external heat source (steam) being dependent on size. The containers then have to be cooled, and remain cool for the duration of the shelf life, which should be in the region of 12–15 months under refrigeration close to 0 °C.

6.2.3   Potential for Microbiological Problems

Hauschild and Hilsheimer (1979)* say

"The pasteurisation treatment received is insufficient to control growth or toxigenesis of *C. botulinum*. This control is generally obtained by a combination of adequately low levels of water activity and pH. Most caviar products contain 0.1% sodium benzoate for control of yeasts and moulds, but it is unlikely that above pH 5.0 it has sufficient effect on clostridia. The safety record of caviar is good, though not without blemish. Imported caviar was incriminated as the cause of one type B botulism outbreak in Japan, and an isolated case of type E botulism from caviar was mentioned by Sebald. In a recent report we listed an incident of suspected botulism in Montreal, Canada, with imported lumpfish caviar as the most likely cause. On the other hand, numerous incidents of botulism have been caused by raw, fermented fish eggs, a traditional Indian food on the North American West coast, and by home-prepared or semi-commercial raw salted fish eggs around the Caspian Sea."

These authors set up a series of experiments to establish the conditions under which growth of *C. botulinum* and toxin production would occur and their results are shown in Table 6.1.

* Hauschild, A.H.W. and Hilsheimer, R., Bureau of Microbial Hazards, Health Protection Branch, Tunney's Pasture, Ottawa, Ontario, Canada K1A 0L2, 1979. Reprinted with permission from *Journal of Food Protection*. Copyright held by the International Association for Food Protection, Des Moines, Iowa, USA.

Table 6.1 *The effect of interactions of sodium chloride concentration and pH on toxin production by strains of C. botulinum inoculated into 50 g jars of caviar*

			Strains							
			A	A	B	B	E	E	A+B	A+B
			Number of weeks the jars were incubated							
Conditions			2	4	2	4	2	4	2	4
pH	a_w	brine (%)	Number of jars (out of 3) showing toxicity							
5.0	0.978	3.95	0	0	0	0	0	0	0	0
5.0	0.986	2.27	0	0	0	0	0	0	0	0
5.2	0.978	3.95	0	0	0	0	ND	ND	0	2
5.2	0.986	2.27	0	0	0	0	0	0	0	0
5.4	0.974	4.67	ND	ND	0	0	ND	ND	0	0
5.4	0.978	3.95	ND	ND	1	1	0	0	2	2
5.4	0.986	2.27	ND	ND	1	2	0	0	0	1
5.6	0.968	5.56	0	ND	0	0	0	0	0	0
5.6	0.974	4.67	0	2	0	0	0	0	0	1
5.6	0.978	3.95	1	3	0	0	0	0	1	1
5.6	0.986	2.27	3	3	1	2	0	0	2	3
5.8	0.959	7.09	0	0	0	0	0	0	0	0
5.8	0.968	5.56	0	0	0	0	0	0	0	0
5.8	0.974	4.67	0	2	0	2	0	0	0	0
5.8	0.978	3.95	3	3	0	0	0	0	1	3
5.8	0.986	2.27	3	3	0	2	0	0	2	3

Notes: pH adjustment by citrate buffers.
Temperature of storage: jars were heated to an internal temperature of 62 °C then cooled, and incubated anaerobically at 30 °C.
Initial inoculum = 50 000 spores per 50 g jar.
Strains:
A A-6
B 13983-IIB (proteolytic)
E Gordon strain
A + B a mixture of A and B strains: A-6, A-62, CK2-A; 13983-II B and 426-B.
ND = not determined.
Source: Hauschild and Hilsheimer, 1979.

When samples of caviar have been analysed considerable variation in their salt contents and pH values have been found, and this, of course, has bearing on the control of micro-organisms contained within the packages. The old standard of the USSR (the 'GOST' standard 7442-79) (Anon, 1979) for canned caviar required sodium chloride concentration to be between 3.5 and 5.0%; transportation at 0 °C to − 3 °C; storage in refrigerated railway wagons to be for a period of less than 7 days, and storage in central railway depots to be at − 2 °C to − 4 °C. It further states that the maximum period for the storage of caviar not containing preservatives should be 2.5 months from the date of manufacture. But analyses in Germany of eight samples of genuine 'low salt' caviar described as originating from Russia, Iran or the Caspian Sea showed the range to be between 1.7% and 5.2% salt, two samples to contain boric acid (0.32 and 0.57%) at a time when that was prohibited in Germany, and one sample to have high levels of microbial contamination while within its stated shelf life period (Brunner, B. *et al.*, 1995). DeSalle and Birstein (1996), working in the USA, found by using the polymerase chain reaction technique (PCR) that the contents of cans of caviar of 23 different commercial samples were not always what they purported to be and they noted that there were also some labelling inconsistencies.

6.3   EXERCISES

According to the Federal Register (of the USA) 60FR 65095 December 18, 1995, "Procedures for the safe and sanitary processing and importing of fish and fishery products: Final Rule" (Anon, 1995) a *"food safety hazard* means any biological, chemical or physical property that may cause a food to be unsafe for human consumption".

This is a specific definition focussing on safety, and today when the hazard analysis critical control point system (HACCP system) is well established as a controlling strategy, many food companies broaden the definition to include micro-organisms responsible for loss of product quality, and implement HACCP with both human safety and product quality in mind.

Secondly the Federal Register (*op. cit.*) describes

"hazard analysis: every producer shall conduct, or have conducted for it, a hazard analysis to determine whether there are food safety hazards that are reasonably likely to occur for each kind of fish and fishery product processed by that processor and to identify the preventative measures that the processor can apply to control those hazards. Such food safety hazards can be introduced both within and outside the processing plant environment, including food safety hazards that occur before, during and after harvest. A food safety hazard that is reasonably likely to occur is one for which a prudent processor would establish controls because experience, illness data, scientific reports, or other information provide a basis to conclude that there is a reasonable possibility that it will occur in the particular type of fish or fishery product being processed in the absence of those controls."

The four exercises (below) are asking you to think about the nature of microbiological hazards and what strategies can be used to control them, and what happens if some, or all of the controls are ineffective. Caviar is a food product which is highly valuable, limiting the possibilities for extensive end product testing, and which, because of its exceptionally high cost, is a luxury not familiar to many people and so limiting our anecdotal knowledge of a product and forcing us to ask some fundamental questions.

Exercise 1. Firstly in the light of the legal and illegal production and distribution of caviar for export as described, and secondly focusing on general microbiological issues, analyse where, when and why the caviar loses quality or becomes unsafe. Then, research the specific microbiological hazards to which caviar is potentially subject.

Exercise 2. Assume yourself to be a caviar importer (intending to buy legally produced product) who repacks it and sells it on to airlines, exclusive restaurants and retail outlets. What strategies could you adopt to ensure that the caviar you sold was of optimal microbiological quality, and safe to eat?

Exercise 3. If a new brand of caviar was already packed into its final container size before the point of import how could the authorities at the first port of entry into the country or into the European Union have confidence it was a safe product to allow in?

Exercise 4. Although on the face of it one might anticipate that blackmarket caviar would be of inferior quality, is that justified when considering only its microbiology?

6.4 COMMENTARY

You need to think logically.

Exercise 1. In the light of the legal and illegal production and distribution of caviar for export as described, and secondly focussing on general microbiological issues, analyse where, when and why the caviar loses quality or becomes unsafe. Then, research the specific microbiological hazards to which caviar is potentially subject.

Set up your own terms of reference.

Hazard

Establish how you are defining the word 'hazard'. Are you limiting its use to pathogens or are you broadening the definition to include spoilage organisms as well?

The process

Systematically determine the process between fishing, through processing, transportation, to re-packing, distribution, sale and use. Using the text a flow diagram (such as shown in Figures 6.1 and 6.2) can be drawn up. The process of doing this will indicate to you both where information is available and where it is lacking. It should stimulate lots of questions. At each step ask yourself whether you have sufficient detail. Don't look at the Figures yet.

The food – caviar

The factors which control microbial growth are primarily the various time/temperature relationships, the pH of the product and its water activity. Other factors contribute to microbial ability to grow and to population increase: the nutrients, microbial inhibitors, the presence of competing or antagonistic organisms, the availability of oxygen, and the gaseous environment.

Thus in order to think about the microbiology of caviar in production you must establish some values for these criteria. These may change throughout the process. Table 6.1 gives some potential values for a_w, pH and salt concentration in samples of caviar. This is a good starting point – although in practice different types of caviar for different markets may be more, or less salty, may or may not contain preservative, could have higher pH value, and so on. But from Table 6.1:

- a_w 0.959–0.986
- NaCl concentration 2.27–7.09% w/v
- pH 5.0–5.8
- presence of preservatives? borax is allowed in the EU but not the USA or Japan. Is sodium benzoate present?
- pasteurised? – this might be undertaken by a re-packer

- vacuum packaged? – similarly when repacking into small containers (30–100 g) it is likely that, after filling and sealing, the pack would be evacuated prior to pasteurising.

Remember that any particular batch of caviar would have specific values for each of the parameters listed – and so in your deliberations you must decide some values for 'your' batch(es) of caviar.

General microbiological issues which can then be addressed are sources of contamination (equipment, water, dirty salt, handling, dirt, other fish materials, detritus from previous fish and so on), exposure to temperatures which allow microbial growth, time periods when organisms have long enough to replicate leading to population increase. Whether growth occurs is of course dependent on whether contaminating organisms can grow in the conditions they find themselves in.

The microflora

As indicated above you need to think about from where the microflora which potentially threatens the safety or quality of the caviar might originate; and then you need to create a list of names of organisms (at a later stage look at Table 6.2).

You then need to decide how each organism might respond to the factors which affect its ability to grow, and its rate of growth. This then enables you determine whether each organism (*i.e.* each hazard) will in fact represent a risk to the safety or quality of the caviar. Clearly predicting how the organisms are likely to respond to particular sets of environmental parameters has to be based in knowledge and in experience (Table 6.3).

If you have access to a software package for predictive microbiology (such as the UK software model 'Food Micromodel') you could evaluate a whole range of combinations of controlling parameters, and see for yourself how the micro-organisms contained might respond.

You can also indicate on your flow diagram the points at which contamination with micro-organisms occurs; you can define values for key parameters such as water activity (a_w), NaCl concentration, pH, temperature values, times at different temperatures, whether the preservative borax is present or not; how the product is transported, the nature of any heat treatment, what sort of packaging is used at that stage, what the expected shelf life will be under defined conditions of storage. There is scope for much variation here – one person's flow diagram will differ from another's – but that will surely reflect the differences which truly exist in the production process.

Outcome

From your analysis of the process you define, you should be able to identify now where the caviar becomes contaminated, and the sorts of controls (within the product and through management of the whole process) which could be implemented to ensure that the microbial hazards cause loss of neither microbiological quality nor safety. You could identify a set of conditions in which the caviar

#	Process	Questions
1	Fishing	
2	Live fish transported to factory, or factory ship	
3	Individual fish measured, weighed and registered	
4	Fish killed	4.–9. What are the temperatures throughout?
5	Oocyte (egg) extraction from fish	5. How? Are the staff hygienic?
6	Eggs washed in water	6. How long? In what? What is the water quality? Is it fresh or salt water?
7	Eggs sieved to separate them from the membranes	7. A separate sieve for each fish? Is the sieve clean? Are the containers used clean?
8	Eggs examined and graded	8. Are the eggs touched? Are they sampled? Is the sampler clean?
9	Stirred in salt, or salt/borax mix	9. How much salt? Borax? Sodium benzoate? What is used to stir the eggs? How long is the process? How are the salted eggs separated from the salt? How long is the exposure to salt? Does it matter? What is the final salt content of the caviar? What is the resulting a_W and pH? temperature? Are samples and/or measurements taken?

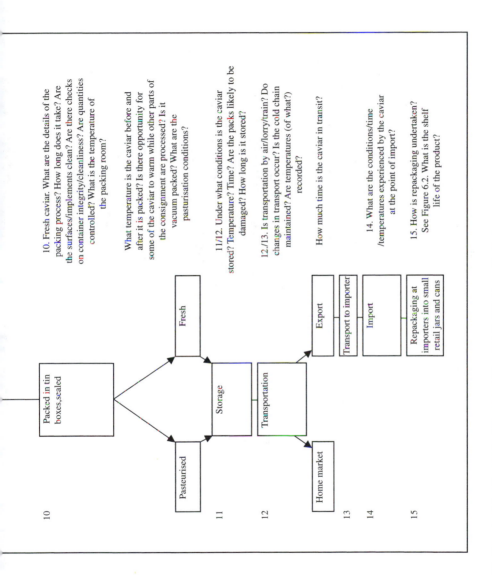

10. Fresh caviar. What are the details of the packing process? How long does it take? Are the surfaces/implements clean? Are there checks on container integrity/cleanliness? Are quantities controlled? What is the temperature of the packing room?

What temperature is the caviar before and after it is packed? Is there opportunity for some of the caviar to warm while other parts of the consignment are processed? What are the pasturisation conditions?

11/12. Under what conditions is the caviar stored? Temperature? Time? Are the packs likely to be damaged? How long is it stored?

12./13. Is transportation by air/lorry/train? Do changes in transport occur? Is the cold chain maintained? Are temperatures (of what?) recorded?

How much time is the caviar in transit?

14. What are the conditions/time /temperatures experienced by the caviar at the point of import?

15. How is repackaging undertaken? See Figure 6.2. What is the shelf life of the product?

Figure 6.1 *Possible flow chart for the production, processing and distribution of caviar*

Table 6.2 Sources of contamination and micro-organisms potentially affecting the quality or safety of caviar

Point of contamination	Reasons for contamination	Micro-organisms affecting quality	Micro-organisms affecting safety
Unsalted caviar (killing, oocyte extraction, washing, grading)	Ovaries will be free from micro-organisms until the fish is opened and oocytes removed. a_w of the oocyte mass will be high, $e.g.$ 0.99, pH at the moment of death close to neutral, $e.g.$ pH 6.8; temperature of the live fish will be that of the sea	Aerobic plate count organisms (bacteria), $e.g.$ $Aeromonas$, $Enterococcus$, $Flavobacterium$, $Lactobacillus$, $Micrococcus$, $Moraxella$, $Pseudomonas$ (Himelbloom and Crapo, 1998)	$Clostridia$: originating from the gut if there is delay between killing and removal of the oocytes; or careless practice in gutting leading to cross contamination. $Staphylococcus\ aureus$ from the handlers. $Vibrios$ from the sea water
	All this changes on death, and the egg mass is then subject to contamination through handling and processing. The microbiological quality of the water used in washing the egg mass is significant, as is cross-contamination from all equipment, and dust	Roe supports the growth of contaminating organisms: $e.g.$ Himelbloom and Crapo (1998) found that aerobic plate counts in roe on a production line varied between 1×10^3 cfu g^{-1} and 1×10^6 cfu g^{-1}, and that APC increased from 5×10^3 cfu g^{-1} to 5×10^6 cfu g^{-1} at 2 °C in 8 days; for further information see also Basby, Jeppesen and Huss, 1998a,b Water quality will affect microbial load	Faecal organisms from sewage polluted sea water, or from insects and rodents infesting the processing area. Examples: coliforms, $E.\ coli$, $Salmonellae$, $Shigella$
			Water could carry faecal pathogens

Salting (sodium chloride); preservatives	Dirty salt could add salt tolerant organisms such as *Micrococcus* and possibly salt tolerant yeasts and moulds. Sodium benzoate inhibits yeasts and moulds	The salt concentration in the salted caviar could range between 2 and 7% w/w lowering a_w in the process. Salt tolerant bacteria, yeasts and moulds organisms which survive and grow will spoil the caviar	Among the more salt tolerant pathogens are *Listeria monocytogenes*; *Vibrio cholerae* and *Vibrio parahaemolyticus*; *Staphylococcus aureus*. For their threshold limits for salt see Table 6.3
Pasteurisation (at whichever processing stage it occurs)	Reduces the numbers of heat sensitive organisms and contributes to keeping quality and shelf life extension	Heat resistant micrococci and sporing organisms liable to remain viable in the product	Survival of pathogens will depend on type and their heat resistance, their numbers and the heat regime applied
Packing for export and repacking once imported	Potentially anaerobic conditions inside the filled containers. Failures to fill containers completely could lead to an air space at the top of the product	Aerobic or anaerobic spoilage organisms could grow favoured by the particular conditions	Potential for *C. botulinum* to grow in anaerobic conditions

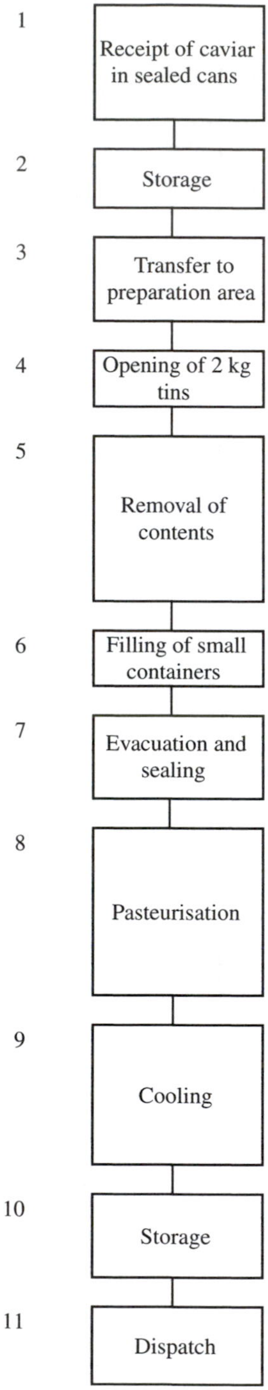
1
Receipt of caviar in sealed cans
2
Storage
3
Transfer to preparation area
4
Opening of 2 kg tins
5
Removal of contents
6
Filling of small containers
7
Evacuation and sealing
8
Pasteurisation
9
Cooling
10
Storage
11
Dispatch

Notes

1.What size cans? 2 kg? Is there a unique number for the fish/batch/factory – traceability? What are the acceptable procedures?

2. Time? Temperatures? Conditions?

3. What is the ambient temperature in the factory? Is the area optimally hygenic?

4. How? Are the staff trained in hygienic procedures?

5. How? Are the implements and utensils clean? Are any samples taken or tests undertaken? What records are kept? How long is the decanted caviar exposed to the air? How is quality judged?

6. How? Are containers checked for cleanliness, integrity?

7. What vacuum is drawn? Is the sealer calibrated? How often? Are samples of the filled cans tested?

8. What time/temperatures settings are used? Are all can sizes subject to the same heating cycle? How many cans are heated at once? Do cans in a batch receive varying treatments? What is the in-can treatment?

9.How are cans cooled? Is water used? Is it of good microbiological quality? Are the cans dried before storage? How? When is the product labelled? Do any cans leak?

10. Where? What temperature? How long? What is the shelf life of the product?

11. Is the cold chain maintained? Could the cans be subject to damage?

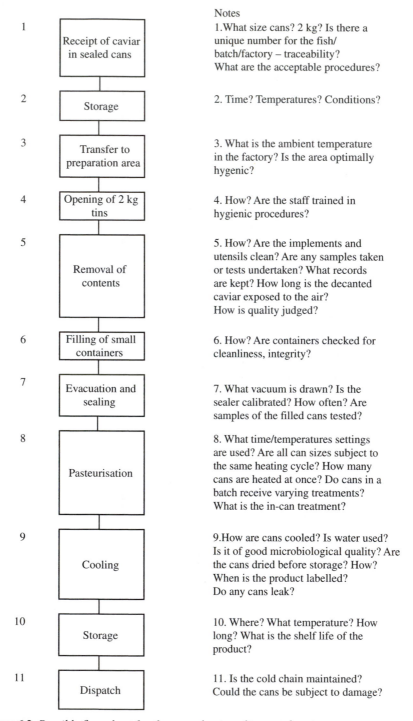

Figure 6.2 *Possible flow chart for the re-packaging of imported caviar*

Table 6.3 *Limits for growth of pathogens which could affect caviar*

	Lower temperature limit (°C)	pH	a_w (reduced by NaCl)	Approximate equivalent NaCl concentration (% w/v)
Listeria monocytogenes	−0.4 °C	4.4–5.0	0.92	13[a]
Vibrio cholerae	10	5.0	0.97	5[a]
Vibrio parahaemolyticus	5	4.8	0.94	10[a]
Staphylococcus aureus (growth)	7	4.0	0.83 (aerobic)	>25[a]
Staphylococcus aureus (toxin production)	10	4.5	0.87 (aerobic)	20[a]
Staphylococcus aureus (growth)			0.90 (anaerobic)	16[a]
Staphylococcus aureus (toxin production)			0.92 (anaerobic)	13[a]
Clostridium botulinum (Group I)	10	4.6	0.94[b]	10.0
Clostridium botulinum (Group II)	3.3	5.2	0.97[b]	5.0

Note: The limits given for any single parameter are valid only when all other growth parameters are optimal. Thus the minimum parameter value for growth tends to be raised when other parameters are less than optimal.

Key: [a]salt concentration interpolated from a_w value.

[b]water activity interpolated from given limiting NaCl concentration.

Sources: ICMSF, 1996; Jay, 1992.

production and export processes are in control and hence the product could have a predicted shelf life; yet you could also identify others where control is lacking.

> **Exercise 2.** Assume yourself to be a caviar importer (intending to buy legally produced product) who repacks it and sells it on to airlines, exclusive restaurants and retail outlets, what strategies could you adopt to ensure that the caviar you sold was of optimal microbiological quality, and safe to eat?

A number of points exemplified below can be considered and researched:

Production and export

- Is the source of the product certain?
- Do you know the producer and have you established a trusting relationship with them?

- Do you know the process used and do you have confidence in it?
- Have you specified the product you wish to buy?
- Do you know that the hygiene certificates which accompany the consignment are valid?
- Do you travel with the consignments from export to import?
- Does someone see them onto and off the planes?
- How is the temperature of the consignment maintained with the cold chain requirements?
- Is the cold chain unbroken?

Receipt of imported product at your repackaging plant

- Do you check your specification against the accompanying documents, and labels on the packages?
- Do you store the cans under refrigeration? At what temperature?
- On opening the cans what tests do you do? – organoleptic, salt concentration? pH? Do you test for (or send away samples for) borates, benzoic acid?
- How do you handle the product? Are the conditions optimally hygienic?
- How do you avoid cross-contamination of the caviar?
- Do you check weigh your repackaged caviar?
- How do you control the times and temperatures used for pasteurisation? How do you know that what you do is effective?
- What are the labelling regulations for the repackaged product?
- How do you know that the 'best before' or 'use by' dates are appropriate?
- Have you considered that your factory should implement a HACCP system?

> **Exercise 3.** If a new brand of caviar was already packed into its final container size before the point of import how could the authorities at the first port of entry into the country or into the European Union have confidence it was a safe product to allow in?

Port authorities at land frontiers, sea and airports in the European Union operate under regulations which ensure that food imported into the EU is of a suitable quality for use in human food, or for direct consumption. Each consignment of food is accompanied by sets of documents – particularly certificates of health verifying that the place of production is certified by the EU and meets the EU hygiene requirements (Anon, 1991).

The port inspectors have the authority also to take samples of imported foods for chemical and microbiological tests, and on the results of the tests the food may be allowed in, returned to its point of origin or destroyed.

In the case of caviar there is a dilemma. It comes in small quantities – perhaps a consignment may comprise eight cans each 1.8 kg in size. Each can is worth perhaps approximately $4000US or more, when divided into 30 g portions and sold retail. Does the inspector take one such valuable can, send it off for testing and then perhaps find that it meets all acceptable criteria, but is wasted in the process? Is the importer going to accept such loss? Or does the inspector not take a sample, and perhaps allow into the country poorly processed caviar and take the risk of it causing illness in consumers?

Exercise 4. Although on the face of it one might anticipate that black-market caviar would be of inferior quality, is that justified when considering only its microbiology?

The quality of caviar is judged organoleptically by the degree of maturation of the eggs, their uniformity of size, by appearance, by taste and by smell. Of course if it was deteriorating microbiologically there could be detectable signs through change in appearance, smell, texture, colour and flavour.

In the case study information provided in Sections 6.1 and 6.2 there are indications that black market caviar is often poorly produced in unhygienic conditions, in poor quality cans, and under poor temperature control. These indicate that there is opportunity for microbial contamination and microbial growth. But consider also that the blackmarketeers are in the game for money. They do not want the product to biodeteriorate – might they not be tempted to add extra salt, or borax, or benzoic acid? If they did that what might the effect on the microbial population be?

6.5 SUMMARY

The eggs of the sturgeon fish are converted into caviar by a salting process. They are very delicate and subject to damage, to quality loss through oxidation of the oils within them and to microbial deterioration.

Processed hygienically, and containing between 2 and 5% salt, held at between -2 and $+1\,°C$, the packaged product can remain of acceptable quality for up to 2.5 months. Pasteurisation extends the shelf life to approximately 12–14 months, again only if the caviar is stored at between -2 and $+1\,°C$.

The microbial hazards potentially affecting the microbial quality and safety of the product contaminate it in its processing and may grow if the salt content, water activity, pH or preservative concentration values alone or in combination are inadequate to prevent it.

The rarity of the sturgeon fish causes caviar to be of high value and this, together with political change in the major producing areas, has led to a black market. While appropriate controls through the fishing, production and distribution chain can assure very high quality product, the lack of those controls leads to the threat of the extinction of many sturgeon species, to quality loss in the caviar and the risk not only of waste, but also of botulism and the presence of other pathogens.

Caviar in the 1990s was a product where the control of safety and quality was subject to great variability, and assurance that much of what was sold as caviar was what it said it was, was safe and good quality, was in fact a game of 'Russian roulette'.

6.6   REFERENCES

Anon, 1979. Canned sturgeon caviar. Technical requirements. Union of Soviet Socialist Republics. Soviet Standard GOST 7442-79.

Anon, 1991. Council Directive 91/493/EEC of 22 July 1991 laying down the health conditions for the production and placing on the market of fishery products.

Anon, 1995. Procedures for the safe and sanitary processing and importing of fish and fishery products; Final Rule 21 CFR 123 and 1240. Federal Register 60FR 65095, December 18. Accessed at: http://vm.cfsan.fda.gov/ ~ lrd/searule3.html

Anon, 1995. *Oxford Concise Dictionary of Current English*, 9th Edition, Oxford University Press, Oxford, UK.

Anon, 1997. What guidance does FDA have for manufacturers of caviar and fish roe? Excerpt from: Requirements of laws and regulations by the US Food and Drugs Administration. Accessed at: http://vm.cfsan.fda.gov/ ~ dms/qa-ind4j.html

Anon, 2000. Sturgeon and paddlefish. Two years after the entry into force of the inclusion in CITES Appendix II. Briefing note from TRAFFIC International, Cambridge, UK. Accessed at http://www.traffic.org/cop11/briefingroom/sturgeon.html

Basby, M., Jeppesen, V.F. and Huss, H.H., 1998a. Spoilage of lightly salted lumpfish (*Cyclopterus lumpus*) roe at 5 °C. *Journal of Aquatic Food Product Technology*, **7** (4), 23–24.

Basby, M., Jeppesen, V.F. and Huss, H.H., 1998b. Characterization of the microflora of lightly salted lumpfish (*Cyclopterus lumpus*) roe at 5 °C. *Journal of Aquatic Food Product Technology*, **7** (4), 35–51.

Birstein, V.J., Bauer, A. and Kaiser-Pohlmann, A. (eds.), 1997. *Sturgeon Stocks and Caviar Trade Workshop*. IUCN, Gland, Switzerland and Cambridge, UK. viii + 88 pp.

Brunner, B., Marx, H. and Stolle, A., 1995. Compositional and hygienic aspects of commercial caviar. *Archiv für Lebensmittelhygiene*, **46** (4), 80, 83–85.

De Meulenaer, T. and Raymakers, C., 1996. *Sturgeons of the Caspian Sea and the International Trade in Caviar*. Traffic International, Cambridge, UK.

DeSalle, R. and Birstein, V.M., 1996. PCR identification of black caviar. *Nature, UK*. Letter to Editor, 16 May, **381**, 197–198.

Hauschild, A.H.W. and Hilsheimer, R., 1979. Effect of salt content and pH on toxigenesis by *Clostridium botulinum* in caviar. *Journal of Food Protection*, **42** (3), 245–248.

Himelbloom, B.H. and Crapo, C.A., 1998. Microbial evaluation of Alaska salmon caviar. *Journal of Food Protection*, **61** (5), 626–628.

ICMSF, 1996. *Micro-organisms in Foods*, Volume 5. *Characteristics of Microbial Pathogens*. Blackie Academic and Professional, London.

Jay, J.M., 1992. *Modern Food Microbiology*, Chapman and Hall, London.

O'Brian, A., 1995. HACCP of pasteurized caviar. MSc project dissertation, South Bank University, London.

Taylor, S., 1997. *The historical development of the caviar trade and the caviar industry*, in *Sturgeon Stocks and Caviar Trade Workshop*. IUCN, Gland, Switzerland and Cambridge, UK, viii + 88 pp.

Further references which may be of use

Anon, 1979. Pressed sturgeon caviar. Technical requirements. Union of Soviet Socialist Republics. Soviet Standard GOST 7368-79.

Anon, 1987. International outbreak of Type E botulism associated with ungutted, salted whitefish. *Morbidity and Mortality Weekly Report Centers for Disease Control and*

Prevention MMWR, **36** (49): 1987 Dec. 18.
Accessed at http://vm.cfsan.fda.gov/~mow/fishbot.html

Anon, 1996. Would you buy a stolen diamond? *Seafood International*, October, p. 39.

Anon, 1999. Salted uneviscerated yellow croaker recalled because of possible health risk. HHS News. Food and Drug Administration, USA.
Accessed at http://vm.cfsan.fda.gov/~lrd/hhscroak.html

Barthelemy, M., Arzouyan, C. and Estienne, J., 1993. Determination of boric acid in caviar by HPLC. *Annales des falsifications, de l'Expertise Chimique et Toxicologique*, **86** (921), 275–282.

Boiko, A.V., Pogorelova, N.P., Zhuravleva., L.A. and Lartseva, L.V., 1993. Microbial content of caviar from sturgeon. *Gigiena i Sanitariya*, **11**, 30–31.

Carlson, M. and Thompson, R.D., 1998. Determination of borates in caviar by ion-exclusion chromatography. *Food Additives and Contaminants*, **15** (8), 898–905.

Chen, I.-C., Chapman, F.A., Wei, C.I. and O'Keefe, S.F., 1996. Preliminary studies on SDS-PAGE and isoelectric focussing identification of sturgeon species of caviar. *Journal of Food Protection*, **61** (3), 533–535, 539.

Christie, T., 1995. Caviar still spells luxury. *Seafood International*, December, pp. 21, 23.

Fukuda, A., Kanzawa, N., Tamiya, T., Seguro, K., Ohtsuka,T. and Tsuchiya, T., 1998. Transglutaminase activity correlates to the chorion hardening of fish eggs. *Journal of Agricultural Chemistry*, **46**, 2151–2152.

Hjul, P., 1996. French farms are to market Fish of Kings. *Seafood International*, May, pp. 60, 61.

Jones, A., 1997. Farming comes to the rescue of caviar. *Seafood International*, October, pp. 24, 27.

Lyhs, U., Hatakka, M., Maki-Petays, N., Hyytia, E. and Korleala, H., 1998. Microbiological quality of Finnish vacuum-packaged fishery products at retail level. *Archiv für Lebensmittelhygiene*, **49** (6), 121–144.

Makarov, A., 1996. Russia sets course to boost fish consumption. *Seafood International*, May, pp. 51, 53.

Pavel'eva, L.G., Ushakova, R.F., Zhungiety, G.I. and Styngan, E.P., 1984. Preserving sturgeon roe. Union of Soviet Socialist Republics. USSR Patent SU 1 064 933 A.

Urch, M., 1999. Caviar crisis as catch controls fall. *Seafood International*, September, pp. 30, 31.

CHAPTER 7

Post-production Product Handling and Acceptability

Key issues
- Vegetables and fruit; canned tomato paste – a high acid food
- Spoilage organisms and *Clostridium botulinum*
- Canning
- Microbial death and commercial sterility
- Microbial growth conditions
- Control of food safety and quality during manufacture, and after
- Product acceptability

Challenge
Canned tomato paste, stored in tropical conditions, is known to be vulnerable to spoilage. The product is liable to deteriorate, the cans to swell and burst.

The case study presents you with some microbiological data for tomato paste and should cause you to think about the meaning of 'product acceptability'. It allows you the opportunity to identify for yourself the key microbiological issues in production which must be addressed to ensure that the product is microbiologically safe and stable in storage. It also enables you to consider whether the market for which a product is destined should be part of that process.

7.1 THE CASE STUDY: THE QUALITY OF TOMATO PASTE AVAILABLE IN NIGERIA

7.1.1 Consider the Following Correspondence:

Document 1: FAX
 To: 'Italian Tomatoes Ltd.', Milan, Italy
 From: 'Nigerian Food Importing Company Ltd.', Lagos, Nigeria
 Date: 1 January 1998

Disclaimer: The situation described in Section 7.1 is entirely fictional, and all company names, names of ships and personal names used in Section 7.1 are from the imagination of the author and do not relate to and are not intended to relate to any real people or companies of the same or similar names.

Sirs,
Re: Consignment of tomato paste ex the Tropical Princess arr. Lagos 24/9/97. dlvd.
our warehouse Lagos, 16.40 23/12/97
Further to our telephone conversations re the above container of canned tomato
paste, and the inspection thereof by Mr Akwere of Lagos Inspection Services
(LIS) on our behalf, we would like to inform you that we will be submitting a
claim to you *via* our insurers Messrs World Wide Insurance for the total loss of
the consignment covered by Bill of Lading ZYXWV246 the lot being unsaleable.
We attach the LIS report for your attention.
Regards,
Kingston Amusu,
Nigerian Food Importing Company Ltd.

Document 1a: Lagos Inspection Services (LIS) – Survey report: Extract
In accordance with instructions from NFIC we attended the warehouse of
NFIC, Lagos on 15 December 1997 in order to be present whilst the container
AABB 500869-9 was destuffed. The doors of the container were opened and it
was immediately apparent that the majority of the cans of tomato paste were
damaged, some merely being swollen, others having already burst and their
contents leaked over the cans in the vicinity. In consultation with the owners it
was decided to reclose the container and to remove it to an area outside the
warehouse, at which the cans could, if necessary, be disposed.
   At that location the container was reopened and with care the first stack of
pallets was removed. Careful examination of the first pallets confirmed the
impression that a number of the cans were a total loss, and that sorting of the
cans on that pallet would have only yielded in the region of 30% cans un-
damaged either by swelling or by leakage. Initially those pallets were put to one
side to allow removal of the second and third stacks of pallets where the same
problem became evident. The same careful examination was undertaken for the
whole consignment and it was clear that all pallets with the exception of one
pallet showed the same minimum damage level of 60–70%. The one pallet
aforementioned showed no visible damage – not one single can showing either
swelling or leakage. However, after discussion a decision was taken to remove all
the pallets from the vicinity and to stack them close to the area of disposal
pending decisions as to disposal.
   It is our opinion that the quality of the tomato paste was different in some way
from previous consignments, problems on this scale not having been encountered before.
   There being the associated danger in the swollen cans of vegetable products
that they may be toxic it is our recommendation that the whole consignment is
disposed of as being unfit for human consumption.
Signed,
A. Akwere
pp. Lagos Inspection Services

Document 2: FAX
From: Italian Tomatoes (Milan) Ltd., Italy
To: Nigerian Food Importing Company Ltd., Lagos, Nigeria.
Date: 5 January 1998

Sirs,
Re: Consignment of tomato paste ex the Tropical Princess arr. Lagos 24/9/97.
We note your letter of the 1/1/98 and advise you that we have put this matter in
the hands of our insurers 'Agenca Roma SRL' and reserve our position in the
meanwhile,
Yours etc,
Antonio di Blanco
Italian Tomatoes (Milan) Ltd.

Document 3: FAX
From: Italian Tomatoes (Milan) Ltd.,
To: Agenca Roma SRL
Date 5/1/98

Sirs,
Re: Consignment of tomato paste ex the Tropical Princess arr. Lagos 24/9/97.
We have received the enclosed notice of claim from our importers in Nigeria (Ref
Docs 1 and 1A) and place the matter in your hands.
Antonio di Blanco
Yours etc,
for and on behalf of Italian Tomatoes (Milan) Ltd.

7.1.2   What Then Happened

In view of the fact that it was undoubtedly the case that the Nigerian consignee,
'Nigerian Food Importers Ltd.' had received a consignment of canned tomato
paste (approximately 100 000 cans each 200 g net weight) which was clearly very
badly damaged and in no way should have been released for sale for human
consumption it was not surprising that the producer's insurers were inclined to
settle the claim. However, they made a counterclaim against the insurers for the
shipping company claiming that the damage clearly occurred while the product
was in transportation from Italy to Nigeria, and also pointing out that the
temperatures inside the metal 20 foot containers can reach, when in the sun, in
excess of 50 °C. This of course triggered a further counter claim from the shippers
against the container yard in Lagos, at which the container resided for three
months after arrival there. That claim was refuted by the yard who argued that
the temperatures and times of storage were not their responsibility. But in the
end the claim between the disputing parties was settled: the importer was fully
compensated, and the insurers for the producer in Italy and the shipping company shared liability.
    However, it was not really the end of the case because the insurers for the

Italian producers had in fact received a number of claims prior to that one and they decided to investigate tomato paste exported to Nigeria. In this process their attention was drawn to some data produced by a group of Nigerian microbiologists who themselves had investigated the quality of tomato paste as sold in Nigeria. The expert for the insurers drew up a report and quoted the following information:*

"Tomato products are of considerable importance world wide (Gould, 1983; Sidhu *et al.*, 1984; Robinson *et al.*, 1994). In Nigeria tomato paste is the most important tomato product because of its widespread use for preparations of various foods/menus.

Until recent years, nearly all the tomato paste consumed in Nigeria was imported from European countries with Italy being the leading country. However, today limited brands are produced locally. Although both imported and the local brands are available on the market, consumers prefer the former in spite of its relatively higher cost.

In general adequate heat processing is given to tomato paste to achieve commercial sterility (Speck, 1984), but subsequent abusive post-process handling/storage may lead to undesirable microbiological changes (Anon, 1980; Speck, 1984). It is public knowledge that cans of tomato paste often show external evidence of spoilage under tropical retail conditions. In addition, and interestingly, these defective products are sold (especially to the less informed) at the same cost as the normal (non-defective) products.

In spite of the widespread occurrence of defective cans and the increasing international concern for food safety (Guardia and Harper, 1990; Motarjemi *et al.*, 1993), published information on the microbial profiles, safety and other quality changes in tomato paste retailed under tropical conditions appears to be lacking. Moreover, the potential hazards associated with imported and local brands are of concern, and such information is therefore needed. Consequently, the objectives of this work were (i) to determine the micro-organisms associated with retailed canned tomato paste (visibly normal and defective), (ii) to identify the isolates from samples (both normal and defective) and the possible health implications and (iii) to evaluate the changes in pH."

Tomato paste samples

"Three popular brands of tomato paste (two imported and one produced locally), designated A, B and C respectively in this work, were purchased from various retail outlets (local markets, local shops/stores and supermarkets) in Port Harcourt, Nigeria. All the brands selected for this study carried comparable expiry dates (*i.e.* 'best before' dates) [. . .]"

The report of the expert for the insurers also highlighted the following points from Efiuvwevwere and Atirike (1998):

* Efiuvwevwere, B.J.O. and Atirike, O.I.E. Microbiological profile and potential hazards associated with imported and local brands of tomato paste in Nigeria. *Journal of Applied Microbiology* (1998), **84**, 409–416. Quoted by permission of Blackwell Science Ltd., Oxford, UK.
Tables 7.1–7.6 are from the above reference and are modified by permission of Blackwell Science Ltd.

"Difficulties were encountered in obtaining these samples since retailers were suspicious of the selective purchase. In contrast visibly normal cans were obtained without difficulty, but the number of normal samples was dictated by the number of defective samples [. . .]"

"Bacteriological and mycological analyses were conducted within two days of sample purchase [. . .] Microbial analysis was carried out by taking small samples with sterile cork borers from the centre of the (opened) can and the immediate vicinity of the side seam [. . .]"

"Temperatures ranging between 24 and 38 °C are often encountered in the retail outlets from which the samples were purchased [. . .]"

"The spoilage in all the brands especially B, showed pronounced swelling [. . .]"

"It was observed that the tomato products did not show any sign (putrid odour) of hydrogen sulphide; rather a sour odour was commonly experienced . . ."

[End of quotes] [Note: references mentioned are included in the References – Section 7.6].

The report of the expert for the insurers also included tables of data based on the published data of Efiuvwevwere and Atirike (1998), and are shown here as Tables 7.1 to 7.6.

These data – although not specifically relating any known particular case – caused the insurers to raise a number of questions with the Italian producers, and also to further investigate practices in Nigeria.

Their fundamental concern was whether the tomato paste should be insured at the premium they were currently charging, or whether they should change their conditions of insurance. In making such a decision they needed to understand better the risk they were insuring.

At the same time, aware of the problem their paste suffered in Nigeria, the concerns of the producers were different. They were making the paste to the buyer's specification (they believed) and it was out of their control how it was handled and sold. But they too had to re-evaluate their processes.

The importers also checked their purchase specification. Yes, it had required the product to be commercially sterile, and so for future consignments they left the specification as it was. However, they also made a decision to buy more locally produced paste to supply their retailers in Nigeria, feeling that this approach would both introduce more control into their supply chain and facilitate discussion should the need arise.

From the standpoint of all these parties – the producers, the insurers, the importers, and also the transportation companies, the retailers and the consumers – the microbiology of the tomato paste was central to both the stability and the safety of the product.

The problem was of course how this would be achieved.

Table 7.1 *Tomato paste: viable counts[a] from the pooled samples obtained from visibly normal and defective cans*

	TVC		Fungal count	
Brands	$cfu\,g^{-1}$ $(log_{10}cfu\,g^{-1})$ Normal	$cfu\,g^{-1}$ $(log_{10}\,cfu\,g^{-1})$ Defective[b]	$cfu\,g^{-1}$ $(og_{10}\,cfu\,g^{-1})$ Normal	$cfu\,g^{-1}$ $(log_{10}\,cfu\,g^{-1})$ Defective[b]
A (13 cans of each type)	1.04×10^8 (8.02)	2.22×10^9 (9.35)	1.17×10^4 (4.07)	2.57×10^5 (5.41)
B (15 cans of each type)	4.26×10^8 (8.63)	1.38×10^{10} (10.14)	3.16×10^4 (4.5)	7.24×10^5 (5.86)
C (14 cans of each type)	6.91×10^6 (6.84)	9.1×10^7 (7.96)	2.57×10^3 (3.41)	6.02×10^4 (4.78)

[a] TVC counts on plate count agar (PCA) incubated at 37 °C/24 h. Fungal counts on malt extract agar, 27 °C/4–5 days. Each value represents the mean of six determinations (*i.e.* duplication analysis of three replicates).
[b] All defective cans were identified by visible swelling.
Source: modified from Efiuvwevwere and Atirike, 1998.

Table 7.2 *Tomato paste: mesophilic and thermophilic microbial populations from pooled samples obtained from visibly normal and defective cans*

	Microbial populations			
	Mesophilic (dextrose tryptone agar 37 °C/24 h; acid producing colonies)		Thermophilic (dextrose tryptone agar 55 °C/72 h; acid and non-acid producing colonies)	
Brands	Normal $cfu\,g^{-1}$ $(log_{10}\,cfu\,g^{-1})$	Defective[a] $cfu\,g^{-1}$ $(log_{10}\,cfu\,g^{-1})$	Normal $cfu\,g^{-1}$ $(log_{10}\,cfu\,g^{-1})$	Defective[a] $cfu\,g^{-1}$ $(log_{10}\,cfu\,g^{-1})$
A (13 cans of each type)	1.3×10^8 (8.11)	1.1×10^9 (9.04)	1.2×10^4 (4.07)	4.7×10^4 (4.67)
B (15 cans of each type)	2.1×10^8 (8.3)	1.2×10^9 (9.079)	5.3×10^4 (4.72)	6.9×10^4 (4.83)
C (14 cans of each type)	1.2×10^8 (8.08)	9.9×10^8 (8.99)	4.5×10^4 (4.65)	6.6×10^4 (4.81)

[a] All defective cans were identified by visible swelling. See note a in Table 7.1.
Source: modified from Efiuvwevwere and Atirike, 1998.

Table 7.3 *Tomato paste: viable spores population from pooled samples obtained*
from visibly normal and defective cans

	Aerobic[a]		Anaerobic[a]	
	Normal cans cfu g^{-1}	Defective cans[b] cfu g^{-1}	Normal cans cfu g^{-1}	Defective cans[b] cfu g^{-1}
Brands	(log_{10} cfu g^{-1})	(log_{10} cfu g^{-1})	(log_{10} cfu g^{-1})	(log_{10} cfu g^{-1})
A (13 cans of each type)	7.6×10^4 (4.88)	4.9×10^5 (5.69)	3.0×10^6 (6.48)	5.7×10^7 (7.75)
B (15 cans of each type)	1.2×10^4 (4.08)	4.3×10^5 (5.63)	1.6×10^7 (7.2)	1.1×10^8 (8.04)
C (14 cans of each type)	8.8×10^4 (4.94)	4.9×10^5 (5.68)	1.1×10^7 (7.04)	7.6×10^7 (7.88)

[a] 0.1 ml samples were exposed to 80 °C/10 min. on PCA and incubated aerobically or
anaerobically at 55 °C/3 days.
[b] All defective cans were identified by visible swelling.
Source: modified from Efiuvwevwere and Atirike, 1998.

7.2 BACKGROUND

Tomatoes are the fruits of the tomato plant, and as they ripen their sugar
content, mainly glucose and fructose, rises and the acidity of their juices reduces.
Citric acid is predominant together with traces of malic, succinic and other
organic acids. The pH of the flesh of the fresh ripe fruit is acidic and normally lies
in the region of pH 4.2–4.9. The actual value depends on how ripe the tomatoes
are, and their variety, and it can vary from batch to batch of fresh fruit and from
season to season. When they are processed into paste the water content is
reduced so its water activity, a reflection of the concentration of solutes (mainly
sugars), will be lower than that of the freshly pulped tomatoes, and thus more
restrictive of microbial growth. The water activity of canned pastes normally lies
between 0.93 and 0.98 a_w, and the pH between 3.5 and 4.7.

Since canned foods are classified into groups based on their pH tomato paste
is viewed as an 'acid food' (Table 7.7).

The pH affects which organisms can grow, whether spores can germinate, and
the heat sensitivity of both vegetative cells and spores. In most nutrient,
anaerobic environments *C. botulinum* cannot grow and produce toxin below pH
4.6, and its spores cannot germinate. So for safety the pH value 4.5 is regarded as
a threshold, and pH values above that are regarded as those in which *C.
botulinum* might grow.

The heat resistance of bacterial spores increases the closer the pH is to
neutrality, but is also affected by the nature of the food in which heating takes
place:

e.g. *C. botulinum* – proteolytic type A strain 62A in

    M/15 phosphate buffer   pH 7.0, $D_{110°C} = 0.88$ min, z = 7.6

    Canned peas   pH 6.0, $D_{110°C} = 1.22$ min, z = 7.5

    Canned peas   pH 5.24, $D_{110°C} = 0.61$ min, z = 7.6

    C. botulinum – proteolytic strain A16037 in

    Phosphate buffer   pH 7.0, $D_{110°C} = 4.29$–4.47 min, z = 9.43

    Tomato juice   pH 4.2, $D_{110°C} = 0.92$–0.98 min, z = 9.43

    (ICMSF, 1996)

These differing heat sensitivities affect the processing needs of foods.

Table 7.4 *Percentages of isolates found in normal and defective[a] tomato paste*

	Normal			Defective[a]		
	A imported	B imported	C Nigerian	A imported	B imported	C Nigerian
B. cereus	10.6	9.5	8.7	11.1	10.7	10.6
B. coagulans	23.2	24.6	30.9	25.6	22.6	26.3
B. subtilis	9.6	9.8	4.3	8.8	9.6	6.4
C. botulinum	9.0	5.9	3.1	9	10.8	5.7
C. pasteurianum	13.9	13.4	11.6	13.2	14.7	14.3
C. thermosaccharolyticum	12.4	10.6	9.4	16.5	15.4	17.5
Lactobacillus spp.	17.6	18.3	25.9	8.3	9.3	11.7
Leuconostoc spp.	6.7	7.4	4.8	6.8	6.1	5.6
Others	0.3	0.5	1.3	0.7	0.8	1.9

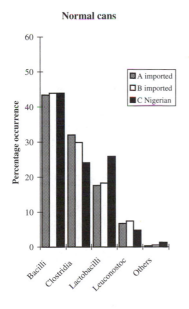
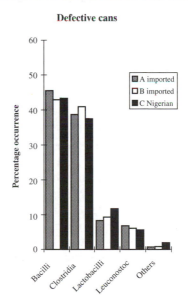

[a]All defective cans were identified by visible swelling.
Source: modified from Efiuvwevwere and Atirike, 1998.

Table 7.5 *Viable moulds found in samples of normal and defective tomato paste*

Absidia
Aspergillus fumigatus
Aspergillus spp.
Fusarium spp.
Rhizopus stolonifer
Saccharomyces spp.
Others

Source: modified from Efiuvwevwere and Atirike, 1998.

Table 7.6 *pH values[a] of imported and Nigerian produced normal and defective samples of tomato paste*

	Normal			Defective[b]		
	A imported	B imported	C Nigerian	A imported	B imported	C Nigerian
pH	4.6	4.3	4.2	4.6	4.8	4.6

[a]Mean of 6 determinations.
[b]All defective cans were identified by visible swelling.
Source: modified from Efiuvwevwere and Atirike, 1998.

Table 7.7 *pH classification of foods*

alkaline	pH > 7.0
neutral	pH 7.0–6.5
low acid	pH 6.5–5.3
medium acid	pH 5.3–4.5
acid	pH 4.5–3.7
high acid	pH < 3.7

In tomato paste a whole range of sporing and non-sporing organisms might occur, derived from the plant material or the environment or soil. Many of these are considerably more heat resistant than *C. botulinum*, and to eradicate them from canned food requires higher temperatures and longer exposure times than are required for 12D processes for *C. botulinum*.

The principal steps in processing tomatoes to tomato paste are shown in Figure 7.1. The goal of the process is to produce a good quality paste in terms of its colour, vitamin content, freedom from pieces of skin, pips or other matter, of smooth texture, safe and microbiologically stable in containers at ambient conditions of storage.

Provided the pH of the paste is below pH 4.5 (just below the limiting value for the growth of *C. botulinum*) the final heating process aims to produce a 'commercially sterile' product – that is one in which the microbial load is reduced; *C. botulinum*, if present is incapable of growing; and other micro-organisms liable to

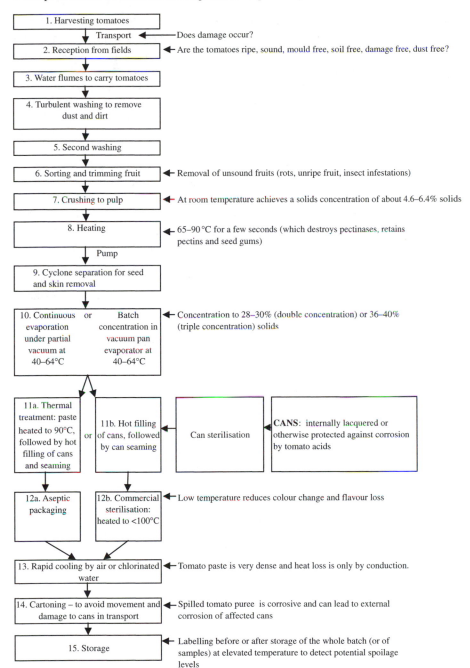

Figure 7.1 *Generic scheme for the production of tomato puree*
(Created from data in Goose and Binstead, 1973)

spoil the product will not be able to grow. It is in effect a heating process which should be adequate to destroy vegetative cells, some bacterial spores, yeasts and heat sensitive moulds and produce a microbiologically stable product. Germination of residual bacterial spores and growth of remaining organisms should be prohibited by the combination of pH value, a_w value, storage temperature value and its duration. At the end of the production line the cans of product will not be truly sterile – *i.e.* free of all viable organisms, but will carry a microbial load incapable of multiplying in the tomato paste unless the conditions in the can become, for some reason, more favourable.

In Italy – where the tomato processing industry is very big – a set of processing parameters have been developed which take account both of the pH and the soluble solids content of the paste. Combinations of the latter control certain sections of the contaminating microbial population while allowing the growth of others. This means that the paste can be processed with target organisms in mind (Figure 7.2) using known D-values for them in those conditions of pH and water activity.

When the combination of pH value and soluble solids concentrations (measured as Brix units) value falls in region A it is essential that processing is designed to at least provide a 12D destruction of *C. botulinum* (which can grow at pH values down to pH 4.6). The 12D treatment is the minimum heat treatment required for safety – but such a heat treatment would not remove other more heat resistant spoilage organisms such as other *Bacilli* or *Clostridia* which, relatively unrestrained by the pH or water activity, could grow and spoil the product. Region A is the most permissive growth environment for a wide range of organisms significant in the processed tomato paste.

In region B, *C. botulinum* cannot grow, so the target organism is either *C. butyricum* or *B. coagulans*, both acid tolerant organisms capable of growth to pH values as low as pH 4.25 under optimum temperature conditions. *C. butyricum* is a mesophilic organism more likely to spoil the tomato paste at storage temperatures below 35 °C, while *B. coagulans* is thermophilic and grows well at 35–55 °C. Since the spores of these organisms have different heat sensitivities processing could be determined using either organism as the target while bearing in mind the future conditions of storage to which the product may be subject.

In region C none of *C. botulinum*, *C. butyricum* or *B. coagulans* can grow. But *C. pasteurianum*, which is acid tolerant and mesophilic, could grow. Thus processing must target the most heat resistant organisms which can grow within the pH and a_w range, which at the Brix values indicated lies above 0.98.

In region D the more acid tolerant lactobacilli, which are also tolerant of reduced water activity caused by the concentration of the tomatoes and their sugars, are the significant spoilage organisms. However, these are non-sporing, heat sensitive organisms, and if processing destroys them then the product will neither suffer spoilage due to sporing organisms nor allow growth of *C. botulinum*. It is products whose parameters fall into this region which need the least processing; and products falling into regions C, B and A which respectively need more severe processing to control both spoilage organisms and *C. botulinum*.

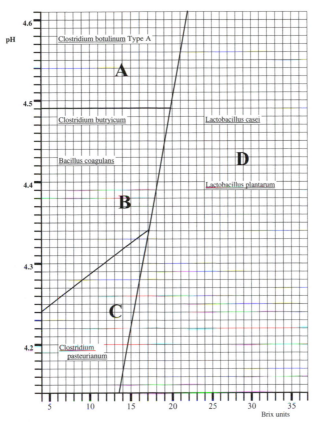

Figure 7.2 *Categories (A–D) of tomato pastes defined by their pH and soluble solids content in Brix units, and the target organisms for processing*

Effective processing has to take into account the initial numbers of organisms, and the acceptable failure rate among the cans. If for example *C. botulinum* occurred in each can at an average concentration of one spore per can, then a 12D process for *C. botulinum* would ensure that the failure rate was 1 in 10^{12} cans. What has to be decided is whether this rate of failure is acceptable or not. In the early 1920s the botulinum cook (the 12D process) was decided upon by the canning industry to assure the safety of low acid canned foods, but its failure rate – *i.e.* how many cans may contain at least one viable spore that could in theory germinate and grow – is dependent on the original concentration of spores. Thus to achieve a failure rate of only 1 in 10^{12} cans the microbial quality of the product before processing must be assured. If the same processing regime is used, and the concentration of *C. botulinum* spores is less than one per can then the failure rate too would be less. The same principle lies behind processing to prevent loss of product by spoilage balanced by a pragmatic assessment of what level of product failure (spoilage) can be accepted. Thus in the case of tomato paste the original microbial quality of the tomatoes dictates the potential failure rate in the canned product if the process is not adjusted to accommodate poor quality ingredients. The combination of pH and water activity to prevent growth of surviving organisms, together with heat processes targeted against key spoilage organisms, and with storage conditions cool enough to prevent growth of residual organisms, are the factors which control the spoilage of tomato paste.

7.3   EXERCISES

Exercise 1. Is it possible to determine from the Efiuvwevwere and Atirike (1998) results what stages in the history of those products may have been responsible for loss of control and which organisms were primarily responsible for the subsequent spoilage? Do you need extra information? What detailed measures could be taken and by whom to avoid or reduce the loss of tomato paste made in, or exported to Nigeria?

Exercise 2. How could a decision be reached as to whether the 'normal' tomato paste as analysed in Nigeria was of safe and acceptable quality? Thus in the case study as described in Section 7.1 would it have been reasonable to have sorted the spoiled from the unspoiled cans, and to have put the normal cans onto retail sale?

Exercise 3. How should the Nigerian health authorities decide whether to allow the entry of the Italian tomato paste? What should their criteria be?

Exercise 4. From the Nigerian consumers' point of view, what needs to change?

Table 7.8 *Summary of microbial quality of normal and defective cans of tomato paste*

	A Imported		B Imported		C Nigerian	
	Normal pH 4.6 cfu g^{-1}	*Defective*[a] pH 4.6 cfu g^{-1}	*Normal* pH 4.3 cfu g^{-1}	*Defective*[a] pH 4.8 cfu g^{-1}	*Normal* pH 4.2 cfu g^{-1}	*Defective*[a] pH 4.6 cfu g^{-1}
TVC	1.04×10^8	2.22×10^9	4.26×10^8	1.38×10^{10}	6.91×10^6	9.1×10^7
Mesophiles	1.3×10^8	1.1×10^9	2.1×10^8	1.2×10^9	1.2×10^8	9.9×10^8
Thermophiles	1.2×10^4	4.7×10^4	5.3×10^4	6.9×10^4	4.5×10^4	6.6×10^4
Aerobic spores	7.6×10^4	4.9×10^5	1.2×10^4	4.3×10^5	8.8×10^4	4.8×10^5
Anaerobic spores	3.0×10^6	5.7×10^7	1.6×10^7	1.1×10^8	1.1×10^7	7.6×10^7
Fungal count	1.17×10^4	2.57×10^5	3.16×10^4	7.24×10^5	2.57×10^3	6.02×10^4

[a] All defective cans were identified by visible swelling.
Source: modified from Efiuvwevwere and Atirike, 1998.

7.4 COMMENTARY

A large UK importer of tomato paste works to the following criteria which they apply to the exporter and to his tomato paste prior to its export to the UK:

- The tomato paste should be commercially sterile – *i.e.* microbiologically stable;
- Samples of cans should have been incubated under elevated standard temperature conditions and times, and zero failure shown.
  'failure' is taken to mean:
  - loss of vacuum demonstrated (the inference of vacuum failure being either can seam leakage, or microbial growth);
  - swelling due to microbial growth and gas production.
- The Howard mould count should normally be less than 40 (this is done to evaluate the quality of the tomato crop);
- The laboratory in which the tests are done should be accredited to a known and acceptable system, *e.g.* UKAS accreditation (see p. 122).

7.4.1 General Principles

This commentary starts by helping you with some general principles, and then later Section 7.4.2 specifically addresses Exercises 1–4.

You should start by asking a number of questions and then assemble the answers into an order which builds an answer to the exercises set. One approach is outlined below.

What is known about the normal and spoiled retail cans of tomato paste in Nigeria?

- The cans were date coded and labelled;
- Swollen cans were easily found on the market;
- They were sold in the same way and for the same price as the normal unswollen cans;
- The retailers were able to sell spoiled cans without apparent restraint;
- Consumers did not differentiate between swollen and unswollen cans;
- Nigeria is a tropical country and conditions of retail sale can often be very hot – above 35 °C for prolonged periods. For example the spoiled imported tomato paste (see Section 7.1) had remained in the heat of the dock side for three months;
- Some tomato paste is made locally in Nigeria.

What might the causes of spoilage be?

From the Efiuvwevwere and Atirike (1998) data it might be helpful to draw up a profile of the microbiology of each of the six types of tomato paste they analysed – the three brands A, B, C, each 'normal' and 'defective' – using data presented in Tables 7.1, 7.2 and 7.3. They are summarised in Table 7.8.

This allows you compare whether there are any clear microbiological differences among them. Secondly consider the types of organisms present, and the proportions of each. Explore what role the various categories of organisms might have had.

Efiuvwevwere and Atirike (1998) demonstrated that swollen (defective) cans of tomato paste, whether Italian or Nigerian made, contained high numbers of bacteria, yeasts and moulds; but they also demonstrated that the non-swollen cans did too, although the bacterial numbers generally were not as great. The demonstrable spoilage – the swollen cans – was due to gas formation inside the can, accompanied with a sour smell, and in these cans there were very high microbial counts. But were the 'normal' cans actually spoiled too? Would you have been happy to eat the paste in the 'normal cans'?

Why were the organisms present?

The organisms may have been present for three reasons:

- Underprocessing: survival of the heating process followed by microbial growth. An inappropriate process may have been used. It may have been insufficiently rigorous for the microflora present which was not subsequently constrained by the acidity and water activity of the paste; or, the microflora was present in much higher numbers that expected and thus a proportion survived the process.
- Post-heat processing contamination at the factory stage followed by growth then or later. Contamination could arise at many points in processing – refer to Figure 7.1. The processing plant can become colonised by spoilage organisms – *Bacillus* for example – which enter product pre-heating, perhaps at the filling stage, or from contaminated water at can cooling.
- Damage to the can at some point between production and retail sale, allowing ingress of organisms which could then grow in the paste. Faulty seams can allow ingress of organisms. This particularly occurs when cans of hot product are cooling and when can surfaces are wet. Equally if cans are roughly handled in transportation [in the movement of pallets, in filling or emptying the 20 or 40 tonnes transportation containers at warehouses or docksides, in rough weather at sea, or travelling over rough roads]. Inadequate protection of the cans may allow perforations to occur and contamination to follow. Damp cardboard or fibre board outer cartons holding 20–24 cans may allow external mould growth and those moulds could penetrate to the product through perforations in the cans.

Why could the organisms grow in the paste?

The parameters of the paste – particularly its pH and water activity provide a specific ecological niche which allows some organisms to grow and prevents the growth of others. The range of organisms able to grow is also further limited by the storage temperature of the product – although the temperature could fluctuate. For example there may have been a period when thermophilic organisms grew (35–55 °C) and their growth was later inhibited by subsequent cool storage.

'Normal' paste had pH values between pH 4.6 and pH 4.2 together with reduced water activity due to the concentration process. The two effects of pH at or below pH4.6 and reduced a_w should have had some effect in restraining microbial growth. The atmosphere inside the cans of processed paste should have been anaerobic – but was there vacuum loss? What was the temperature of the storage of the cans, and how long had that temperature been maintained? Had the tropical conditions exposed the cans to temperatures at which thermophilic organisms had grown? There are data in Table 7.9 which should aid the process of analysis of what may have happened, or could happen.

Thus the organisms which were found in high numbers were those which were tolerant of the actual conditions in the cans which had not controlled their growth.

What numbers and types of viable organisms occurred in the tomato paste in retail sale?

The 'best before' date on cans of tomato paste sold in the UK – a temperate climate – allow for a period of storage between 18 months and two years before the product may show signs of deterioration. These signs may not relate to change in microbiological status, but possibly to colour change, loss of texture and so on. The product is expected to be microbiologically stable. Organisms are not expected to increase. Thus the number of organisms at the beginning of the product's life should be similar to that at the end.

If a paste has low pH, for example pH 4.2, and its water activity is below 0.98 (Brix units would be greater than 20) then viable spores of most *Bacillus* and most *Clostridia* would not be able to germinate in those conditions, and the vegetative cells of those species would not be able to multiply. But bacteriological tests on samples taken immediately post-processing would remove the organisms from those restraining environments and in non-restraining laboratory tests show the actual numbers of viable spores and vegetative organisms. Such tests would indicate the microbiological quality of paste post-processing, and it would be these organisms which might grow in the paste if the storage temperature changed and became more permissive. Growth of some organisms could change the environment in the paste allowing the growth of other members of the flora.

'Normal paste' – paste in cans not visibly changed or swollen – had counts in the region of $10^7 \, g^{-1}$ and had various populations of organisms including heat sensitive lactobacilli, yeasts and moulds. Clearly these counts, composed of such a range of organisms, are only acceptable if they are constantly constrained by the pH, water activity and temperature of the paste. But the numbers in the 'normal cans' at the time of analysis may have been be rising steadily and it may only be a matter of time before they became either toxic or swollen or both.

The organisms in the tomato paste could have originated from the tomatoes and their agricultural environment and from soil, and from the processing factory. Among the organisms recorded is *C. botulinum*, demonstrated present in all pooled samples regardless of the pH or 'normal' or spoiled appearance.

Table 7.9 *Characteristics of some sporing organisms significant in tomato paste*

Organism	Spores produced	Lower pH limit for growth	Temperature preference	Growth conditions	Significant products	Gas production
C. botulinum ('mesophilic anaerobe')	Spores	4.6	Mesophilic	Anaerobic	Neurotoxins	Possible below 30 °C, toxins significant before gas evident. Putrefactive
C. thermosaccharolyticum ('thermophilic anaerobe')	Spores	4.6	Thermophilic	Anaerobic	Acids	Hydrogen, carbon dioxide
C. pasteurianum (acid tolerant, thermophilic anaerobe)	Spores	4.0	Thermophile	Anaerobic	Butyric acid	Hydrogen, carbon dioxide
C. butyricum (butyric anaerobe)	Spores	4.2	Mesophilic	Anaerobic	Butyric acid	Hydrogen, carbon dioxide
B. cereus (mesophilic flat souring, aerobe)	Spores	4.9	Mesophilic	Aerobic/facultative anaerobic	Acids – flat sours	None
B. coagulans (acid tolerant, flat souring, thermophilic, aerobe)	Spores	4.2	Thermophilic	Aerobic/facultative anaerobic	Acids – flat sours, phenolic odours	None

Sources: ICMSF, 1980; National Canners Research Association, 1968.

C. botulinum grows and produces toxin at pH values above pH 4.6. Canned foods whose pH is higher than 4.5 are normally given a 'botulinum cook'. Tomato paste is not given a botulinum cook if its pH is below 4.5, relying on the low pH to prohibit spore germination, and vegetative growth.

Thus botulinum toxin could have been present with lethal consequences in 'normal' cans of paste whose pH was above or at one time had risen above 4.5. Efiuvwevwere and Atirike (1998) record that all the defective cans had pH values above 4.5. Equally could the pH have moved in the opposite direction: could the pH of the 'normal cans' have once been higher, and have been reduced by acid production during the growth of other organisms present? Had botulinum toxin been produced would it become inactivated or remain potent?

Clostridium pasteurianum is an anaerobic, sporing, spoilage organism capable of growth down to pH 4.0 producing souring, butyric acid particularly, and gases (hydrogen and carbon dioxide). This is an organism which can lower pH.

Clostridium thermosaccharolyticum is anaerobic, sporing, and obligately thermophilic and capable of producing through fermentation of sugars considerable quantities of hydrogen, carbon dioxide and acid.

Bacillus coagulans – is a soil organism, producing heat resistant spores and capable of growing in tomato paste under microaerophilic conditions, and is tolerant of pH values down to around pH 4.0. It is also an organism which is thermophilic, growing well at temperatures above 40 °C, and whose spores only germinate above 40 °C. Its growth in the paste would cause 'flat sours' – that is acidic spoilage without gas formation.

Bacillus subtilis and *Bacillus cereus* are also spore formers, capable of microaerophilic growth and acid without gas production.

Lactobacillus and *Leuconostoc* species are strongly associated with tomatoes, are acid tolerant and acid producing in their growth and some strains produce gas. They are heat sensitive, destroyed at temperatures well below 100 °C, and they do not form spores.

Saccharomyces species are yeasts, capable of growing in acid conditions well below pH 4.0, and in their growth fermenting sugars to acids and carbon dioxide. They are heat sensitive and readily destroyed by pasteurisation processes.

Moulds are generally aerobic, acid tolerant and heat sensitive, but some produce heat resistant ascospores which may germinate in the product and spoil it slowly.

What heat process would have destroyed the organisms?

The heat processing used should have taken into account both the pH of the paste (to which, by law, acid may not be added), and its Brix value. Brix units relate to the refractive index of solutions, are a measure of the dissolved solids in a solution and are particularly measured to evaluate overall sugar content of mixtures. Brix values rise as the tomatoes ripen. Tomatoes with high Brix values are high in sugar. They will produce sweeter tomato pastes, and the sugars will contribute to the reduction of the water activity. Thus there is an inverse relationship between Brix value and water activity [as Brix units (high soluble solids concentrations) increase water activity becomes lower].

Figure 7.2 shows the organisms of concern in pastes of different combinations of pH and Brix units. Processing has to be adjusted to ensure that organisms capable of growing in the paste under those combinations of parameters are eliminated. The other organisms may not be killed but should be incapable of growth. Thermal death time curves for the range of organisms in Figure 7.2 may have to be consulted. The processes need to be geared to the concentrations of organisms found in the untreated paste, and have to be hotter or longer if cell concentrations are higher. This implies that the manufacturer should have known the microbial quality of the crushed raw tomato and have been particularly aware of the presence of thermophilic sporers, bearing in mind the product was for the Nigerian market. Secondly all aspects of the processing plant needed be of the highest hygienic standard to avoid contamination moving from equipment into the paste. Underprocessing could arise if unexpected organisms were present, microbial counts were too high for the process used, or the process used was inadequate for pH and Brix combination of the paste.

The outcome of processing should have been a commercially sterile paste. But thermophilic organisms would become significant for product likely to be exposed to tropical ambient temperatures. Was processing sufficient for that export market? Were storage tests done prior to export?

7.4.2 Exercises

> **Exercise 1.** Is it possible to determine from the Efiuvwevwere and Atirike (1998) results what stages in the history of those products may have been responsible for loss of control and which organisms were primarily responsible for the subsequent spoilage? Do you need extra information? What detailed measures could be taken and by whom to avoid or reduce the loss of tomato paste made in, or exported to Nigeria?

The presence of such a range of sporing organisms suggests under-processing, while the presence of heat sensitive organisms suggests post-processing contamination – although gross under-processing is still a possibility. Either of these conditions would have been exacerbated by poor handling, and the high temperature storage likely to be experienced in Nigeria. From the evidence available it is not possible to attribute full blame. Extra information is needed relating to details of the processing history, transportation history and storage times and temperatures. Furthermore information about the original microbiological specification, whether it was met and whether it was appropriate for the market, is also needed for evaluation. It is pertinent to ask too whether the consignments from which the samples originated had also been subjected to prior incubation tests at their point of origin in order to evaluate how robust the material was in relation to tropical temperature tolerance.

The measures which would prevent such high counts in the tomato paste would extend throughout the entire chain of production ensuring the highest quality, lowest practicable count ingredients were used, that processing to eliminate *Clostridium botulinum* occurred or that the microbial count in the paste was very low, and held by the combination of high acidity and reduced water activity. No viable heat sensitive organisms should have been detectable in the paste unless the paste was of a very low pH/high Brix combination which would restrain the growth of sporers, *Lactobacilli* and *Leuconostocs* alike.

Exercise 2. How could a decision be reached as to whether the 'normal' tomato paste as analysed in Nigeria was of safe and acceptable quality? Thus in the case study as described in Section 7.1 would it have been reasonable to have sorted the spoiled from the unspoiled cans, and to have put the normal cans onto retail sale?

The microbiology of the 'normal' cans was in terms of viable cell counts remarkably similar to the defective cans. Had similar results been found in the UK, the products would not have been acceptable under UK food law because of the suspected presence of *C. botulinum* and the clear potential for selling a product of such inferior microbiological quality infringing consumer quality expectations and rights under the Food Safety Act (1990). In Nigerian terms the data raise the question of the potential risk for consumers associated with tomato paste, and indicate that Nigerian processed paste should be processed to destroy greater numbers of organisms approaching true sterility, or that a more acid product than is currently produced should be developed to combat the potential for thermophilic spoilage under hot storage conditions.

In the case study described in Section 7.1 and in the absence of microbiological analyses the fact that 60–70% of the cans showed defects would not be acceptable. Had the failure rate in the consignment of 100 000 cans been at a level of 1% or less then sorting might have been a reasonable approach, if all the cans were of the same date code. Even if sorted it would be wise to retain the cans for a further time period (one month) to evaluate whether they were stable under the tropical heat, or whether further failures became evident. In the absence of evident swell it would still be appropriate to sample cans for spoilage by flat souring organisms (see Section 7.4.3). If positive cans were found it could be assumed that the whole consignment was suffering from incipient spoilage, and a much safer approach would be to return the whole consignment to the producer, or to dispose of it.

Exercise 3. How should the Nigerian health authorities decide whether to allow the entry of the Italian tomato paste? What should their criteria be?

The normal procedure at customs borders in a country is not to examine every consignment of food – that would be too time consuming and highly impractical, but to rely on an examination of the accompanying paperwork. Canned tomato paste is considered to be a low risk food because of its acidity and reduced water activity, and it is probable that provided the paperwork was in order – describing the origins of the product and to whom it was consigned – it would have been allowed through without inspection. However, it is the prerogative of inspectors at the borders of countries to inspect food – and had they become aware that the tomato paste consignment was spoiling they would have been within their rights to return it to Italy, or to order its disposal on the basis of sampling as indicated earlier in Section 7.2.

> **Exercise 4.** From the Nigerian consumers' point of view, what needs to change?

The information of Efiuvwevwere and Atirike (1998) indicates that Nigerian consumers are not as well protected from food risk as they should be, nor are some aware of the risks associated with the use of swollen cans of food. While the processing of the tomato paste produced both in Nigeria and Italy needs to be appropriate to the storage of the product in a hot climate, control of the quality of what is on sale also needs to be implemented together with a programme of public education.

7.4.3 Additional Information

Testing tomato paste for its microbial quality and stability

Samples of the production line should be taken at the beginning of production to test for the efficacy of cleaning and the possible build up of contaminating organisms in the plant (see National Canners Association, 1968; Department of Health, 1994). In addition to the regular monitoring of the production plant for its hygiene, regular line samples should be removed during production and tested for the presence of organisms which could give rise to spoilage. In principle cans are taken and incubated at both mesophilic temperatures (30–35 °C/14–21 days), or thermophilic temperatures (50–55 °C/7–10 days) and examined daily for the presence of swells. At the end of the incubation period flat cans are tested for their retention of vacuum, and then opened and the pH measured to detect souring. Cans which show change can be further examined for viable organisms by cultural methods in order to identify the types of organism present. Microscopic examination of stained smears may indicate the presence of bacteria and their Gram's reaction, bacterial spores, mould and yeasts.

Howard mould count

In 1964, in the UK, a principle was established in court that if a product is made from unsound fruit the product itself is deemed unsound, although it may have been sterilised. The Howard mould count was accepted as evidence for that unsoundness (Anon, 1964). The count is based on the principle of viewing successive smears of sample food (tomato paste) and establishing in what percentage of fields mould is found to be present. Mould mycelia can be seen regardless of whether the mould is viable or not. In the test court case referred to above (Anon, 1964) 90–100% of fields were positive, and in the view of the public analyst concerned this indicated that the concentrated tomato paste had been made from partly rotten tomatoes. The Howard mould count, established in 1911, has been the subject of much controversy and demonstration of how little correlation there can be between the counts obtained and the quality of the original fruit (see Dakin, 1964). It seems to be the case that low counts are indicative of good quality fruit, but that high counts are not necessarily indicative of poor quality fruit, rather a reflection of the fibrous, clumping structure of moulds and the possibility of localised mould spoilage. It is no longer a legal standard in the UK for imported tomato products, but still low Howard mould counts (<40%) in processed pastes are used by the food industry as a measure of high fruit quality.

7.5 SUMMARY

- The processing of canned food is to some extent dictated by its acidity. Tomato paste is an acidic product, its pH normally being below pH 4.5.
- *Clostridum botulinum* does not grow, nor do its spores germinate below pH 4.6.
- Processing of tomato paste therefore can be less than is required to eliminate *C. botulinum*, but care also has to be taken to eliminate or control other spoilage organisms which are acid tolerant.
- The range of spoilage organisms commonly associated with tomato paste are both heat resistant, acid tolerant sporing clostridia and bacilli, as well as non heat resistant, acid tolerant lactobacilli, yeasts and moulds.
- Heat resistant, thermophilic sporing organisms can be problematic in processed tomato paste stored at high temperatures, such as in tropical climates particularly if the pH of the product is above pH 4.2–4.3.
- The risk of such spoilage is reduced if several control strategies are implemented, including the reduction of contamination of raw ingredients, optimal plant hygiene avoiding plant colonisation by sporing organisms, processing which takes account both of the microflora and the end destination of the product, export of the more acidic varieties of tomato paste to hotter countries, end product sterility tests before export and clear product specification. Additionally care should be taken at all stages of the canned product to avoid can damage and post-processing seam leakage, together with cool storage, appropriate can labelling and setting shelf life appropriate to the expected market conditions. The intention is that the canning process is the critical control point (CCP) for the product, but if the product is sold to an unexpected or unsuitable market then the CCP is no longer effective, and control needs re-thinking.
- The standards for the microbiological quality of the paste need to be agreed between producer and buyer.

7.6 REFERENCES

Anon, 1964. Imported tomato products. Recent test case results. *Food Trade Review*, July, 48, 51.

Anon, 1980. *Microbial Ecology of Foods*, Volume II. *Food Commodities*. Academic Press, New York.

Anon, 1990. Food Safety Act, HMSO, London.

Dakin, J.C., 1964. The validity of the Howard Mould Count as a means of assessing the quality of tomato purees – a review. *Food Trade Review*, June, 41–44, 69.

Department of Health, UK, 1994. Guidelines for the Safe Production of Heat Preserved Foods, HMSO, London.

Efiuvwevwere, B.J.O. and Atirike, O.I.E., 1998. Microbiological profile and potential hazards associated with imported and local brands of tomato paste in Nigeria. *Journal of Applied Microbiology*, 84, 409–416.

Feld, H.N., 1964. The decline and fall of the Italian tomato puree empire. *Food Trade Review*, July, 49–51.

Goose, P.G. and Binstead, R., 1973. *Tomato Paste and Other Tomato Products*. 2nd edition. Food Trade Press Ltd., London.

Gould, W.A., 1983. *Tomato Production, Processing and Quality Evaluation*, 2nd edition. Van Nostrand Reinbold/AVI Publishing Co., New York.

Guardia, E.J. and Hopper, P.A., 1990. Food safety – an international concern for industry leadership in producing safe food. *Food Microbiology*, 7, 69–72.

ICMSF, 1980. *Microbial Ecology of Foods*. Volume 1. *Factors Affecting Life and Death of Micro-organisms*. Academic Press Inc., London.

ICMSF, 1996. *Micro-organisms in Foods*. Volume 5. *Characteristics of Microbial Pathogens*. Blackie Academic and Professional, London.

Motarjemi, Y., Kaferstein, F., Moy, G. and Quevedo, F., 1993. Health and development aspects of food safety. *Archiv für Ledensmittelhygiene*, 44, 25–56.

National Canners Association Research Laboratories, 1968. *Laboratory Manual for Food Canners and Processors*. Volume 1. *Microbiology and Processing*. The AVI Publishing Company, Inc., Westport, CT, USA.

Robinson, T.P., Wimpenny, J.W.T. and Earnshaw, R.C., 1994. Modelling the growth of *Clostridium sporogenes* in tomato juice contaminated with mould. *Letters in Applied Microbiology*, 19, 129–133.

Sidhu, J.S., Bhumbla, V.K. and Joshi, B.C., 1984. Preservation of tomato juice under acid conditions. *Journal of the Science of Food and Agriculture*, 35, 345–352.

Speck, M.E., 1984. *Compendium of Methods for the Microbiological Examination of Foods*, 2nd edition. American Public Health Association, Washington, DC.

CHAPTER 8

HACCP and the Responsibilities of the Food Producer

Key issues
- Shell eggs and chocolate mousse
- *Salmonella*
- Cross contamination
- HACCP

Challenge
Shell eggs, often used in catering, can contain *Salmonella*. The main case study describes an outbreak of salmonellosis which arose from a chocolate mousse dessert made with raw shell eggs. The mousse was made by a caterer and supplied to a party. In reading this, and the other two case studies, you should be evaluating what controls are potentially available to reduce risk of salmonellosis and food poisoning in the catering production of lightly cooked, chilled foods. But you should also be considering why a HACCP system properly implemented would lead to a level of control where risk of salmonellosis was negligible.

8.1 THE CASE STUDIES: RAW EGGS AND CHILLED CHOCOLATE MOUSSE DESSERT

8.1.1 "Chef must serve time after wedding treat turned sour

(From *The Times*, London, 29 May 1997, p. 3, by Lin Jenkins. © Times Newspapers Limited, 1997)

It was a perfect wedding reception. The sun shone as 300 guests dined in a poolside marquee at the country home of the groom, a property developer. The chef was a childhood friend who worked unpaid as a wedding gift.

The guests enjoyed themselves into the night dancing to a live jazz band. Within hours the groom, Neil Morgan, 35, was on a transatlantic flight on his honeymoon to the Bahamas with his bride Fiona, when he suddenly doubled up with pain. In Britain, three out of four of the guests were struck down with stomach cramps, diarrhoea and vomiting. Six were so ill with salmonella poisoning that they were admitted to hospital.

Yesterday the chef, Alain Baxter, 33, was jailed for four months for breaching the Food Safety Act in his celebratory meal for his old friend. Tunbridge Wells magistrates had been told that as temperatures soared to 85 °F, food had been left standing in dishes covered with cling film inside the marquee for up to four hours.

The menus included seafood, followed by roast turkey, ribs of beef and chicken. Baxter made the mayonnaise by hand using raw eggs and olive oil, and used it with the prawns in the first course and to make coleslaw. A former chef to the Saudi Royal family, he was in charge of catering. He was paid £4785 for the food, but donated his labour.

Peter Blackwell, chairman of the bench told him: "You have experience of 15 years catering and had a duty of care to provide food fit for human consumption. Quite evidently it was not." Kuldeep Clare for the defence said: "The extreme heat increased the rate at which the food was contaminated. He has lost two stone in weight since the incident because of stress."

The Morgans, who had to cut short their honeymoon, were not in court yesterday. Elizabeth Johnston, the bride's mother said the sentence saddened her. "I am really sorry for the man. He did not mean to poison us all."

Justine Bard, Environmental Health Officer for Tunbridge Wells Borough Council said: "We want this case to stand as a warning to others to take precautions."

The main problems were lack of refrigeration, the undercooking of the poultry, and the use of raw eggs in the mayonnaise which should always be avoided at large functions such as this."

[end of quote]

8.1.2 "D-day veterans protected from dangers of soft eggs

From *The Times*, London, 7 August 1997, p. 3. © Times Newspapers Limited (1997)

Second World War veterans were left shell shocked at their retirement home yesterday after environmental health officers banned their favourite breakfast of soft boiled eggs. The owners of Albury Park Hotel in Albury, Surrey had been instructed to stop serving lightly boiled eggs, or using raw eggs in cooking because of the risk of salmonella poisoning.

Major John Howard who led airbourne forces at the capture of Pegasus Bridge during D-day landings has turned to bananas. The 84 year old now tucks into fruit every morning. "Every time you read the papers some scientist is saying such and such a thing is bad for you. If you followed all their advice there wouldn't be anything left to eat. I enjoy my eggs on the softer side and to be told I can't have them like that is a bit of a blow. It's a good job I have my bananas."

Fellow resident Group Captain Peter Johnson, 88, one of the most decorated pilots in Bomber Command, added: "I was so furious about the ban I wrote to my local paper. I and most other people staying here have enjoyed waking up to a good soft boiled egg and I don't see why we should stop now. This could lead to some of us sneaking over the border into the next county, where people are still allowed the pleasures of a nice omelette or meringue."

Lalage Glaister who runs the hotel with her husband said: "The fact is we've been told that in the event of any of our guests falling ill with salmonella by eating undercooked eggs we would be liable to prosecution."
[end of quote]

8.1.3   The Chocolate Mousse Outbreak

(From *The Times*, London, 26 April 1997, p. 5, by Russell Jenkins. © Times Newspapers Limited, 1997)

"Passover guests hit by food poisoning

Thirty people have been struck down by food poisoning after celebrating Passover, and many more cases are expected among the Jewish community in north Manchester.

The victims of the salmonella bug all come from Whitefield and attended a celebration on Monday evening when the eight day festival began. Three people, including a girl aged nine, are in hospital. Four members of one family have been bedridden; a woman, her father, her husband, and her daughter were all in serious discomfort until they began to feel slightly better yesterday. The woman said she was awaiting results of tests.

All are believed to have eaten chocolate mousse served by an outside organisation, and began to suffer discomfort and diarrhoea within 48 hours.

Environmental Health Officers at Bury Council believe there could be as many as 400 cases. Samples are being analysed to determine the source of the outbreak."
[end of quote]

Laboratory analysis of faecal samples from affected people established that they were all positive for a strain of *Salmonella* – *Salmonella enteritidis* PT6 (source: Bury Metropolitan Council).

"Carving knives were drawn again yesterday in the row that has engulfed celebrity chefs"
From the *Guardian*, 4 February 1999, page 7. © The Guardian.

Once again the target was the doyenne of traditional cooking, Delia Smith, while her accuser was Antony Worrall Thompson, star of the TV programmes "Ready, Steady, Cook" and "Food and Drink".

Appearing as an expert witness in a court case, Mr Worrall Thompson said the runny omelette Ms. Smith made on her latest television show "How to cook", [. . .] would not have reached the temperature to kill bacteria [. . .]

Mr Worrall Thompson yesterday was giving evidence before Bury Magistrates in Greater Manchester on behalf of caterer Marc Cooper, who is charged with five offences after an outbreak of salmonella poisoning in 1997. One hundred people in the Prestwich area were affected, four of whom were detained in hospital, the court heard.

At the heart of the allegations was a chocolate mousse sold for family parties at the Jewish Passover. Mr Worrall Thompson told the court: "I think its a farcical situation where the chef is blamed if he gets a faulty egg."

He said he, like other chefs, had received government warnings in the late 1980's that pasteurised eggs should be used instead of fresh ones. But in his own kitchen he preferred to use fresh ones. "Pasteurisation cannot enhance the flavour."

He admitted having "poisoned" customers himself in the past – with oysters. Cooper, who runs Marc Cooper Catering based in Whitefield, has pleaded not guilty to five offences of selling chocolate mousse unfit for human consumption. If convicted he could be fined up to £20,000 or jailed for six months for each offence."
[end of quote]

From *Environmental Health News*, 26 February 1999 (Volume 14, part 8, page 3):*

"A verdict has been delayed in the case of Marc Cooper v Bury MDC. Mr. Cooper pleaded not guilty to five offences, following a serious outbreak of salmonella food poisoning in 1997 caused by eggs."
[end of quote]

From *Environmental Health News*, 5 March 1999 (Volume 14, part 9, page 5):*

"Eggs case causes confusion

A court victory for a caterer who used shell eggs has left the food safety enforcement position confused and vulnerable to challenge. It also undermines the status of the Chief Medical Officer's (CMO) advice that caterers should only use pasteurised egg for uncooked foods. Marc Cooper, whose chocolate mousses were associated with a salmonella outbreak involving at least 100 cases in Bury in April 1997, was cleared last week of blame.

Stipendary magistrate Christopher Darnton, from Bury Magistrates Court, accepted a scientific argument by food safety consultant Dr. Richard North demonstrating that Mr. Cooper had been sold refrigerated eggs instead of the fresh ones demanded by law for caterers. A fresh shell egg, used under correct HACCP, was not a risk, he said. Neither was it illegal.

Judgement
However, there was no suggestion from the court that Bury MDC, which had brought the prosecution, had acted hastily or improperly. No costs were awarded against the council.

Mr. Cooper had prepared the mousses for Passover celebration and his

* With permission of the copyright holders, Chadwick House Group Ltd.

defence argued that old stock had been cleared out and the kitchen cleaned from top to bottom beforehand, in accordance with Jewish kosher rules. Dr. North's arguments convinced the court that the mousses had been safely prepared and transported, and that salmonella found in samples of the mousse could only have come from an egg held for weeks in refrigeration by the producer. Egg producers, he explained, could be paid up to 60p a dozen for fresh Grade A eggs. Once refrigerated, these become Grade B and the price drops to 13p a dozen. Bury's commercial section manager, Andrew Johnson, told EHN that the council's solicitors were examining the judgement: "We don't even know at this stage whether there are grounds for appeal."

As EHN went to press, the council was still waiting for a return phone call from the Department of Health."
[end of quote]

From *Environmental Health News*, **5 March 1999** (Volume 14, part 9, page 9):*

"Egg case leaves bad taste

If the court decision in the Bury salmonella outbreak involving a caterer's shell egg mousse remains unchallenged, then advice from the Chief Medical Officer can be ignored with legal impunity.

The decision to hold the egg producer and not the caterer responsible for the illness that struck down more that 100 people was based on a scientifically convincing argument. However, stipendary magistrate Christopher Darnton's judgement kicks away a prosecution prop relied upon by local authorities in these cases.

Ever since the 'salmonella in eggs' scare of the late 1980's, successive governments have shied away from declaring the use by caterers of raw, shell egg in uncooked products illegal. Instead, they have relied on caterers following, and local authorities taking into account, the CMO's advice that caterers should only use pasteurised eggs for these foods.

Up to now, if a caterer ignored this advice and used shell-eggs in a food linked later to a food poisoning outbreak, this fact would weigh heavily in court. In 1996, Reading District Council took a successful case against a school caterer on this basis.

Government must sort this mess out. Either ban caterers from shell-egg use, or concede a right to use it under strict safety conditions. It needs, too, to look at an egg-producing industry beset by peaks and troughs, and by foreign imports that may not be the fresh, Grade A eggs they purport to be."
[end of quote]

8.2   BACKGROUND

8.2.1   *Salmonella* spp. and Salmonellosis

The salmonella group of organisms, members of the family Enterobacteriaceae,

* With permission of the copyright holders, Chadwick House Group Ltd.

are Gram negative, non-sporing, rod shaped bacteria which are facultatively anaerobic. The salmonella group contains over 2200 species identified serologically by combinations of the somatic (O), flagellar (H) and capsular (Vi) antigens they possess. Additionally some strains may be further identified through phage typing (PT); for example, there are over 30 phage types of *Salmonella enteritidis* of which *Salmonella enteritidis* PT4 is an example. Salmonellae are ubiquitous, occurring in the gut contents of many animals. Many of the salmonellae are pathogenic for human beings (D'Aoust, 1991), largely affecting the gastro-intestinal tract and causing gastro-enteritis, although some strains are also invasive leading to generalised infections.

Illness is caused through the ingestion of the organisms contained in food or in water, followed by their survival of the acid environment of the stomach in sufficient numbers to colonise and multiply in the large intestine. The infected person sheds the organism in their faeces during infection and also for a while after recovery but eventually they eliminate the organism, rarely retaining the carrier state for more than three months.

The *Salmonella* reservoir is the gastro-intestinal tract of a wide range of animals (Table 8.1) many of which are used for human food; others have access to human food, and act as sources of cross-infection.

The numbers of salmonella required to cause infection vary, but can be as low as 1000 organisms (refer to Table 5.7). After an incubation period of 12–72 hours the symptoms of illness occur when the numbers of the multiplying salmonellae in the gut have risen to such levels that they cause inflammation, followed by diarrhoea, nausea, abdominal pain, fever, and chills. Salmonellosis infection can be fatal, particularly to the debilitated, the immuno-suppressed, the elderly and to young children and babies. Statistics for the incidence of food associated salmonellosis in the UK, in Europe, USA and other developed countries show a rising trend over the last decades of the 20th century, with particular serotypes tending to predominate (Table 8.2).

8.2.2 *Salmonella* spp. and Infection of Eggs

In the UK in the period 1981–91 an increase of 170% in cases of gastro-enteritis caused by *Salmonella enteritidis* was observed. Around the same period a rise in the incidence of *S. enteritidis* gastro-enteritis was seen in other countries too, sufficient to question whether it could be considered a new pandemic (Rodrigue *et al.*, 1990). It was known that *S. enteritidis* strains occurred in poultry and in poultry meat, but it also had become evident that the organism was occurring in some eggs, but to an unknown extent. An investigation into this was commissioned by the UK Government, and a report published in 1993 (ACMSF, 1993). The investigation clearly established that while it had long been known that the gastro-intestinal tract of chickens could be infected with many strains of salmonella, and that eggs could sometimes become contaminated through wet or dirty shells, one strain – *S. enteritidis* PT4 was particularly invasive (Cox, 1995) (although other strains such as *S. enteritidis* PT6 and PT8 are also invasive) and could also colonise the reproductive tract of laying hens. While the investigation

Table 8.1 *Salmonella strains and their reservoirs*

Reservoir	Commonly associated species	Cross-infection from:	Cross-infection to:
Man – infected by wide range of *Salmonella* spp.	ST: *S. typhi, S. paratyphi A, S. paratyphi C, S. sondai*	From man: human faeces, sewage	To man from infected foods, infected water; direct contact from infected source; person to person
	GE: *S. typhimurium, S. enteritidis, S. heidelberg, S. agona, S. newport, S. infantis, S. panama, S. saint-paul, S. weltrevreden.* Many other strains also incriminated	From man: human faeces, sewage	To man: as above
Poultry – infected by a wide range of *Salmonella* spp. which may not cause clinical symptoms	*S. pullorum, S. gallinarum.* Both host adapted and cause widespread illness in breeding flocks and chicks	From bird to bird through faeces, feathers, dusts, infected feeds, poultry house to other birds	To man through infected meat and eggs
Eggs	*S. enteritidis* strains, *e.g. S. enteritidis* PT4	From breeding environment to chicks, and ovarian and oviduct transmission to eggs	To man from external contamination on eggs; and from strains within the eggs – particularly *S. enteritidis*
Pigs	*S. cholerae-suis, S. typhi-suis* are host adapted. Pigs may be infected by other species too	From pig to pig in the piggery through mud, faeces, feed	To man: through faecally contaminated meat and meat products; factory environment
Cattle – infected by a wide range of *Salmonella* spp. which may not cause clinical symptoms in the cattle. Milk and cream	*S. typhimurium* commonly isolated from cattle, pasture land, raw milk, and milking parlours	From cow to cow through the farming environment and feeds. Milk can be infected on farm	To man through meats, and milk contaminated during production, and consumption of unpasteurised products, or cross-contaminated pasteurised products

| Goats, sheep, horses, cats, dogs, camels, buffaloes, elephants, kangaroos, hares, minks, rabbits, bats, whales, dolphins, rats, mice, guinea pigs | All these species have been shown to be able to carry *Salmonella* whether or not they display clinical symptoms | Cross infection arises through consumption of infected foods and through contact with infected materials such as faeces, and may enter the animal not just through eating but through grooming or by other routes | To man through faeces, meats, direct contact (vermin, insects, pets); cross contamination to foods and/or food preparation surfaces, handlers' hands |
| Sparrows, seagulls, pigeons, ducks, geese
Turtles, tortoises, snakes, lizards, crocodiles
Frogs, toads
Shrimp, cockroaches
Catfish, fish and other aquatic species from faecally polluted waters
Flies | All these species have been shown to be able to carry *Salmonella* whether or not they display clinical symptoms | Cross infection arises through consumption of infected foods and through contact with infected materials such as faeces, and may enter the animal not just through eating but through grooming or by other routes | To man through faeces, meats, direct contact (vermin, insects, pets); cross contamination to foods and/or food preparation surfaces, handlers' hands |

Key: ST = septicaemia-typhoid syndrome; GE = gastro-enteritis.
Source: ICMSF, 1996.

Table 8.2 *Salmonella in humans: faecal and unknown reports excluding S. typhi and S. paratyphi, England and Wales, 1981–1999*

Year	S. typhimurium	S. enteritidis (all)	S. enteritidis PT4	Other serotypes	Total salmonellae
1981	3992	1087	395	5172	10 251
1982	6089	1101	413	5132	12 322
1983	7785	1774	823	5596	15 155
1984	7264	2071	1362	5392	14 727
1985	5478	3095	1771	4757	13 330
1986	7094	4771	2971	5111	16 976
1987	7660	6858	4962	6014	20 532
1988	6444	15 427	12 522	5607	27 478
1989	7306	15 773	12 931	6919	29 998
1990	5451	18 840	16 151	5821	30 112
1991	5331	17 460	14 693	4902	27 693
1992	5401	20 094	16 987	5860	31 355
1993	4778	20 254	17 257	5618	30 650
1994	5522	17 371	13 782	7518	30 411
1995	6743	16 044	12 482	6527	29 314
1996	5542	18 256	13 127	5185	28 983
1997	4778	23 008	15 266	4810	32 596
1998	3039	16 397	10 288	4292	23 728
1999[a]	2402	10 596	6645	4253	17 251

[a]Provisional data.

Source: PHLS Laboratory of Enteric Pathogens 1981–1991; *Salmonella* Dataset 1992 onwards. Updated to 22 March 2000.
Accessed through: http://www.phls.co.uk/facts/Gastro/SalmHum.Ann/htm

concluded that nationally the proportion of infected eggs was likely to be small,* never the less consumption of raw or lightly cooked eggs could lead and had led to outbreaks of serious gastro-intestinal illness.

An early response to the 'salmonella in egg' problem was that the Chief Medical Officer (CMO) issued advice (Anon, 1988) that pasteurised (and by implication salmonella free), rather than raw egg should be used in catering. Furthermore soft boiled eggs should not be eaten, particularly by the vulnerable, the elderly or by children. This did not produce the wanted results.

Subsequent research (Gast and Holt, 1999) has shown that one day old chicks, experimentally orally infected, can remain infected to maturity, and that some of those infected chickens will lay infected eggs. Whereas eggs have been shown responsible for some large salmonella outbreaks, infected eggs may also be responsible for sporadic outbreaks in which single individuals become ill (Delarocque-Astagneau *et al.*, 1998).

In the UK, following the 1993 investigation indicated earlier, a number of measures have been implemented to reduce or eradicate *S. enteritidis* from the laying flocks (Table 8.3). However, an early policy of slaughter of positive testing

* Not all flocks were infected. It was subsequently estimated that approximately 0.6% of eggs from known infected flocks were infected (Humphrey, 1994).

Table 8.3 *Control measures to reduce or eliminate Salmonella in laying hens, broilers and their products*

Feeds	Adequate heat treatment (or other process) of feed to eliminate contained *Salmonella*
	Testing for *Salmonella*
	Regular feed plant inspections
	Rigorous import controls
Flocks	Control and monitoring on farm for *Salmonella*
	Improved hatchery hygiene to prevent transmission from hen to chick, or from environment to chick
	Chlorination of chicken drinking water on farm, or hatchery
	Use of 'competitive exclusion' feeding of chicks to prevent early gut colonisation by *Salmonella*
	Chicken vaccination against *Salmonella enteritidis* PT4*
Eggs	Shell eggs: 21 day 'best before' date on shells and packs. No retail sale of cracked eggs
	Egg products: heat treatment to destroy *Salmonella*
Poultry meats	Improved processing hygiene
Home	Improved public education in hygiene

Source: based on *Eggs and Salmonella – the facts*, 1999. British Egg Information Service at http://www.britegg.co.uk, and *Environmental Health Journal*, 1999, **107** (01), 65.

flocks was contentious and was stopped two years after it started. To date the overall programme is on-going and it cannot be said that it has fully achieved its aim, although it is evident that reported cases of salmonellosis in people are reducing (see Table 8.2). What produced a downturn in the *S. enteritidis* PT4 food poisoning statistics in the UK was the fact that in 1994 the chicken industry introduced the practical measure of a vaccination programme for chickens against the organism.

But in the view of Hayes *et al.* (1999) the risk of illness associated with consumption of infected eggs remained high [in 1999]. Indeed in 1999 a gastro-enteritis outbreak affecting 73 staff out of 2700 in a food factory was caused by *S. enteritidis* PT4, and the probable source of organisms was raw shell egg used in making an uncooked dessert (Wilson *et al.*, 1999).

8.2.3   *Salmonella* spp. – Its Growth, Survival and Death

As with all organisms the rate of growth *Salmonella* shows in environments nutritionally capable of supporting growth is dependent on the actual values of all the influencing factors and their combined effect – particularly the temperature, the pH and the water activity. The limits for the growth of *Salmonella* are detailed in Table 8.4, and examples of growth rates at 8–30 °C are shown in Table 8.5.

Table 8.4 *Salmonella: characteristics and growth ranges*

	Temp. (°C)	pH	Water activity (a_w)
Range	7–46.2	3.8–9.5	0.94–>0.99
Optimum	35–43	7–7.5	0.99

Note: a few serotypes can grow slowly below 7 °C down to 5.2 °C.
Source: ICMSF, 1996.

Table 8.5 *Salmonella spp. growth in meat systems*

	Temp. (°C)	pH	Approximate generation time (hours)
Beef slices[1]	8	5.6–6.3	35
Minced meats[2]	10	5.4–6.3	9.6–15.2
Minced meats[2]	30	5.4–6.3	0.49–0.56
Beef slices[1]	30	5.6–6.3	0.9

[1] Goepfert, J.M., Iskander, I.K. and Amundson, C.H. (1970). Relationship of the heat resistance of *Salmonella* to the water activity of the environment. *Applied Microbiology*, **19**, 429–433.
[2] Elliott, P. and Heiniger, P.K., (1965). Improved temperature-gradient incubator and maximal growth temperature and heat resistance of Salmonella. *Applied Microbiology*, **13**, 73–76.

But when computer models, *e.g.* Food MicroModel, or other models (see Section 16.4) are used to predict whether growth may occur under a range of defined parameters it is important to remember that growth stimulants may be present. For example Radford and Board (1995) found in experiments that 1–2% sodium chloride stimulated growth of *S. enteritidis* PT4, and this would not necessarily have been predicted by computer models.

Furthermore the growth rates in egg are affected not only by the temperature, the pH and the a_w but also by the age of the egg and whether the natural protective factors which exist in fresh eggs are still functioning effectively.

Secondly although the growth of *Salmonella* is prohibited outside the ranges shown in Table 8.4 it has become evident that it can frequently survive in non-growth conditions. For example *Salmonella* can survive in dried milk powders, dried egg, and dried animal feedstuffs ($a_w < 0.6$). It can multiply once the dried protective medium is rehydrated as occurs when dried milk or egg is reconstituted, or feeds are consumed by animals. It can also persist in mayonnaise, a phenomenon responsible for a number of outbreaks of egg-associated salmonellosis (Lock and Board, 1995). It can also survive in other products of low water activity and high fat content such as salami and chocolate. When these foods are eaten, the organisms seem protected against the stomach acid by the fats, and they then multiply in the large intestine.

In eggs infected before they are laid the *Salmonella* seem to be primarily located on the membrane surrounding the yolk, or within the yolk. Investigations have shown that when eggs are laid infected the actual number of *Salmonella* cells per egg is small – ranging between 10 and 1000 per egg (Humphrey *et al.*, 1994). This range is at the lower limit of numbers able to cause infection, and gastro-intestinal disease is probably only possible at this dose in the most vulnerable people in society. Normally in healthy adults a greater number is needed to cause infection. Thus for egg to be a cause of salmonellosis, in most cases growth of the initial infecting population is needed. This will happen inside the egg when the egg ages – which happens during storage, and is faster at ambient temperatures than under refrigeration. Investigations have shown that the natural inhibitory factors in eggs, such as ovotransferrin, lysozyme and avidin in the egg white (the albumen), become less effective during the period when the water moves from the white to the yolk, and the pH of the white rises towards pH 9.5 (Board, 1994), while at the same time nutrients (especially iron) leach out from the yolk and stimulate salmonella growth. These processes take time and it is known that a significant increase in salmonella numbers only occurs in storage at 20 °C after a period of around 21 days. Refrigeration of eggs is one way of reducing the risk from the growth of *Salmonella* in eggs. But in long term refrigerated storage of eggs unless the temperature is very cool, slow loss of protective factors still occurs, and slow multiplication of *Salmonella* is possible above 5.2 °C (see Table 8.4). Thus the safety of refrigerated eggs also diminishes with their increasing age.

Consideration of this factor is implicit in the grading of eggs (Table 8.6). But there is also the need to observe the on-pack advice given to consumers "to refrigerate [Grade A] eggs once purchased", and to observe the 'use by dates' to minimise risk from lightly cooked egg (see also Box 8.1).

Salmonellae are non-sporing organisms, and high proportions of their populations can be destroyed by exposure to temperatures below 100 °C. Pasteurisation regimes applied to foods are expected to effectively eliminate viable cells of *Salmonella*, and normally achieve this unless there are abnormally high populations. But the D-values are affected by the nature of the food system because some are more protective that others (Table 8.7).

The pasteurisation of liquid egg products,* in order to make them safe from *Salmonella*, uses a time/temperature regime adequate to ensure that viable *Salmonella* cells are not demonstrably present from test samples (usually 25 g), while at the same time ensuring that the egg remains liquid. It is thus a delicate balance between these two constraints. In commercial pasteurisation a controlled process can be achieved by the pumping of raw liquid egg in a thin film through a heat exchanger; but in domestic or catering circumstances such fine control is impossible. Achieving the effective pasteurisation of egg while cooking an omelette is at best a matter of chance. If a shell egg is infected with any *Salmonella* strain, soft boiling the egg (not causing the yolk to coagulate) yet

---

* UK legislation: The Egg Products Regulations, 1993

Whole egg or yolk – must be treated at a minimum 64.4 °C for a minimum 2 minutes 30 seconds (or equivalent process) followed by cooling as quickly as possible to below 4 °C. Pasteurised egg should pass a specified α-amylase test.

Albumen – should be heat treated by a process designed to take account of the likely microbial contamination so as to meet specified microbiological criteria.

Table 8.6 *Eggs – gradings defined in the European Union*

Council Regulation (EEC) No. 1907/90	Class A	Class B	Class C	Eggs not complying with Grades A, B or C
Council Regulation (EEC) No. 1907/90: general description	Fresh eggs	Second quality or preserved eggs	Down graded intended for the food industry in accordance with Directive 89/437/EEC	Industrial eggs
Commission Regulation (EEC) No. 1274/91 which provides detailed rules for implementing Regulation (EEC) No. 1907/90	Grade A: May NOT be commercially refrigerated (except under specified conditions of short duration) for the purpose of preservation	Grade B: Can be refrigerated below +5 °C for the purpose of preservation; can be preserved by other means	Grade C: Eggs not meeting the specification for Grade A or Grade B	
For human consumption?	Yes	Yes	Yes	No
Shell	Normal, clean, intact	Normal, undamaged; does not have to be clean	May be visibly cracked but not broken	
Air cell	Not more than 6 mm depth	Not more than 9 mm depth		
White	Clear, translucent, gelatinous, free of foreign substances	Clear, translucent, free of foreign substances		
Yolk	Stationary; visible as a shadow under candling; free of foreign substances	Stationary; visible as a shadow under candling; free of foreign substances		

Germ cell	Imperceptible development	Imperceptible development	Imperceptible development
Odour	No odour	No odour	No odour
Shell washing allowed	No	Yes	
Comments	Generally sold for breaking out and subsequent use as eggs products	In the UK may be sold for use as egg products as long as they comply with the Egg Product Regulations (SI 1993/1520)	Used for non-food products

Note: after retail sale consumers are normally advised to refrigerate eggs.

Table 8.7 *Salmonella: D values (in minutes) in various food systems*

Organism	Food system	D-values (minutes) at °C												z(°C)	Ref.
		51.6	52.5	54	55	56	57.2	60	62.8	64*	65	70	80		
Salmonella	Ground beef	62													1
Salmonella (three strains mixed)	Egg white (pH 9.5)			1.1			4.2		0.7					4.8	2
Salmonella anatum	Milk								0.085 [5.1 s]		0.023 [1.4 s]				3
Salmonella enteritidis PT4	Whole egg				6.4 ± 0.6			0.44 ± 0.1		0.22 ± 0.02				4	
Salmonella enteritidis PT4	Egg yolk				21.0 ± 1.5			1.1 ± 0.2							4
Salmonella enteritidis PT4	Albumen				1.5 ± 0.2			0.2 ± 0.1							4
Salmonella other than *Salmonella senftenberg* 775W	Milk										0.02–0.25			4.4–5.5	5
Salmonella senftenberg	Milk chocolate											360			5
Salmonella senftenberg	Milk chocolate												108		5
Salmonella senftenberg 775W	Whole egg				34.3 ± 1.2			5.6 ± 0.1		2.80 ± 0.2					4
Salmonella senftenberg 775W	Egg white, (pH 5.3) (adjusted with HCl)		105												6

Salmonella senftenberg 775W	Egg white, (pH 7.0) (adjusted with HCl)	24						6
Salmonella senftenberg 775W	Egg white, (pH 8.1) (adjusted with HCl)	9.5						6
Salmonella senftenberg 775W	Egg white, (pH 9.1) (adjusted with 28% NH₃ (aq))	3.5						6
Salmonella typhimurium	Whole egg (pH 5.5)		17	5.5			4.2	7
Salmonella typhimurium	Egg yolk		9.5	3	0.4		4.4	6
Salmonella typhimurium	Egg white (pH 9.2)		1				4.2	6
Salmonella weltrevreden	Flour, a_w 0.5					29 (at 63–65°C)		1
Salmonella weltrevreden	Flour, a_w 0.4				875 (at 60–62°C)			1

[1] Archer, J., Jervis, E.T., Bird, J. and Gaze, J.E., 1968. Heat resistance of Salmonella weltrevreden in low moisture environments. Journal of Food Protection, **61**(8), 969–973.
[2] Kohl, W.F.,1971. A new process for pasteurising egg white. Food Technology, **25**, 1176–1184
[3] Read, R.B. Jr., Bradshaw, R.W., Dickerson, R.W. Jr. and Peeler, J.T., 1968. Thermal resistance of salmonellae isolated from dry milk. Applied Microbiology, **16**, 998–1001.
[4] Humphrey, T.J., Chapman, P.A., Rowe, B. and Gilbert, R.J., 1990. A comparative study of the heat resistance of salmonellas in homogenised egg, egg yolk, or albumen. Epidemiology and Infection, **104**, 237–241.
[5] Goepfert, J.M. and Biggie, R.A., 1968. Heat resistance of Salmonella typhimurium and Salmonella senftenberg 775W in milk chocolate. Applied Microbiology, **16**, 1939–1940.
[6] Garibaldi, J.A., Straka, R.P., and Ijichi, K., 1969. Heat resistance of Salmonella in various egg products. Applied Microbiology, **17**, 491–496.
[7] Lategan, P.M., and Vaughn, R.H., 1964. The influence of chemical additives on the heat resistance of Salmonella typhimurium in liquid whole egg. Journal of Food Science, **29**, 339–344.
* Pasteurisation of whole egg (UK requirement) 64.4°C/2.5 minutes minimum. Expected population reduction factor 1–10000, strain and egg characteristics dependent.

BOX 8.1 *EU control of shell eggs and egg products*

The quality and safety of eggs marketed in the European Union are controlled under national legislation which implements EU Decisions and Regulations.

Because pathogenic and spoilage organisms can enter eggs through cracks in the shells cracked eggs may not generally enter retail sale;[1] rather they are directed to other, controlled uses.

Certain UK legislation[2] implements the public health conditions in relation to the sale of shell eggs and to the use of eggs in catering kitchens laid down by the Council (of the EU) Decision 94/371/EC in Articles 2, 3(1), 3(2) 3(4), and 5. Article 2 requires eggs to be kept dry, out of direct sunlight and stored and transported at a constant temperature. Article 3(1) requires eggs to be delivered to the consumer within a maximum time limit of 21 days of laying; Article 3(2) requires the respective sell-by date to correspond to the date of minimum durability less 7 days; and Article 4 requires the date of minimum durability to be clearly indicated on the egg packaging.

Article 5 says "Only eggs packaged in small or large packs in accordance with the requirements of Regulation (EEC) No. 1907/90 and (EEC) No. 1274/91 or egg products in accordance with Directive 89/437/EEC may be used in catering kitchens including restaurants and for the preparation of non-industrial scale egg products, or products containing eggs."

The same UK legislation[2] implements the detailed rules defined in Regulation (EEC) No. 1907/90 which classifies shell eggs into grades A, B and C (Table 8.6).

Thirdly other regulations[3] define the conditions under which egg products, such as fresh or frozen liquid egg, intended for human consumption must be produced, and define the parameters for pasteurisation, and microbiological and other quality criteria.

[1] Ungraded Eggs (Hygiene) Regulations 1990 (SI 1990/1323)
[2] Eggs (Marketing Standards) Regulations 1995 (SI 1995/1544)
[3] Egg Products Regulations 1993 (SI 1993/1520) and its amendments

---

achieving adequate central temperature rise to destroy the organisms is probably impossible, at least under normal kitchen conditions.

8.3   EXERCISES

In the case study described in Section 8.1.3 *Salmonella* organisms were found in the chocolate mousse (which was made using shell eggs) and over 100 people became ill. It was argued successfully in court that the caterer was not to blame. In the words of the expert (reported in *The Times* newspaper) "a fresh shell-egg, used under correct HACCP, was not a risk".

Exercise 1. Consider what steps would have been taken by the relevant authorities to investigate the chocolate mousse outbreak described in Section 8.1.3 and to determine the probable cause.

Exercise 2. Ex-students of mine will recognise the following:
"The entrepreneurial cook
A semi-retired school cook has always made excellent chocolate mousse at home. She makes it from real chocolate, fresh eggs, and double cream. She never uses a recipe, for it does not take much effort – all she does is simply melt the chocolate in a saucepan over a low heat and stir in the beaten eggs and cream, and mix it all carefully. Then she pours it into little pots and lets it set. She has always had a great success with this when she has made it for parties at home. But now, no longer working at the school, she has had the great idea of going into business. She is going to make chocolate mousses for some of the local restaurants. She reckons she can get a tray of product into the boot of her car, and she could easily deliver once a week here and there, and if things go well she could expand making strawberry desserts, apricot ones, some she could decorate with cream . . . her imagination is fired and she will have a wonderful productive, profitable time . . ."

Using the example above, and the case study described in Section 8.1.3 identify the similar microbiological issues between the two, while also clarifying the differences in microbiological safety management of production of chocolate mousse in a domestic environment and on a commercial scale?

Exercise 3. Bearing in mind the case study described in Section 8.1.3 evaluate the commercial catering production of chocolate mousse and identify where and how a HACCP system would minimise the risk to consumers from *Salmonella enteritidis,* and from *Salmonella* generally.

Determine whether there are any critical control points and decide what is necessary in order for a producer be able to argue that all due care has been taken.

Would the HACCP system you have defined for the commercial catering production of chocolate mousse control the risk from all food pathogens?

Exercise 4. Should caterers be allowed to use raw eggs for lightly cooked foods?

8.4   COMMENTARY

In Europe – as around the world – there are legal obligations on commercial food producers to produce safe food. In fact the obligation extends right through the food chain from primary sources of food materials through manufacture to the point of retail sale. In the UK the relevant legislation is embodied in The Food Safety Act 1990 (the 'FSA'). The Food Safety (General Food Hygiene) Regulations 1995, made under the FSA define how food shall be produced hygienically and also require all food premises to have in place a hazard analysis and control system, although not necessarily a fully documented HACCP system.

Exercise 1. Consider what steps would have been taken by the relevant authorities to investigate the chocolate mousse outbreak described in Section 8.1.3 and to determine the probable cause.

It is necessary for you to consider how an incident like this comes to light, and then to consider what needs to be done in order to find out what is causing it and from where the infection originates (refer to Case Study 2).

The actual background to the outbreak is as follows:

Investigation of this outbreak started when two separate families reported to their local doctor with stomach pain, diarrhoea and other symptoms of gastro-intestinal illness. In conversation with their doctor they each mentioned they had eaten chocolate mousse. The Environmental Health Services were alerted by the doctor, and they initiated investigations. These quickly revealed that other families who had been participating in the same Passover celebration were affected. Furthermore the families who were ill only bought chocolate mousse from the caterer and prepared other foods themselves. Very early in the investigation it became obvious that chocolate mousse was the source as all the people who were ill had eaten mousse, and some had eaten only mousse. The mousse was the only product supplied for the celebration by the outside caterer. He did produce other foods but not for the families who were ill. Samples of chocolate mousse from families who had some left over, including one sample which had not been touched were taken by the EHOs and were sent to a laboratory for analysis. The analysis established that they were all positive for a strain of *Salmonella enteritidis* – *S. enteritidis* PT6. Faecal samples were also taken from 100 people and of these 80 were positive for the same strain of organism. It was not established from where the *S. enteritidis* PT6 had originated because although samples of chocolate and of eggs taken from the producers premises were analysed, all of these were negative. In fact the *Salmonella* was only found in the chocolate mousse.

   The principal steps taken in the investigation are those shown in Figure 2.4. and the cases would have been recorded in the national statistics. It is also of note that *S. enteritidis* PT6 is strongly associated with eggs (Table 8.8).*

Exercise 2. Ex-students of mine will recognise the following:

"*The entrepreneurial cook*
A semi-retired school cook has always made excellent chocolate mousse at home. She makes it from real chocolate, fresh eggs, and double cream. She never uses a recipe for it does not take much effort – all she does is simply melt the chocolate in a saucepan over a low heat and stir in the beaten eggs and cream, and mix it all carefully. Then she pours it into little pots and lets it set. She has always had a great success with this when she has made it for parties at home. But now, no longer working at the school, she has had the great idea of going into business. She is going to make chocolate mousses for some of the local restaurants. She reckons she can get a tray of product into the boot of her car, and she could easily deliver once a week here and there, and if things go well she could expand making strawberry desserts, apricot ones, some she could decorate with cream . . . her imagination is fired and she will have a wonderful productive, profitable time . . .."

Using the example above, and the case study described in Section 8.1.3 identify the similar microbiological issues between the two, while also clarifying the differences in microbiological safety management of production of chocolate mousse in a domestic environment and on a commercial scale?

The preparation of food, whether in the home or in commercial premises, carries with it the moral responsibility to prepare it to the highest safety standards possible. Since food can very often support the survival and growth of pathogenic organisms, the health of family members – babies, children, old people, as well as healthy teenagers and adults – is potentially at risk from food prepared at home. There is a growing suspicion in the UK and elsewhere (Motarjemi and Kaferstein, 1999) that the incidence of food poisoning is much higher than official statistics show because, it is suspected, many cases occur at home yet are not reported. People do not necessarily consult their doctors when they suffer the symptoms of gastro-enteritis, particularly if the symptoms are mild. (See Ryan, Wall *et al.*, 1996.)

Similarities

Those who prepare food, on whatever scale, have a responsibilty to produce microbiologically safe food, that is to say food in which, under the anticipated conditions of storage and use, pathogen contamination and growth are minimised. But most food can support the growth of pathogens (and, if not growth, their survival), so in food preparation the methods of handling and preparation must prevent pathogens getting into or growing in it, and those people who

*Acknowledgement is made to Bury Metropolitan Council (Environmental Services), UK for background information in the public domain concerning the outbreak described in Section 8.1.3.

Table 8.8 *General outbreaks of Salmonella enteritidis PT6 infection: January 1995 to May 2000*

Year	Health authority	Place of outbreak	Number ill	Cases positive	Suspect vehicle	Evidence
1995	Oxfordshire	School	3	3	Raw fairy cake mixture	S
1996	North Cumbria	Restaurant	8	8	Chicken nuggets	M
1996	South Staffordshire	Nursery	16	5	None	–
1996	Coventry	Residential institution	4	2	Chicken sandwich	S
1996	Tees	Public house	46	12	Egg sandwiches, egg and bacon quiche, and cheese and onion quiche	S
1996	Barnet	Private house	30	24	Ice cream made with raw shell eggs	S
1996	Newcastle	Restaurant	8	4	Tiramisu made with raw shell eggs	D
1997	Bro Taf	Residential institution	7	4	Puréed food	D
1997	Bury	Caterer	>100	90	Chocolate mousse made with raw shell eggs	D
1997	Bro Taf	Canteen	5	3	Fish cakes bound with raw shell eggs	D
1997*	North West Anglia	Restaurant	12	8	Shell eggs	M
1997	Wiltshire and Bath	Private house	2	2	None	–
1997	Southampton and SW Hampshire	Private house	7	3	None	–
1997	Wirral	Residential institution	11	7	Cheese on toast bound with raw shell egg	S
1997	NW Lancashire	Residential institution	4	4	None	–
1997	Cornwall and Isles of Scilly	Private house	10	6	Lemon mousse made with raw shell eggs	M
1997	Merton, Sutton and Wandsworth	Nursery	20	11	None	–
1997	South Staffordshire	Restaurant	13	8	Mayonnaise made with raw shell eggs	S

Year	Location	Setting			Vehicle	Evidence
1997	Avon	Residential institution	38	21	None	–
1997	NW Lancashire	Restaurant	6	5	Tiramisu made with raw shell eggs	S
1997	Bro Taf	Restaurant	8	7	Tiramisu and cassata both made with raw shell eggs	M + S
1998	South and West Devon	Residential institution	16	6	None	–
1998	East London	Hall	45	15	Chocolate mousse made with raw shell eggs	D
1999	Dorset	Retailer	35	35	Eggs	M
1999	West Surrey	Retailer	61	21	Profiteroles with parveh cream	S
1999	Somerset	Restaurant	3	3	Honeycomb ice cream made with raw shell egg	M
2000	West Kent	School	11	4	Fresh cream cake made with raw shell eggs	D

*S. enteritidis PT21B also isolated.

D (descriptive): other evidence, usually descriptive, reported by local investigators as indicating the suspect vehicle.

M (microbiological): identification of an organism of the same type from cases and in the suspect vehicle or vehicle ingredient(s), or detection of toxin in faeces or food.

S (statistical): a significant statistical association between consumption of the suspected vehicle(s) and being a case.

Source: Anon, 2000. *Communicable Disease Report*, **10**(28), 251. Reproduced with permission of the PHLS Communicable Disease Surveillance Centre, Colindale, London.

prepare food must know how these hygienic measures are achieved (Anon, 2000).

Thus the similarities between home preparation and commercial preparation reside in the fact that foods may be sources of pathogenic organisms and may support their growth. Chicken may contain *Salmonella* and *Campylobacter*; vegetables may contain *Listeria*; meat may contain *Clostridium perfringens*, and so on. These foods may themselves introduce their pathogens into the kitchen or commercial premises environment, which in turn act as sources of contamination. Once in the foods, the pathogens may grow and their population increase. If this occurs in foods which are not further processed – such as dessert products, cooked meat pies, pâtés, sliced cooked meats and so on – they may cause illness in the consumers. There are errors in common between domestic and commercial cooking processes – poor personal hygiene, dirty equipment which allows cross contamination, preparation of foods in such a way that raw foods contaminate cooked food, allowing cooked foods to stand around so that microbial growth occurs, allowing cooked foods to remain at temperatures and/or for time periods at which pathogen multiplication is possible and significant.

All of these things can lead to food becoming dangerous to eat, and to food poisoning arising in the consumers whether they are family members or the general public.

Thus issues in common between domestic food preparation and commercial food preparation are:

- knowledge of sources of pathogenic organisms,
- the risk of cross contamination,
- potential for growth, survival and death of pathogenic organisms,
- the time limits and storage conditions within which foods are more likely to remain safe.

Differences

The primary differences between domestic food preparation and commercial production relate to the difference in scale, and responsibility. A single error of hygiene – any of the things mentioned above has the potential on a commercial scale to lead to great numbers of people becoming ill. This is because microorganisms are self replicating, and in commerce food quantities produced are large. The introduction of a small number of organisms at the initial error leads through growth and replication to large numbers. If mixing or filling operations distribute those organisms into a large number of containers of food, they can often continue to grow there too. An organisation producing food for sale has not only a moral responsibility to produce safe food, but also a legal one as indicated in the introduction to Section 8.4. A commercial food producer must be aware of the relevant food law and codes of practice and abide by them.

Commercial food production is complex. Many staff may be engaged in the production process, and those staff may change day by day, week by week. Worldwide the food industry is notorious for employing large numbers of low paid, under-skilled staff. Yet those staff could well be at the front line of production and their failures could compromise safety. Thus, in contrast to the domestic environment, commercial food production needs formalised systems of safety management.

These systems, grouped together under the name 'Good Manufacturing Practice' (GMP), often operating to internationally recognised standards such as ISO 9000 series, include systems for recognising and managing the risk associated with the presence of microbial hazards. The Hazard Analysis Critical Control Point System (HACCP) is a component part of GMP. The management of food safety with respect to micro-organisms requires food ingredients of the optimal microbial quality, and that their storage prior to use is managed both in terms of stock rotation, time, temperature and humidity such that the microbial population increases minimally. Of course ingredients of different qualities can be purchased, and a commercial company will source their ingredients from suppliers who can provide the quality and quantity to the microbial specification agreed. In a sense domestic purchasers do this too – except they buy on price and presentation, and rely on the retail outlet from which they buy for assurance of microbial quality. They also rely on the retailer complying with legal requirements in respect of microbial safety. This is particularly significant in regard to prepared ready-to-eat foods, which require no further cooking.

Commercial food manufacture is complex, and the sequence of production events to produce it has to be thought out and the space in which it is undertaken designed to minimise the risk of cross contamination while creating an efficient work flow (see Case Studies 11 and 14). Catering too demands many activities – peeling potatoes, chopping raw meat, icing cakes, whipping eggs, making desserts – which may not be readily compatible in terms of the associated microbiological risk, yet which in practice may take place in poorly designed premises. For example Duguid and North (1991) commented on the fact that the old system of catering management in which different food types were prepared in different areas ('parties') was tending to be replaced by 'general parties' in which both raw and cooked meats and other foods are prepared. In such environments people as well as food materials may be moving around and the risk of cross-contamination of foods and surfaces can be high. Good management analyses these risks and redesigns processes, work flows and peoples' activities to minimise them.

Commercial products are packaged and labelled with codes (batch codes, dates of production, shelf life) which facilitate traceability should that be necessary, and guide use. Commercially produced food is often made in one place and eaten elsewhere. But the hygiene of transport, its temperature, humidity and duration all significantly affect the level of risk associated with the food. Notice that the 'entrepreneurial cook' was happy to transport her chocolate mousses in the boot of her car because in her ignorance she proposed no better. What about the hygiene of such an environment? Was the dog in there the day before? And what about the temperature? Is the boot of a car refrigerated? (see Case Study 13).

Commercial food production – on whatever scale – demands of the food producer an understanding of the sources of microbial pathogens, how they can be disseminated, the fact that they can grow in many foods, the cooking or processing conditions when they may be killed or controlled, and the food preparation and storage conditions under which they increase and/or cross-contamination occurs. It also demands sufficient knowledge to produce food safely and the implementation of production management systems to ensure that happens. This is not necessarily easily achieved, yet the HACCP system, which is a proactive preventative system of food production management, is increasingly being demanded worldwide. Modern food safety management requires the understanding of HACCP (Table 8.10).

Exercise 3. Bearing in mind the case study described in Section 8.1.3 evaluate the commercial catering production of chocolate mousse and identify where and how a HACCP system would minimise the risk to consumers from *Salmonella enteritidis*, and from *Salmonella* generally.

Determine whether there are any critical control points and decide what is necessary in order for a producer to be able to argue that all due care has been taken.

Would the HACCP system you have defined for the commercial catering production of chocolate mousse control the risk from all food pathogens?

The first point to note in thinking about the chocolate mousse outbreak (Section 8.1.3) is this: the case study itself presents you with retrospective knowledge. You know that the problem proved to be the occurrence of *Salmonella enteritidis* PT6, found both in some of the chocolate mousse and in the stools (faeces) of many of those who presented with symptoms of gastro-enteritis. But in food safety management HACCP (Table 8.10) looks to the future – what might occur one day? The caterer concerned suffered a problem – but could he have anticipated it? Could he have taken specific actions to avoid *S. enteritidis* PT6? What about *S. enteritidis* PT4? Are the preventative actions for the two the same? What about all the other food poisoning pathogens?

To understand how HACCP can help to manage the risk from such pathogens it is appropriate to analyse where the pathogens – in this case *S. enteritidis* PT6 – might have entered the food, and where control(s) for it existed. Secondly it is appropriate to evaluate whether those controls would be equally effective against other pathogens.

Keep these points in mind.

Where could the *S. enteritidis* PT6 come from? In Court, although the association between *Salmonella enteritidis* was debated, it is useful for you to step back from that knowledge. In making a product like chocolate mousse organisms could enter the product through many routes including from its ingredients. Recipes for chocolate mousse indicate that they can contain combinations of egg, chocolate, cream, and gelatin. Thus the first question is – what is the recipe for the mousse and which of those ingredients could introduce the pathogen? Shell egg is known to sometimes carry salmonella (Section 8.2) but outbreaks of salmonellosis have also been associated with chocolate (Table 8.9) although when that was first identified as a source it was a bit of a surprise. At the time salmonellae were known to be strongly associated with animal products, but few people suspected that chocolate – ambient stable and fatty – could also be a vehicle. The chocolate had been contaminated in manufacture, and the organisms were not destroyed by processing and had survived in low numbers while the product was stored. Subsequently when the chocolate was eaten the fat protected the *Salmonella* from the acids in the stomach which later multiplied in the gut causing illness. Cream and milk products have also been identified as vehicles for *Salmonella* and associated with causing food poisoning.

So in principle any of the ingredients mentioned could introduce the *Salmonella* into the mousse. Suppose one of them does. Is it not possible that the way the mousse is made and cooked would destroy the *Salmonella* – they are after all heat sensitive organisms. But is the mousse cooked in any way? This fact needs to be established, and if it is – what process is used?

Equally at this stage of analysis you will see that the ingredients may not be responsible. The possibility exists that the mousse may be contaminated from an external source – other food, raw meat, catering equipment, people's hands . . . before or after it is fully made, as possibly happened in the wedding incident described in the Case Study 8.1.1. Assuming the contamination affects the bulk product – thought then needs to be given to the level of contamination introduced at that moment, and whether any subsequent conditions allow microbial growth. Chocolate mousse is a moist product of high water activity, approximately neutral pH and a supportive nutritive medium for microbial growth, unlikely to be containing preservatives (is this allowed?). So if *Salmonella* is present only rapid cooling and storage under refrigeration (product temperature below 5 °C), or frozen (-20 °C) would prevent growth. But remember also that these low temperatures would also preserve any viable *Salmonella* if they are present.

You need also to keep in your mind the minimal dose of organisms needed to cause illness, while also remembering that the greater the number the more likely illness is, and also consider the quantity of mousse eaten which might carry that loading. Furthermore consideration needs to be given to who the consumers are – healthy adults, or young children, old people (as in Case Study 8.1.2), immuno-compromised people . . .

There are a lot of possibilities here and these are part of the thinking process in understanding how a HACCP system (see Table 8.10) can lead to management and reduction of food safety risk. Thus consider

- What is the food made of?
- How is it processed, packaged and distributed?
- What is it for? who is the consumer? how long is its shelf life?
- Is the process from ingredient to consumer adequately controlled to ensure a safe product?

The flow chart

Figure 8.1 shows a possible flow chart and it also identifies where microbial hazards could enter the process. Check this through and ask yourself whether

Table 8.9 *References relating to Salmonella and cocoa and chocolate*

S. anatum in milk chocolate	Barrile, J.C., Cone, J.F. and Keeney, P.G., 1970. Effect of added moisture on the heat resistance of *Salmonella anatum* in milk chocolate. *Applied Microbiology*, **19** (1), 177–178
S. eastbourne in chocolate	Craven, P.C., Mackel, D.C., Baine, W.B. *et al.*, 1975. International outbreak of *S. eastbourne* infection traced to contaminated chocolate. *Lancet*, **i**, 788–793
Review	D'Aoust, J.-Y., 1977. *Salmonella* and the chocolate industry. A review. *Journal of Food Protection*, **40**, 718–727
S. napoli in chocolate bars	Greenwood, M.H. and Hooper, W.L., 1983. Chocolate bars contaminated with *Salmonella napoli*: an infectivity study. *British Medical Journal*, **286**, 1394
S. nima in Belgian chocolate	Hockin, J.C., D'Aoust, J.J., Bowering, D. *et al.*, (1989). An international outbreak of *Salmonella nima* from imported chocolate. *Journal of Food Protection*, **52**, 51–54

this diagram would be adequate for a real process? Where would it need expansion or alteration?

In reality a HACCP system has to apply to a real, individual food producer and process. Specific detail relevant to that commercial concern is needed. Figure 8.1 is a generic diagram – illustrating principles only.

Control

Where could controls exist? On the diagram potential control points are marked with numbers. In evaluating how the chocolate mousse might be infected with *S. enteritidis* PT6 questions have to be raised about how effective the controls at the various points marked could be.

Refrigeration (points *13,*14, and *15) would limit microbial growth, and would be an effective way of preventing increased risk, but would not prevent salmonellosis occurring if sufficient organisms were present in a 'portion' particularly if the consumer was in a vulnerable group. So although a good control, it is not a critical control in this process because the risk is not reduced to an acceptable level.

A better control is needed.

Cooking the mousse mixture (*6) should raise the temperature adequately to destroy salmonellae, but the product which results might not have the texture and flavour characteristics sought. Secondly when a product is mixed in a batch great care has to be taken to ensure that all parts of the mix receive the adequate minimum heat treatment. This is very difficult to achieve in a catering environment – the edges, and the mixing implements particularly, can be areas of low heat treatment and cross-contamination. So if the cooking method dictates that pasteurisation cannot be effectively achieved in the sense of every part of the mixture receiving a minimum heat treatment, or that cross-contamination from undercooked mixture to heat treated mixture might arise, that point is not in practice a critical control point although it has the potential to be.

Thus control lies in two areas – prevention of entry of the *S. enteritidis* PT6

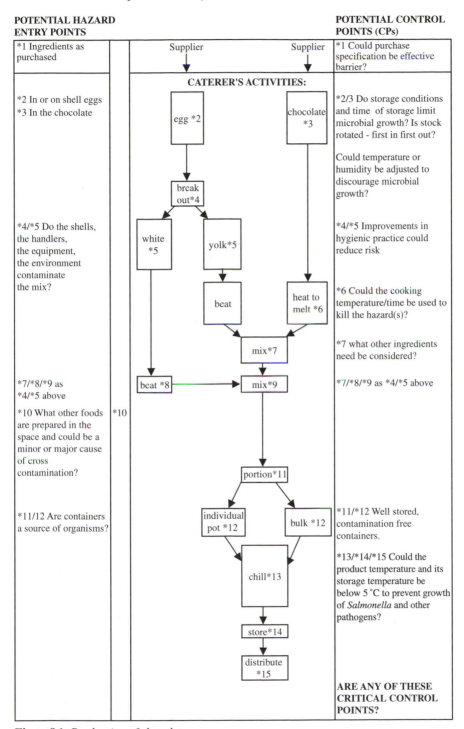

POTENTIAL HAZARD ENTRY POINTS

*1 Ingredients as purchased

*2 In or on shell eggs
*3 In the chocolate

*4/*5 Do the shells, the handlers, the equipment, the environment contaminate the mix?

*7/*8/*9 as *4/*5 above

*10 What other foods are prepared in the space and could be a minor or major cause of cross contamination?

*11/12 Are containers a source of organisms?

POTENTIAL CONTROL POINTS (CPs)

*1 Could purchase specification be effective barrier?

*2/3 Do storage conditions and time of storage limit microbial growth? Is stock rotated - first in first out?

Could temperature or humidity be adjusted to discourage microbial growth?

*4/*5 Improvements in hygienic practice could reduce risk

*6 Could the cooking temperature/time be used to kill the hazard(s)?

*7 what other ingredients need be considered?

*7/*8/*9 as *4/*5 above

*11/*12 Well stored, contamination free containers.

*13/*14/*15 Could the product temperature and its storage temperature be below 5 °C to prevent growth of *Salmonella* and other pathogens?

ARE ANY OF THESE CRITICAL CONTROL POINTS?

CATERER'S ACTIVITIES:

Supplier — Supplier

egg *2 — chocolate *3

break out*4

white *5 — yolk*5

beat — heat to melt *6

mix*7

beat *8 — mix*9

*10

portion*11

individual pot *12 — bulk *12

chill*13

store*14

distribute *15

Figure 8.1 *Production of chocolate mousse*

Table 8.10 *HACCP principles and aims*

HACCP Principle		Aim
1. Conduct a hazard analysis	**Hazard:** a biological, chemical or physical agent in or property of food that may have an adverse health effect (WHO, 1995) **Analysis:** a scientific process to differentiate between control and critical points through a risk assessment evaluation of hazards	Risk identification
2. Identify critical control points (CCPs)	**Critical:** a few points (the 'vital few') in a specific food system where loss of control may result in a HIGH probability of a health risk (Bryan, 1988; Humber, 1992) **Control:** many points (the 'trivial many') in a specific food system where loss of control may result in a LOW probability of a health risk but may result in an economic or quality defect (Bryan, 1988; Humber, 1992) **Point:** an operation (practice, procedure, process) at which control must be exercised (Notermans *et al.*, 1995)	
3. Establish preventative measures with critical limits for each CCP. For a cooked food, for example, this might include setting the minimum cooking temperature and time required to ensure the elimination of any microbes		
4. Establish CCP monitoring procedures		Risk management
5. Establish corrective actions to be taken when monitoring shows that a critical limit has not been met		

6. Establish procedures to verify that the system is working properly

7. Establish effective records keeping to document the HACCP system. This would include records of hazards and their control methods, the monitoring of safety requirements and action taken to correct potential problems

Documentation control

Adapted from Panisello, P.J. and Quantick, P.C., 1998. HACCP and its implementation, *Food Science and Technology Today*, **13** (3), 130–133.

References

Bryan, F.L., 1988. Risks associated with vehicles of foodborne pathogens and toxins. *Journal of Food Protection*, **51**, 498–508.

Humber, J.Y., 1992. *Control Points and Critical Points*. In Pierson, M.D. and Corlett, D.A. (eds.), *HACCP Principles and Applications*, Chapman and Hall, New York.

Notermans, S., Gallhoff, G., Zwietering, M.H. and Mead, G.C., 1995. Identification of critical control points in the HACCP system with a quantitative effect on the safety of food products. *Food Microbiology*, **12**, 93–98.

WHO (World Health Organisation). Applications of risk analysis to food standards issues. WHO/FNU/FOS/95.3 Geneva, Switzerland, 13–17 March, 1995.

from the catering environment and from the ingredients used. Both need to be controlled.

Environmental contamination can be controlled by the use of space, equipment and staff dedicated to the preparation of 'high risk' foods (ready-to-eat foods, which are not subject to further processing), and the whole process managed in a way that other sources of contamination including insects, rodents, and dust are eliminated.

But all that is ineffective if the ingredients are themselves contaminated. A critical control point in the chocolate mousse is the control of the quality of the incoming eggs and chocolate.

The purchasing specifications for the chocolate should specify absence of *Salmonella* in (at least) a sample size of 25 g, and be verifiable. That for the eggs should define 'Grade A' eggs. A caterer would then have reasonable confidence in their quality – but only of course if the eggs themselves came from chickens free from salmonella. But the egg management post-receipt is an issue too. If after receipt they are then stored for an extended period, even under refrigeration, the possibility exists for *Salmonella*, if present (and thus not from a guaranteed *Salmonella* free source), to multiply slowly in the stored eggs.

The caterer has to trust the supplier of the eggs ensuring that the supplier is a reliable source in whom he has confidence. Assuming both the critical controls and the controls are in place then it should be possible to argue that all due care had been taken in production – that is to say due diligence had been observed.

In the Case Study (8.1.3) the Court debated the known association between shell eggs and the potential for the presence in them of *Salmonella enteritidis*. Grade A shell eggs should be free of this risk. Thus to evaluate the robustness of that control it is also necessary to define what is the level of risk which is acceptable, whether that risk can be tolerated, and if not to institute another critical control point. If that cannot be achieved then a caterer has the choice of either not making products which do not receive pasteurising heat treatment, or of using pasteurised eggs.

Since HACCP is a knowledge based system, thought would have to be given to whether the controls in place were appropriate to other pathogens entering with the ingredients or from elsewhere. Of particular note would be those which are heat resistant, which multiply under refrigeration, which are infective at very low dose. It might well be the case that controls effective for *Salmonella* would not be effective for other pathogens. An analysis specific to all relevant pathogens should be done and the existing controls should not be relied on to control the risk from all hazards.

This is the problem with HACCP – it requires knowledge of micro-organisms which individuals in the catering and restaurant industry may not have. Generic systems of HACCP have now been evolved, which instead of focusing on particular named microbial hazards, identify hazards as for example 'potential for microbial growth', and 'microbial contamination'. Perceiving microbial hazards in this light is more manageable in a catering environment. A careful analysis of the process is still required, and appropriate controls need to be put in place.

> **Exercise 4.** Should caterers be allowed to use raw eggs for lightly cooked foods?

In UK law caterers are allowed to use raw eggs for uncooked or lightly cooked foods but they are advised not to, and to use pasteurised egg instead. It is thus a matter for the caterer to implement a control system for the catering process which ensures that raw egg containing products are managed in such a way that they do not cause illness.

8.5 SUMMARY

This chapter has described outbreaks of food poisoning in which raw eggs feared to be carrying *Salmonella*, might have been eaten undercooked by vulnerable old people, or were used in foods which actually did cause food poisoning.

The problem with *Salmonella* is that although a large dose of them is likely to cause illness, a small dose of them – perhaps as few as 500 cells, can in some circumstances also cause illness. Thus although all good hygienic practices may limit cross-contamination from source (in this case study from raw eggs) to other foods, and may prevent their multiplication (good temperature control and effective refrigeration of prepared foods), in the case of raw eggs used in lightly cooked foods the risk taken is that they might be present in the raw material, and not be destroyed by any subsequent process. Thus the responsibility of the caterer is to accept that this risk exists and to manage it. As explained in the text various options exist, and the implementation of an effective HACCP system which extends to the source of the eggs and the other raw ingredients will allow that risk to be evaluated and an appropriate course of controlling action taken.

8.6 REFERENCES

ACMSF, 1993. Report on *Salmonella* in eggs. HMSO, London.

Anon, 1988. Raw shell eggs. EL/88/P136, Department of Health, UK.

Anon, 2000. *Keeping it safe*, Institute of Food Science and Technology, UK Advisory statement. *Food Science and Technology Today*, **14** (2), 102–105.

Archer, J., Jervis, E.T., Bird, J. and Gaze, J.E., 1998. Heat resistance of *Salmonella weltevreden* in low-moisture environments. *Journal of Food Protection*, **61** (8), 969–973.

Board, R.G., 1994. Chapter 3: *The egg: a compartmentalized, aseptically packaged food*, in *Microbiology of the Avian Egg*, ed. Board, R.G. and Fuller, R. Chapman and Hall, London.

Cox, J.M., 1995. *Salmonella enteritidis*: virulence factors and invasive infection in poultry. *Trends in Food Science and Technology*, **6**, 407–410.

D'Aoust, J.-Y., 1991. Pathogenicity of food borne *Salmonellae*. *International Journal of Food Microbiology*, **12**, 17–40.

Delarocque-Astagneau, E., Desenclos, J.C., Bouvet, P. and Grimont, P.A.D., 1998. Risk factors for the occurrence of sporadic *Salmonella enterica* serotype *enteritidis* infections in children in France: a national case-control study. *Epidemiology and Infection*, **121** (3), 561–567.

Duguid, J.P. and North, R.A.E., 1991. Eggs and salmonella food poisoning: an evaluation. *Journal of Medical Microbiology*, **34**, 65–71.

Gast, R.K. and Holt, P.S., 1999. Persistence of *Salmonella enteritidis* from one day of age until maturity in experimentally infected layer chickens. *Poultry Science*, **77** (12), 1759–1762.

Hayes, S., Nylen, G., Smith, R. *et.al.*, 1999. Undercooked hens eggs remain a risk factor for sporadic *Salmonella enteritidis* infection. *Communicable Disease and Public Health*, **2** (1), 66–67.

Humphrey, T.J., Chapman, P.A., Rowe, B. and Gilbert, R.J., 1990. A comparative study of the heat resistance of salmonellas in homogenized whole egg, egg yolk or albumen. *Epidemiology and Infection*, **104**, 237–241.

Humphrey, T.J., 1994. Chapter 5: *Contamination of eggs with potential human pathogens*, in *Microbiology of the Avian Egg*, ed. Board, R.G. and Fuller, R. Chapman and Hall, London.

ICMSF, 1980. *Microbial Ecology of Foods*. Volume II. *Food Commodities*. Academic Press, London.

ICMSF, 1996. *Micro-organisms in Foods*. Volume 5. *Characteristics of Microbial Pathogens*. Blackie Academic and Professional, London.

Lock, J.L. and Board, R.G., 1995. The fate of *Salmonella enteritidis* PT4 in home-made mayonnaise prepared from artificially inoculated eggs. *Food Microbiology*, **12**, 181–186.

Motarjemi, Y. and Kaferstein, F., 1999. Food safety, hazard analysis and critical control point and the increase in foodborne disease: a paradox? *Food Control*, **10**, 325–333.

Panisello, P.J. and Quantick, P.C., 1998. HACCP and its implementation. *Food Science and Technology Today*, **13** (3), 130–133.

PHLS Laboratory of Enteric Pathogens, 1981–1991; Salmonella dataset 1992 onwards. Accessed at website June 2000.
http://www.phls.co.uk/facts/Gastro/SalmHum.Ann.htm

Radford, S.A. and Board, R.G., 1995. The influence of sodium chloride and pH on the growth of *Salmonella enteritidis* PT4. *Letters in Applied Microbiology*, **20**, 11–13.

Rodrigue, D.C., Tauxe, R.V. and Rowe, B., 1990. International increase in *Salmonella enteritidis*: a new pandemic? *Epidemiology and Infection*, **105**, 21–27.

Ryan, M.J., Wall, P.G., Gilbert, R.J., Griffen, M. and Rowe, B., 1996. Risk factors for outbreaks of infectious intestinal disease linked to domestic catering. *Communicable Disease Report*, **6** (13), 6 December, R179–R183.

Wilson, D., Patterson, W.J., Hollyoak, V. and Oldridge, S., 1999. Common source outbreak of salmonellosis in a food factory. *Communicable Disease and Public Health*, **2** (1), 32–34.

Product Formulation and Control

Key issues
- Yoghurt
- *C. botulinum*
- The 'combined treatment' approach to control
- HACCP

Challenge

Yoghurt, a fermented milk product, has widely been considered to be a 'safe' product. Yet this case study describes a situation which should lead you to an exploration of when the formulation of a product can, but may not, control the potential hazards within it.

9.1 THE CASE STUDY: BOTULISM AND HAZELNUT YOGHURT, 1989

From the UK newspaper the *Guardian*, 13 June 1989. © The Guardian

"Botulism alert over yoghurts
Tom Sharratt

A national alert against eating any brand of hazelnut yoghurt was issued by the Department of Health yesterday after it was disclosed that 10 people, five of them children, were seriously ill with suspected botulism in north-west England.

All the victims are believed to have eaten hazelnut yoghurt made by two small Lancashire dairies. The Department of Health says the use of a hazelnut puree from a firm in Kent may be linked to the outbreak. But it warned that hazelnut puree from a third firm might be implicated in another suspected case of botulism and advised the public to avoid all brands of hazelnut yoghurt.

The Department emphasised that although botulism had been diagnosed it had yet to be confirmed. The suspected victims are all showing symptoms of botulism including difficulty in breathing and swallowing. Botulism attacks the nerves in the head, causing paralysis of the eyes and the throat. There have been 21 cases this century with 12 deaths in seven different incidents, most recently in 1978 from contaminated tinned salmon.

Two men in their twenties, one from Preston, the other from Blackpool, in intensive care at the Royal Preston Hospital, were said to be "very poorly". At Blackpool Victoria Hospital, a man aged 26, a woman 44 and a boy of 13 all from the Fylde coast area of Lancashire, were in a stable condition. A young Blackpool man admitted to hospital yesterday was believed to be suffering from the same infection. The four other patients were all children at Booth Hall Children's hospital in Manchester. A 14-month-old boy from the Blackpool area was said to be in a satisfactory condition and his three year old sister was stable but on a ventilator. A boy of six from the Rochdale area was "improving" and a 14-year-old boy from the Oldham area was said to be "poorly but stable".

Dr. Richard Newton, a consultant neurologist at Booth Hall, said yesterday he would not be surprised if there were more victims as botulism has an incubation period of up to a week.

The Department of Health said the hazelnut puree was manufactured by the 'XYZ Fruit Company', of Folkestone, and was used for yoghurt by two dairies in the north west – 'Producer 1' of Bilsborrow, and 'Producer 2' of Longridge, both near Preston.* All the hazelnut yoghurt produced by the two companies has been withdrawn from sale. A partner in Producer 1 said the company had suspended production until the cause of the outbreak was determined. The firm supplies shops within a radius of 15 or 20 miles. He said: "I believe we are looking at one batch of between 300 or 350 pots." A spokeswoman for 'Producer 2' said there had been no problems with any of their yoghurts, but they had withdrawn hazelnut.

Eden Vale, which makes Ski yoghurts, said it would temporarily cease production of its hazelnut brand although it was made from nut pieces rather than from puree and the suppliers were not implicated in the current alert.

St. Ivel, the other big yoghurt maker, was still considering what to do last night. Last night the Co-op, Sainsbury's, Safeway and Asda[†] had removed hazelnut yoghurt from their shelves."
[End of quote]

Three days later the *Guardian* (16 June, 1989) contained the following report:

"Four more yoghurt botulism poison cases in North-west

Four more people were suffering from botulism poisoning yesterday after eating hazelnut yoghurt. Three are being treated in hospital. The fourth, a 74 year old woman, suffered only mildly and was not admitted to hospital. They bring the number of known cases to 22. All live in the North-west. Seven of the 22 are aged 25–44, two are children aged four and under, and three are over 65.

Medical teams said yesterday there was no evidence to suggest a particular age group was at risk from botulism poisoning. Four children are making progress in Manchester's Booth Hall children's hospital, a spokesman said yesterday. The most seriously ill [a girl] aged three from Fleetwood, Lancashire, was "still

---

* Names have been changed to be consistent with data on pages 209–211.
† Large retail food supermarkets.

poorly but improving slowly". Warnings not to buy hazelnut yoghourt were issued when bacteria were identified in the hazelnut puree used in the manufacture of the yoghurt. Samples of a yoghurt brought to hospital by one of the poison patients have been sent to the Central Public Health Laboratory in Colindale, north London.

Health officials in Folkestone, Kent, yesterday said that inadequacies in the heat treatment caused the outbreak of botulism in hazelnut puree supplied by the XYZ Fruit Company, a factory in the town. The manufacture of all low-acid food products at the factory has been halted."
[End of quote]

In the preliminary investigation of the outbreak in June 1989 [O'Mahony *et al.* (1990)] it was established that the hazelnut yoghurt made by Producer 1 was the only food consumed in common by the first patients in the week before symptoms became evident. Of these eight, seven could be interviewed but one who was too ill and could not be. Once the common link was established the authorities – the Department of Health – first stopped production of the implicated brand of hazelnut yoghurt made by Producer 1, and the next day, June 12, advised the public not to eat any brand of hazelnut yoghurt after Producer 2, who also used the same make of hazelnut puree as Producer 1 had stopped production.

Some while later, in October 1989, a paper was published in the *Lancet* describing medical aspects of the June outbreak and reported that 27 people were suspected to be affected, of whom one died (Critchley, Hayes and Isaacs, 1989). 25 of the 27 had eaten hazelnut yoghurt in the week before the onset of symptoms, and a few had noted an abnormal taste or appearance of the yoghurt, but others had not. In one family two children who shared a carton were affected, yet their cat who had eaten the remainder of the carton of yoghurt was not.

Diagnosis for *C. botulinum* type B intoxication rested on identification of toxin in one tin of hazelnut puree, in the stool specimen of one patient, and in two cartons of hazelnut yoghurt obtained from patients' homes.

In the investigation (O'Mahony *et al.*, 1990) the following types of samples were sent to the Public Health Laboratory Service (PHLS) Food Hygiene Laboratory (FHL) in Colindale, London:

- serum and faecal specimens from suspected cases for detection of *C. botulinum* toxin,
- 21 cans of implicated hazelnut conserve,
- opened and unopened cartons of hazelnut yoghurt (Producer 1),
- two samples of hazelnut yoghurt from another producer (Producer 3).

The tests carried out at the FHL or other specialist laboratories included:

- injection into mice of sera, food and faecal extracts,
- protection tests (of mice) with polyvalent and monovalent antisera against extracts of hazelnut conserve from a blown can,
- culture for *C. botulinum* from faecal and food samples,
- confirmation tests of identity of *C. botulinum* on suspected isolates,

- tests on sera and food samples for the presence of botulinum toxins using an amplified ELISA procedure.

The outcomes of these analyses reported by O'Mahony *et al.* (1990) were:

Cans of hazel nut puree:

- One can positive for *C. botulinum* B toxin at a concentration of $600-1800 \, MLD \, ml^{-1}$
  (Note 1: this can was badly blown, and it came from the premises of Producer 1. Note 2: 'MLD' = 'mouse lethal dose').
- pH:
  15 of 17 cans – between 5.0 and 5.5
  One can pH 4.5
  One can pH 4.7

Cartons of Producer 1's hazelnut yoghurt:

- Two opened cartons from patients homes and 15 unopened cartons were positive for *C. botulinum* B toxin (by mouse bioassay). Concentration of toxin varied between $14-30 \, MLD \, ml^{-1}$. Additionally viable cells of *C. botulinum* B were found to be present.
- All 17 cartons had sell-by date 13 June 1989.

Cartons of Producer 3's yoghurt:

- Two cartons tested and were negative for *C. botulinum* toxin.

Patients faecal specimens:

- One sample positive for *C. botulinum* B toxin by mouse bioassay test. Culture tests showed viable cells of *C. botulinum type B* (*proteolytic*) present.
- Eight samples negative.

Patients sera

- All 23 samples negative.

Furthermore, during the epidemiological investigation the premises of Producers 1, 2 and 3 were inspected and those of Producer 1 were found to be of an adequate standard for the production of yoghurt, and those of Producers 2 and 3 satisfactory. In the inspection process information was also gained about how Producer 1 made the yoghourt.

The hazelnut puree manufacturing process at the XYZ Fruit Company*

The official investigation team (O'Mahoney *et al.*, 1990) determined from the hazelnut puree processing company (the XYZ Fruit Company) how the product was made.

---

*Disclaimer. This is not the real name of the company concerned and has no relation to any real company of that name, should there be one.

- *Ingredients*: pre-roasted hazelnuts, water, starch and other ingredients.
- *Method*: these ingredients were mixed and heated in a steam jacketed vat with half a ton capacity to a temperature of 90 °C for 10 minutes. It was then pumped into metal cans which were closed at the top using a manually operated seamer, and then placed in a retort of boiling water for a minimum of 20 minutes.

Most of the hazel nut conserve made contained sugar but 76 cans of conserve sweetened with aspartame rather than sugar were manufactured in July 1988. 36 cans of this consignment were received by Producer 1 in November 1988 and stored at ambient temperature. A further 36 cans of this consignment had been delivered to another firm, Producer 2, in July and August 1988. Two of the remaining four cans were still held by the manufacturer, and it was believed that two had been used for testing purposes some time previously. Producer 3 had received a supply of this manufacturer's hazelnut conserve on 28 June 1988, which Producer 3 believed had been used soon after delivery, but records were not able to confirm this. Since that time they had received supplies of hazelnut puree from another firm. Customers reported to the conserve manufacturer that cans of the aspartame sweetened preparation had blown and following this, in October 1988, potassium sorbate was introduced into the mixture in an attempt to control yeasts.

The hazelnut yoghurt production process at Producer 1

The premises of Producer 1 were inspected by the outbreak investigating team and were considered to be of adequate standard for the production of yoghurt. The yoghurt was produced from a mixture of pasteurised milk, skimmed milk powder and starch, which was heated to 82 °C, had sugar added and was held for 30 minutes. The mixture was cooled, pumped to an inoculating tank, starter culture added and mixed for 20 minutes, following which it was poured into 10 gallon churns and placed in an incubator for 2–3 hours until 'set'. Hazelnut conserve was then mixed with the yoghurt, before dispensing into 360 cartons, each with a 'sell-by' date 25 days after production. Advice was given on the cartons that the yoghurt should be refrigerated and consumed within two days of purchase.

The problem for the outbreak investigating teams was to identify how such an incident of botulism came about in order both to limit the numbers of cases at the time and secondly to learn lessons from the outbreak in order to prevent a similar outbreak happening again. The outbreak had serious implications for those who suffered it. As already indicated at least 25 of the 27 suspected cases definitely suffered botulism, and unfortunately one old person died. The remaining 24 cases recovered, but with their future health and well-being compromised. The adverse publicity severely affected the national overall sales of hazelnut yoghurts regardless of by whom they were produced. Later, in 1993, the fruit puree producer [the XYZ Fruit Company] went out of business, as did Producer 1 in 1994, both several years after the botulism outbreak.

9.2   BACKGROUND

This case study relates to two very different types of products – fruit puree and yoghurt – but which often have in common a low pH. However, they come from very different sources of raw materials – from plants on the one hand and from an animal source on the other. Thus the natural flora of the raw materials is associated with their different origins and will tend to reflect that. But the flora of the products is also associated with how they have been produced, handled and stored. When mixed together to make a fruit yoghurt the flora of one product is introduced into the other and the combined effects of composition and environment, which are expected to protect the final product from quality loss or even ensure its safety, are challenged and may be insufficiently robust to achieve those ends.

9.2.1   Fruit Purees and Their Preservation

As fruits ripen both their sugar content and their pH rise, while acidity reduces. pH values seen across a wide variety of fruits may extend from around 2.2 (lemon), to around 6.0 (melon), due to the presence and concentrations of a range of organic acids (Table 9.1).

Ripe fruits also contain a high concentration of water, *i.e.* have high a_w. The flora they carry is a result of a combination of factors – the flora which naturally develops on the fruit; flora associated with the agricultural processes in their production; soil, dust and handling contamination, as well as the development of post-harvest spoilage organisms. Because of their acidity and sugar content the spoilage of fruits is often fermentative and is due to acidophilic or aciduric bacteria, yeasts and moulds. For preservation of the fruit that flora must be controlled so that further microbial growth does not reduce either the safety or quality of the product. The combined effects of the actual sugar and acid concentrations, the nature of the organic acids in the fruits, the pH and a_w values as well as the temperature determine the type of microbial spoilage to which raw and processed fruit is subject. Preservation techniques may well change the balance of these factors so that fruit products provide microbial growth environments different from the raw fruit but which may nevertheless be subject to microbial spoilage. Equally fruits may carry organisms among their flora which do not grow but merely survive on the fruit which then acts as a vector to other foods. Significant in this context is the presence of faecal pathogens originating from sewage used as a field fertiliser or from direct contamination by human, animal, bird or insect faeces. A number of outbreaks of food-borne diseases have been connected with contaminated fruits (refer also to Case Studies 1 and 19). Published data demonstrate that pathogens have varying rates of survival in fruits (Table 9.2) illustrating that provided sufficient numbers of pathogens are present, even if the population as a whole is dying, acidic foods such as fruits have the potential to act as vectors of food-borne disease organisms.

Survival of pathogens in fruits is generally longer at refrigeration temperatures (5–10 °C) than at ambient (20–30 °C) and in some fruits of low acidity pathogens

Table 9.1 *Approximate pH values of some foods*

lemons	2.2
vinegar	2.9
gooseberries	3
prunes, apples, grapefruit	3.1
rhubarb	3.2
apricots	3.3
strawberries	3.4
peaches	3.5
raspberries	3.6
oranges	3.7
cherries	3.8
pears	3.9
tomatoes	4.2
bananas	4.6
egg albumin	4.6
carrots	5
cucumbers	5.1
cabbage	5.2
bread	5.4
meat, ripened	5.8
tuna fish	6
potatoes	6.1
peas	6.2
egg yolk	6.4
milk, fresh	6.6
shrimp	6.9
meat, unripened	7
egg white	7–9

Source: Parry and Pawsey, 1984.

may actually grow on the cut surfaces. Thus when fruits are preserved the process must take account of the potential nature of the microbial load as well as the total numbers of organisms present (refer to Case Study 7).

Frozen fruit puree will not spoil while frozen, but will retain most of its viable microbial load which will be able to multiply once the product is thawed. Dried fruits have lowered water content and lowered water activity and may therefore be spoiled by osmophilic yeasts and moulds. Frequently this is prevented by the use of permitted preservatives such as sulfur dioxide (Table 9.3). The very short shelf life of fruit juices can be extended by refrigeration, or further lengthened by pasteurisation and chilling. Further extension of shelf life can be achieved by sterilisation and/or protected by the addition of preservatives.

However, if a fruit concentrate or jam is to be produced then the process will reduce the water and raise the sugar concentrations. It is very common to add more sugar to obtain a sweeter product than the fruit alone would provide, but the maximum sucrose concentration cannot be more than 66 g/100 g because beyond that the solution is saturated and the sucrose will crystallise out. At that concentration the water activity is approximately 0.88 (at 37 °C). Including some lower molecular weight sugar such as glucose and adjusting the proportions of

Table 9.2 Survival of pathogens in fruits and acid systems

	Temp. (°C)	pH	Organism	Effect	Time	Ref.
Orange serum broth	30	5.0 (HCl)	Listeria monocytogenes	decrease 10^6 to 10^2	8 days	1
Orange serum broth	4	5.0–4.8 (HCl)	Listeria monocytogenes	decrease 10^6 to 10^2	25 days	1
Orange serum broth	4	4.6 (HCl)	Listeria monocytogenes	decrease 10^6 to <25	81 days	1
Orange serum broth	4	4.0 (HCl)	Listeria monocytogenes	decrease 10^6 to <25	43 days	1
Orange serum broth	30	3.6 (HCl)	Listeria monocytogenes	decrease 10^6 to <25	5 days	1
Orange serum broth	−4	3.6 (HCl)	Listeria monocytogenes	decrease 10^6 to <25	25 days	1
Frozen orange juice	−17	3	Shigella flexneri	survival	1–2 days	2
Orange juice	−4	3.5	Shigella dysenteriae	survival	<170 h	3
White grape juices	20	3	Shigella dysenteriae	survival	2–7 h	4
Red grape juices	20	3	Shigella dysenteriae	survival	4–28 h	4
Citrus	2–4	not stated	Vibrio cholerae	survival	<1 day	5
Citrus	28–30	not stated	Vibrio cholerae	survival	<1 day	5
Other fruits	2–4	not stated	Vibrio cholerae	survival	1–5 days	5
Other fruits	28–30	not stated	Vibrio cholerae	survival	3–11 days	5
Dried fruits	2–4	not stated	Vibrio cholerae	survival	1–4 days	5
Dried fruits	28–30	not stated	Vibrio cholerae	survival	<1 day	5
Apple cider	8	3.9–4.0	E. coli O157:H7	decrease from 10^5 to not detectable	16 days	6
Apple cider	25	3.9–4.0	E. coli O157:H7	decrease from 10^5 to not detectable	3–6 days	6
Apple cider	refrigerated	3.3–3.5	Salmonella typhimurium DT104	Survived one week	Undetectable in 2 weeks	7

Sources: 1. Parish, M.E. and Higgins, D.P. (1989); 2. Hahn, S.S. and Appleman, M.D., (1952); 3. Beard, P.J. and Cleary, J.P. (1932); 4. Rochaix., A. and Jacqueson, R. (1938); 5. ICMSF, 1996; 6. Zhao et al., (1993); 7, Roering et al., (1999).

Table 9.3 *Preservatives effective in fruit products*

Preservative	Source	Action	Comments	Foods used in
Sorbic acid and sodium, potassium, and calcium sorbate	Occur naturally in some fruits	Very effective against moulds and yeasts at pH 4.0–6.0. Effective against some bacteria, and prevent spore outgrowth	Break down at high temperature	Fermented milks, yoghurts, dried fruits, fruit salads, wines and cider, cheeses, some other foods
Benzoic acid, and sodium, potassium and calcium benzoates	Occur naturally in some fruits	Inhibits yeasts, moulds and bacteria in an acid medium		Jams, fruit pulps and purees, fruit juices and other foods
Ethyl, methyl and propyl 4-hydroxybenzoates, and their sodium salts	Derived from benzoic acid	Inhibits yeasts, moulds and bacteria in an acid medium		Jams, fruit pulps and purees, fruit juices and other foods
Sulfur dioxide and sulfites	Oxidised sulfur	Inhibits moulds, and bacteria	Most of the added sulfur dioxide is released on cooking	Very widely used in the protection of fruits and fruit products from spoilage. Used also in many other foods

sucrose and glucose could permit a further lowering of the water activity while reducing the risk of crystallisation. Equally the same sweetness could be achieved by reducing the total sugar content and adding low calorie sweetener, but that would result in a water activity value in the product not as low as 0.88. Many spoilage micro-organisms, and all pathogenic bacteria, are incapable of growth below 0.88 a_w, but among the osmophilic yeasts and moulds there are some which can grow at a_w values as low as 0.60 a_w.

The boiling point of sugar solutions is dependent on their sugar concentration, and the higher the concentration the higher the boiling point above 100 °C. In jam making using the open pan method the boiling point is around 104 °C, but use of that temperature can have adverse effects on quality. In commercial practice an alternative is to boil the mixture under vacuum which lowers the boiling temperature to as little as 71 °C while improving the colour of the jam. However, the process of boiling fruit to make concentrates and jam should kill all heat sensitive bacteria, moulds and yeasts, and leave only the heat resistant bacterial spores whose heat resistance tends to increase the lower the water activity of the heating menstruum. The subsequent microbial stability of fruit concentrates and jams thus depends on a number of factors: reduction of the microbial load in processing, elimination of those organisms which can grow in the product in its final condition (of water activity, pH, organic acid concentrations and types, and storage temperature), use of packaging methods which do not introduce organisms and packaging which retains its integrity during storage. Where processing is mild and insufficient to destroy the populations of spoilage organisms producers may sometimes also resort to the addition of permitted chemical preservatives (within maximum permitted concentrations) to retard the increase of the target populations.

9.2.2 Nuts and Their Preservation

Botanically nuts are also fruits or parts of fruits. While developing they are protected by their shells, and as mature edible nuts they lose moisture and become drier, water activity reduces, the oilyness becomes more pronounced, while pH remains near neutral. In storage they are largely spoiled by moulds tolerant of the low water activity and as such may suffer from the growth of mycotoxin producing moulds. Contamination by other organisms (bacteria and yeasts) originating from the plants, soil, dust and the environment can occur in the handling and shelling processes. Many nuts are sold roasted and the fierce heat of this process has the potential to destroy the microbial flora of the nuts rendering them effectively sterile. The reliability of the process in this respect depends on how the process is implemented and whether the nuts are exposed on all sides, for how long and what temperatures their surfaces achieve. Additionally, as with all other foods, unless they are handled hygienically post-processing, they can become re-contaminated.

9.2.3 Yoghurt and Its Manufacture

Through the controlled growth of starter organisms in liquid whole, skimmed, fortified or sweetened milk lactic acid and flavour compounds are produced resulting in the gel known as yoghurt.

The draft Codex Alimentarius (2000) standard for fermented milk describes yoghurt as "symbiotic cultures of *Streptococcus thermophilus* and *Lactobacillus delbruekii* subsp. *bulgaricus*".

Normally the yoghurt maker chooses the starter culture used so that yoghurt of the desired flavour and texture characteristics is produced. Before the addition of the starters the milk is heat treated under a regime usually more severe than that used for drinking milk pasteurisation. Heating at around 80 °C for 30 minutes both destroys the heat sensitive contaminating flora, leaving only the most heat resistant of organisms, and also facilitates changes to the milk proteins which favour starter growth, and milk protein gel formation. Milk casein gels at around pH 4.8 and this pH drop is achieved by the growth of the starter over a period of 3–4 hours of incubation at 41–42 °C. The formed yoghurt is then cooled rapidly to below 4 °C – which stops the further growth of the starter organisms and ensures that the continued rate of pH drop over the product's storage life is then very slow.

Traditional yoghurt starter culture is a 50:50 combination of *Streptococcus thermophilus* and *Lactobacillus delbruekii* subsp. *bulgaricus*, added at about 2% by volume to the milk. The optimum temperature for growth for *Strep. thermophilus* is 37–42 °C, and higher for the lactobacillus at 45–47 °C. The initial population of around 10^6 cfu ml^{-1}, achieved after inoculation of the milk, rises during incubation to approximately 10^8 cfu ml^{-1} by which time lactic acid has accumulated, the buffering power of the milk is overcome and the pH drops rapidly to about 4.5 when the iso-electric point of the casein micelles is reached and they coagulate. In the freshly inoculated milk the *Streptococcus* tends to grow faster than the *Lactobacillus* but as the pH drops its growth rate diminishes and ceases at around pH 4.2 while the population of the *Lactobacillus* continues to develop. The final acidity in the yoghurt varies according to the product being made – mildly acidic products having a lactic acid concentration 0.8–0.9%, and sharper flavoured, more acidic products having 1.0–1.2% lactic acid.

Under optimal conditions the rapid growth of the starter cultures ensures that their population dominates, and that contaminating organisms, should they be present, have to compete for available resources in a changing environment in which lactic acid (and lower quantities of other organic acids) accumulates and the pH drops. However, should any growth inhibitory factors occur in the milk, such as traces of antibiotics resulting from veterinary treatment of the cows, or should the temperature of the milk be too cool for the designated process then the starter organisms will grow more slowly. Slower starter growth will be evident to the yoghurt maker either because measurements of rate of acid development will indicate it, or because observation shows that the gel is late in forming, the gel is weak, lacks viscosity or the yoghurt has an unexpected odour. These conditions may afford the opportunity for spoilage organisms, or even pathogenic organisms if present, to increase in number. Advisory standards for yoghurt at the point of consumption recommend that coliforms (as indicators of the hygiene of production) should not be detectable in one gram, and that less than one mould and less than 10 yeasts are detectable per gram (Adams and Moss, 2000). As would be expected surveys of commercially available yoghurt products often show that good quality has been achieved (Nogueira *et al.*, 1998) but sometimes demonstrate the reverse (Fleet, 1990; Abd-El-Ghani *et al.*, 1998).

Finally flavoured yoghurt can be made from the fermentation of flavoured milks, or through stirring fruit or other flavours into the formed gel. This latter method was what was used in the hazelnut yoghurt/botulism outbreak being considered in this chapter. Many strains of aciduric yeasts and moulds are major causes of yoghurt spoilage, being able to endure the acid conditions and grow during storage. They enter the product from the fruits or from the production environment, tolerate the acid conditions in the yoghurt and grow, causing spoilage which is detectable through changes in appearance, smell, colour or texture. For example, yeasts at populations above 10^5 ml^{-1} above can produce unacceptable fizzy flavours and gassiness (Fleet, 1990).

9.3   EXERCISES

Exercise 1. Was it just bad luck that one tin of the **XYZ Fruit Company**'s hazelnut conserve caused botulism? In thinking about this, analyse the evidence presented to you in the case, and debate your views using other evidence if necessary.

Exercise 2. Managing the safety and quality of fruit yoghurt: if the yoghurt producer, referred to by O'Mahony *et al.* (1990) as **Producer 1**, had purchased their hazelnut conserve from a different supplier do you think their yoghurt would have been consistently free from quality and safety problems?

9.4 COMMENTARY

Exercise 1. Was it just bad luck that one tin of the **XYZ Fruit Company**'s hazelnut conserve caused botulism? In thinking about this, analyse the evidence presented to you in the case, and debate your views using other evidence if necessary.

In order to come to a view on this question you need, as usual, to ask yourself what you need to know and whether information supplied in Sections 9.1 and 9.2 or elsewhere is enough to help you.

A number of aspects of the overall production and distribution process need to be considered:

- What was their recipe for making hazelnut puree?
- What was the microbial status of the mixed raw ingredients?
- What was the process for heat treating the hazelnut puree?
- How would this process affect the microflora?
- Would every batch receive adequate treatment?
- What was the expected shelf life of the product?
- How was the process managed, and was a HACCP system in place?

O'Mahony *et al.* (1990) indicate that three types of puree were made:
- the original puree containing pre-roasted hazelnuts, water, starch and other ingredients not specified, but including sugar;
- one in which the sugar was substituted by aspartame; and
- one in which the aspartame was retained but sorbic acid was added to control yeast growth.

The *C. botulinum* spores must have come from these ingredients unless they came from contamination from the factory, its equipment or the packaging.

Because hazelnuts are plant material it is conceivable that they were contaminated with soil containing the spores. As far as the puree producer, the XYZ Fruit Company, was concerned it is possible that the organisms came from the pre-roasted hazelnuts bought in, in spite of the fact that roasting is a process which involves exposing the nuts to very high temperatures, well above 100 °C, a process which should kill all micro-organisms and their spores. However, if the roasted nuts were either improperly roasted or were subsequently contaminated after roasting, and also exposed to humid atmospheres in storage causing their water content and hence water activity to rise, if any residual viable spores of *C. botulinum* were present could they have germinated and grown, meaning that the hazelnuts themselves became toxic before use in the puree? There is no evidence presented in Section 9.1 to indicate that this was so, but it is a possibility which merits consideration. The other ingredients – the starch and the sugar – are less likely to have contained the organism, although, as with the water used in the process, it is possible.

We do not know the exact proportions of the ingredients used but they would

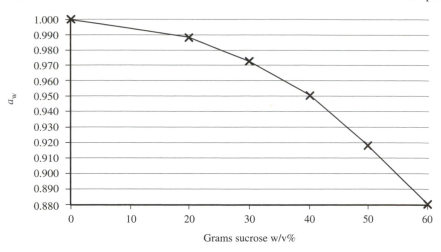

Figure 9.1 *Isotherm for sucrose/water solution at 37 °C*

have been important in determining the eventual water activity of the processed puree. The main ingredient determining the water activity of the mixture would have been the sugar (sucrose and/or glucose) (Figure 9.1).

It seems that after mixing the raw ingredients, batches of up to 500 kg were heated to a temperature of 90 °C for 10 minutes at a time in a steam jacketed vat. Exposure to this time/temperature treatment would have destroyed heat sensitive bacteria, yeasts and moulds, would have destroyed any pre-formed botulinum toxin, and would have left most bacterial spores viable. However, the efficacy of the treatment for the batch would depend on good mixing – ensuring that all of the batch (every gram of the 500 000 g present) received that treatment.

The subsequent pumping into cans could have re-contaminated it with a variety of organisms from the pipes, pumps and fillers, as would any exposure to the air. The can seamer too might have contaminated the product, or may have made imperfect seams through which contamination was later drawn as the product cooled. Thus the second heat treatment – the in-can treatment – should have been an important process in killing viable cells and spores present, while the soundness of the seaming process should have protected the product from later re-contamination. From the description provided by O'Mahony *et al.* (1990) in Section 9.1, it appears that the maximum temperature used was that of boiling water – 100 °C, although the time of exposure seems to have been variable ("for a minimum of 20 minutes"). Of course what would have been critical would have been the time and temperature of exposure in the centre of each can and whether it was enough to reduce the population of spores to zero per can. Being thickened by starch there would have been little convection inside the can, so considerable care would have been needed to ensure that every can of puree was adequately heat-treated. The company would have (or should have) ascertained the point in the cans of the greatest heat lag, and secondly how long it took for the temperature to rise to a stable value, so that they could then determine the heat treatment required. The time that it took would also have depended on the can geometry, each can's position in the retort and on the uniformity or otherwise of distribution of heat in the retort. The other factor to consider is how long the puree was expected to remain of good quality under the expected conditions of storage. There is no statement of what the expected shelf life of the canned puree was, but it was handled and stored at ambient temperature and, at least in the case of Producer 1, used 11 months later. Thus the shelf-life expectation was at least 11 months at UK ambient ($20 \pm 5\,°C$).

So what does 'adequate heat treatment' mean in this case study? (refer for comparison to Case Study 7).

If the processor considered the possible presence of *C. botulinum* spores in the unprocessed puree he should also have considered what constraints would have existed in the product to prevent outgrowth of spores should they have remained viable. Firstly the heat treatment itself should have reduced the overall population of organisms – yeasts, moulds and heat sensitive bacteria. However, either of the two heating steps might have had the effect of activating the residual bacterial spores, thus stimulating them into germination.

The most effective constraint in addition to the heat treatments could have been the manipulation of the water activity through the concentration of sugar, ensuring that the water activity was low enough to prevent spore outgrowth and/or multiplication of vegetative cells. The limiting water activity was dependent on the combined values of other parameters of the food system – particularly the pH and the temperature, and thus would have been higher in some combinations, lower in others. Minima of pH 4.6 and a_w 0.93 are quoted as growth threshold values for mesophilic strains of *C. botulinum* (ACMSF, 1992) when the other parameters are at optimal values. Combinations of reducing the pH, the a_w and/or the temperature have the effect of raising the minimal value at which organism growth – in this instance *C. botulinum* – is possible.

Thus only if the puree had a sugar content of greater than 46% sucrose would the a_w have been lower than 0.93 (refer again to Figure 9.1). Reduction of the pH of the puree might have introduced another constraint but it is not stated that any action was taken to lower the pH of the puree nor is the actual pH of the newly made product given. But if the pH and/or the water activity values were above the threshold values at which spore germination could occur and viable cells could grow, the heat treatment should have been a 'botulinum cook' *i.e.* a 12D process for heat resistant *C. botulinum*. If the pH and/or water activity could have prevented even slow growth of *C. botulinum* then the heat treatment could have been milder and geared to destroying any problematic spoilage organisms (refer to Case Study 7). An adequate heat treatment would have destroyed organisms which could have grown in the puree (*i.e.* tolerating its a_w, its pH and storage temperature). But because of the radical difference in a_w between the sugar sweetened product and the aspartame sweetened product any heat treatment which was adequate for the former would not have been adequate for the latter.

There is no evidence presented that spoilage of cans of the puree sweetened with sugar occurred, indicating that in that product the combined effects of heat treatment and low a_w actually did constrain microbial growth in storage.

But when the sugar was substituted by aspartame problems with quality did occur. Customers reported that some of the cans already received had blown. The texture of the aspartame sweetened product may have been different from the sugar sweetened product, and maybe the producer adjusted the added starch content to accommodate this. But even so, increasing the starch content would have had little effect on lowering the water activity. Starch molecules are of high molecular weight; when heated they would hydrolyse to some extent and some low molecular weight sugars would be present and would marginally reduce the water activity. But the reduction would not be equivalent to the initial addition of sucrose or glucose. Thus the aspartame sweetened product, while possibly thick in texture, would have had a much higher water activity than the sucrose sweetened product, and its a_w would have been insufficiently low to prevent growth of yeasts. It is evident now that bacteria (*e.g. C. botulinum*) as well as other organisms were indeed able to grow and were not completely inhibited by reduced a_w.

The XYZ Fruit Company sought in October 1988, 9 months before the botulism incident, to solve the spoilage of the aspartame sweetened product by adding the preservative potassium sorbate rather than by changing the heating processes to more stringent regimes or by preventing the entry of the yeasts they saw to be problematic.

The problem of the cans blowing may have been one of contamination after the first heat processing stage, and before final can sealing. But if the heating at the second stage had been adequate yeasts – which are heat sensitive – would have been destroyed by the heating process long before any level of security regarding bacterial spores was achieved. Thus the problem of the cans blowing due to yeast growth (if indeed that is what it was) needed to be seen in the context of the process given. Account needed to be taken of the fact that the hazelnut puree would have been considerably less acidic than many other fruit products. Evidence in Section 9.1 indicates that the pH of most of the cans of the hazelnut puree was above 5.0. While a process for a fruit product with a low pH, well below 4.6, can be relatively mild, that will not be adequate for a product whose pH is above 4.6.

You need also to consider the addition of potassium sorbate. This hazelnut puree was produced in a fruit-processing factory, an environment where there could well have been many yeasts prevalent in the air of the processing facility. In some contexts sorbic acid (and sorbates) is a useful preservative which is most effective at pH 4.0–6.0. Not only do sorbates inhibit yeast growth, but also inhibit the outgrowth of bacterial spores, provided that the concentration is adequate for the microbial load. Lund *et al.* (1990) reported, in some experiments conducted at pH 5.5 and over a period of 14 days, a probability of between 1:100 000 and 1:1 000 000 of the outgrowth of a single spore of proteolyic *C. botulinum* in the presence of 280 mg undissociated sorbic acid/litre at 30 °C. But in the absence of sorbic acid the probability increased to between 1:1 and 1:10.

However, the efficiency of sorbate as a preservative depends on the concentration of undissociated acid, itself dependent on the pH because the pK_a of sorbic acid is 4.76 at 25 °C. So only if an adequate concentration of undissociated sorbic acid had been present for the combination of the pH of the product and the

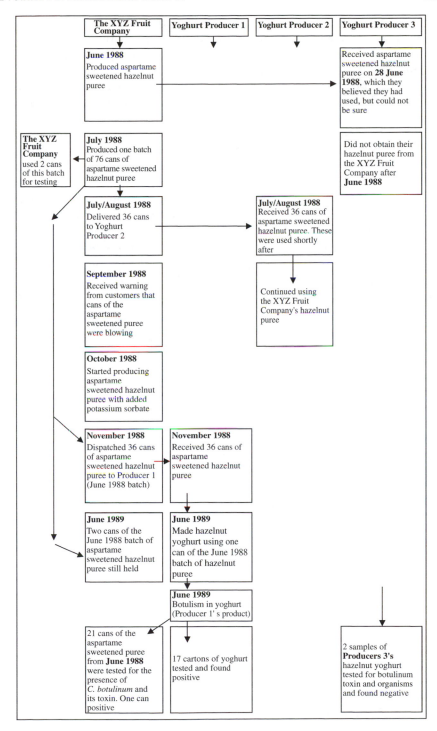

Figure 9.2 *Hazelnut puree production and distribution*
(Based on data in O'Mahony *et al.*, 1990)

contaminating cell concentration would added sorbate have had any significant effect. From the evidence presented in Section 9.1 the pH of the hazelnut puree as made is not given, so consider the options concerning heating regimes and hygiene management that have been implemented before sorbate was added to control contaminating organisms (refer to Table 9.3). Sorbates are not stable in heating – they break down – so that the concentration present after heating would be very much less than that added. In the UK Miscellaneous Food Additives Regulations, 1995 (Anon, 1995) sorbic acid and sorbates are only allowed in listed foods, so check whether canned and bottled products are among them and evaluate this strategy of adding sorbate to the spoiling cans of aspartame sweetened hazelnut puree. But what is not clear from the evidence presented in this case study is whether the cans of puree which were analysed by the authorities after the botulism outbreak contained sorbic acid or not. The evidence indicates that the changes were made after the particular batch of aspartame sweetened puree which caused the botulism was manufactured.

Turning to the process of production and the distribution of canned product, Figure 9.2 gives an analysis, based on the evidence of O'Mahony *et al.* (1990) given in Section 9.1, of when the different types of puree were produced and used by customers.

The total batch of 76 cans of aspartame sweetened puree were made in July 1988 and 36 cans were used by Producer 2 in July and August 1988 (4–8 weeks after production). Another 36 cans were delivered to Producer 1 in the same period but stored by him until the next year. It was in June 1989 – eleven months later – that the outbreak of botulism arose. That period of eleven months of storage at ambient temperature (perhaps at temperatures ranging between 15 and 25 °C allowing for seasonal variations in temperature in the UK) of the infected hazelnut puree allowed the bacterial spores present to germinate, grow and produce toxin or spoilage in the puree. Producer 1 said that probably only one batch of yoghurt of 350 cartons was involved. The hazelnut puree needed for such a small batch cannot have amounted to much more than one can. Producer 1 also had another can of the same aspartame sweetened puree which was badly blown and which proved on testing to be positive for botulinum toxin. Another 17 cans were tested and although they were not reported to be positive for botulinum toxin, two cans had pH values close to 4.6 (4.5 and 4.7) while others had pH values of between 5.0 and 5.5. pH 4.6 is the threshold for the growth of *C. botulinum*. But in the two most acidic cans had the pH changed from a higher value to a lower one due to microbial growth which could have included *C. botulinum*? If the puree's pH was normally above pH 4.6 it should have received a 'botulinum cook' or its combined pH and a_w should have been capable of preventing the growth of the organism for the period and conditions of storage anticipated.

To conclude, was it just bad luck that the XYZ Fruit Company's hazelnut conserve caused botulism? You need to weigh the evidence and the information you have amassed to come to a view on the assumptions made about the processing of the product and its stability which allowed contaminating organisms, including *C. botulinum*, to grow in it.

By adjusting different parameters in the product such as the pH, the a_w, the addition of preservatives and, if necessary, the temperature of storage so that singly or combined they restrain microbial growth can be an effective strategy for managing the safety and quality of food products. But such a strategy has to be implemented on the basis of sound understanding of what is needed to achieve those ends. Predictive computer models can be used to evaluate the possible adjustments which can be made in each of the parameters when targeting the control of individual microbial hazards, but once a combination treatment which will control all the targetted hazards has been decided, it can only be effective if implemented within a process which is well managed.

Exercise 2. Managing the safety and quality of fruit yoghurt: if the yoghurt producer, referred to by O'Mahony *et al.* (1990) as **Producer 1** had purchased their hazelnut conserve from a different supplier do you think their yoghurt would have been consistently free from quality and safety problems?

With the growing market for yoghurt seen in Western Europe and America in the late 1960s and early 1970s a number of workers investigated the possibilities of the growth or survival of pathogens in yoghurt. Mocquet and Hurel (1970) actually reported an outbreak in which raspberry flavoured yoghurt was responsible for causing food poisoning at a military hospital in Strasbourg in 1964. Although 10^8 *Staphylococcus* g^{-1} were found in the yoghurt, how this situation arose could not be determined and they speculated that there was a lactic acid starter failure. Unless the starter is specifically selected, starters, in contrast to contaminating *Staphylococcus*, may be slowed by the presence of sugar in the milk – and this may have happened in the above case.

Other workers in the early 1970s studied the growth or survival of *Staph. aureus, Salmonella typhimurium, Yersinia enterocolitica*, coliforms, and *E. coli* in yoghurt and other milks, and some of their results are included in Table 9.4.

It was observed that inhibition of contaminating organisms might be attributable not only to the antimicrobial effects of the lactic acid (and other organic acids if present), or to the pH drop, but also to antimicrobial compounds produced by the starter organisms. The ability to produce bacteriocins, of which a number are recognised and some chemically defined, is possessed by some members of the lactic acid bacteria used as starters in yoghurt making and in other dairy fermentations. However, they are produced in varying quantities and have a range of antimicrobial effects (Sahl and Bierbaum, 1998). For example studies have indicated that there is potential to exploit the ability to produce bacteriocins of lactic cultures commonly used in various fermented milks. Apella *et al.* (1992) showed in *in vitro* experiments that the strains of *Lactobacillus acidophilus* and *Lactobacillus casei* produced (unidentified) bacteriocins capable of inhibiting the growth of *Shigella sonnei* and *Shigella flexneri* and these authors suggest that there is scope for the development of fermented milk products for use in the prevention and treatment of infantile diarrhoea caused by strains of *Shigella*.

Nisin is the bacteriocin which has been studied most extensively. It is produced by *Lactococcus lactis*, is a polypeptide and is capable both of inhibiting growth of Gram positive bacteria through membrane damage and also of preventing the outgrowth of bacterial spores. Because nisin resists degradation during heating, particularly at acid pH values, it can be used to prevent spore outgrowth in heat-processed foods. It also occurs in fermented milk products and contributes to the control of the population there. It has been suggested as having use in cheeses where it is envisaged to be of value in the control of Gram positive competing organisms as well as pathogens such as species of Listeria (also Gram positive). However, it is possible that the addition of nisin to cheese milk could be counterproductive in the ripening stages when the development of flavour occurs. There is potential for the genes encoding for nisin production to be introduced by genetic manipulation techniques into yoghurt starter cultures to cause them to produce nisin and thus control the growth of contaminating organisms. Of course it is also necessary that the nisin does not also affect the starter cultures' own ability to produce enough acid to coagulate the milk proteins. Additionally under some circumstances in which the integrity of the cell membrane has been damaged nisin can also be inhibitory to Gram negative organisms (see for example Elliason and Tatini, 1999; Gaenzle *et al.*, 1999).

However, it has become clear that although there are factors in yoghurt which normally prevent the growth of contaminating organisms, there are also occasions when pathogens can grow in the fermenting milk prior to the formation of the yoghurt gel, and also it is known that under many conditions of storage of the formed yoghurt pathogens die, but over differing time periods which sometimes extend into days. The controlling factors which normally protect yoghurt (if infected) from transmitting pathogens to consumers are its low pH, its lactic and other organic acid content, the possible presence of bacteriocins, and its storage at refrigeration temperatures of 5 °C or below. In practice moreover there are very few reported outbreaks of food poisoning associated with yoghurt. In 1964 there was the *Staphylococcus* case described earlier; in 1989 there was of course the botulism case which forms the case study in this chapter, and in 1991 there was an outbreak of *E. coli* 0157:H7 PT49 infection affecting 16 people, and in which yoghurt was the suspected source of infection (Morgan *et al.*, 1993). The literature generally indicates that while fermented milk drinks may sometimes be associated with the transmission or even growth of pathogens, yoghurt itself is rarely incriminated. But with the proliferation of types of fermented milk products, and with the boundaries of definition between them blurring as new products are developed, it is not easy for producers to predict what could grow or survive in them. Producers may still be faced with the unanswered question of whether, for their particular yoghurts, made in their particular production conditions and eaten at any time over the duration of the products expected shelf life, pathogens could survive (Table 9.4) or grow there (Table 9.5).

Thus for protection of public health and for the well-being of their businesses it is necessary that the control of production processes, through good manufacturing practice and the application of HACCP, ensures that such contaminating organisms either do not enter the process or are eliminated or controlled within it.

Confidence in Producer 1's processes, judged by the visiting team of inspectors as satisfactory for the production of yoghurt, is affected by the fact that the company used a can of aspartame sweetened hazelnut puree in their yoghurt when at the same time they had in their store another can which was blown (see Figure 9.2). How did they judge that the can they did use was of a suitable quality and safe to use? A few consumers of the affected yoghurt reported that it had an abnormal taste – a view which could be attributable to their not having eaten low calorie hazelnut yoghurt, or not even having eaten any hazelnut yoghurt before, as well as the possibility that the puree itself was 'off'. Was Producer 1 one of the XYZ Fruit Company's customers who back in October 1988 had complained (to the XYZ Fruit Company) that some cans of puree had blown? But should the fact that one can in Producer 1's store was blown at the very time in June 1989 when the aspartame sweetened hazelnut yoghurt was to be made have warned them that the 'good can' was a risk? How robust were the lines of communication within the factory between the store and the production line?

To be consistently free from quality and safety problems Producer 1 actually needed a management system in place incorporating HACCP which defined the yoghurt production process. A well designed HACCP system accounting for the difference between sugar and aspartame sweetened hazelnut puree would have alerted the producer to the potential for problems and would certainly have started alarm bells ringing when one can (not used) of puree was evidently spoiled. Under a good HACCP system any other can of the same production date code – even if it showed no evident signs of spoilage – would not have been used.

Thus in forming your view that they would have been consistently free from quality and safety problems even if they had never purchased puree from the XYZ Fruit Company you need to consider how the company managed risk in relation to production, and where its management strengths and weaknesses were.

9.5 SUMMARY

The successful use of combined treatments as a control strategy for the quality and safety of foods depends on understanding that is what is being done. A producer needs to understand the growth restraining factors which exist, those he is introducing and those which may no longer be effective should any change in product formulation or product processing occur. If, for example, the product should be stored refrigerated yet is in practice stored at ambient, or if it is stored for a period longer than the anticipated shelf life, then he should understand that the controls which exist under one set of conditions may no longer exist under another. Predictive microbiology based on computer models tries to support product design, but is only useful if the product is then manufactured and used within the controlling factor specifications (see Shadbolt, Ross and McMeekin, 2001).

Unless steps are taken to conserve them low pH, high-acid fruits are spoiled by aciduric organisms. However, it is a mistake to imagine that all fruits are high in

Table 9.4 Selected examples of survival of pathogens in fermented milk products

Product	Organism	Temp. (°C)	Initial pH	Initial population/g ($log_{10} N$)	Final population/g ($log_{10} N$)						Notes	Ref.
					Day 1	Day 2	Day 3	Day 4	Day 7	Day 8		
Yoghurt	Campylobacter jejuni/coli (10 strains)	4	4.4–5.4	2.5 log reduction in 0.11–0.31 hours								1
Yoghurt (Brand 'C')	Escherichia coli H52	7.2	4.2	3.04	2.4	1.36	<1	<0				2
Skimmed milk fermented by homofermentative LA	E. coli strains (toxigenic, invasive, and non-pathogenic strains)	7	4.5	2.01	Longest survival time observed = 17 days							3
Yoghurt	E. coli O157:H7 (added at beginning of yoghurt fermentation)	4	4.5	3	3.3	2.93	2.86	2.83	2.72			3
Cultured skimmed milk containing Strep. thermophilus and L. bulgaricus	Salmonella typhimurium	11 (storage temp.)[a]	6.50 dropping to 4.50 due to growth of LA organisms	Approx. 3.0 (before co-culture with LA organisms)						>1.0 to <2.0	Rate of decline dependent on the lactic culture	4

Product	Organism								Ref.
Yoghurt (Brand 'S')	*Staphylococcus aureus*	7	3.7	5	3.69	2.8	2	<1.0	5
Yoghurt	*Shigella sonnei* (1 strain)	4	Not stated			Survival time: 6–7 days			6
Yoghurt	*Shigella flexneri* (1 strain)	4	Not stated			Survival time: 1–2 days			6
Yoghurt	*Yersinia enterocolitica* E657	5	4.2	7		6	5		7
Yoghurt	*Yersinia enterocolitica* 2635	5	4.2	6		5	4		7

[a]The salmonella were co-cultured with various LA organisms at 20–30 °C, after which they were stored at 11 °C for a number of days.

References: 1. Cuk *et al.*, 1987; 2. Goel *et al.*, 1971; 3. Frank and Marth, 1977; 4. Park and Marth, 1972; 5. Minor and Marth, 1972; 6. Mihajlovic *et al.*, 1980; 7. Ahmed *et al.*, 1986.

Table 9.5 *Limits for growth for a range of food poisoning organisms*

Organism	Lower temperature limit (°C)	pH	a_w (NaCl)	[NaCl]
Aeromonas spp.	0–4	<4.5		
B. cereus	4	5	0.93	
C. perfringens	12	5.5–6.0	0.97	
E. coli (intestinally pathogenic strains)	7–8	4.4	0.95	
Salmonella spp.	5.2	3.8	0.94	
Shigella sonnei	6.1	4.9		
Shigella flexneri	7.9	5		
Yersinia	−1.3	4.2–9.6		5–7%

Note: Each factor considered alone under otherwise optimal conditions. See also Tables 4.3 and 6.3.

Source: ICMSF, 1996.

acid. The successful preservation of fruits has to take into account the actual pH of the product, its acids profile and its water activity in order to process them in such a way that while the desired product is achieved the microbial flora is controlled, by either its destruction or its inability to grow.

The 'combined treatments' approach to the fruit purees relies on the successful combination of factors such as the pH, the water activity, and the temperature at values which together limit or completely inhibit microbial growth, where used singly they would not have been capable of so doing.

In the case of the hazelnut purees the manufacturers relied on heat processing, combined with the low water activity produced by the added sugars, to control the growth of acid tolerant organisms. But when they changed the sugar for aspartame, the water activity was not low, was permissive to microbial growth and yeasts were able to grow. But sporing bacteria including *C. botulinum*, not inhibited by the water activity, could also grow. There was insufficient constraint in the system to inhibit growth. When they added a different constraint – sorbic acid – to control yeast growth they may also have been successful in inhibiting *C. botulinum* – but it was a policy which needed very careful consideration. Whereas sorbic acid may have controlled the yeasts, since the pH of the hazelnut puree must have been greater than 4.6, *C. botulinum* might have been present; the pH might not have allowed sufficient undissociated sorbic acid to be present for effective inhibition of spore germination. In the case of low acid foods, whose pH is permissive for *C. botulinum* growth, adequate heat processing is the primary barrier. In this case clearly that barrier was inadequate.

Yoghurt, being an acid environment, is subject to spoilage by aciduric organisms. But while many fermented milk products can induce the loss of viability of some pathogenic organisms which put the product at risk, this depends on both the type of infecting organism and the specific nature of the product. Thus some constraints to growth of both spoilage organisms and pathogens exist in yoghurt too. In this outbreak experience had shown that the puree could go off, and that would indicate that the puree might also contain pathogens. Secondly, yoghurt itself does not necessarily constrain pathogen survival. Yoghurt is not automatically a safe food.

Experience is showing that, while no food can be assumed to be 'risk free', the approach of control through well designed and implemented HACCP, taking proper account of the limitations of constraints introduced to inhibit microbial growth, brings foods closer to that ideal.

9.6 REFERENCES

Abd-El-Ghani, S., Sadek, Z.I., Fathi, F.A., 1998. Reliability of coliforms as an indicator for post processing contamination in yoghurt manufacture. *Annals of Agricultural Science, Ain Shams University*, **43** (1), 221–230.

ACMSF, 1992. Report on vacuum packaging and associated processes, HMSO, London, UK.

Adams, M.R. and Moss, M.O., 2000. Chapter 9 *Fermented and Microbial Foods*. In *Food Microbiology*, 2nd Edition. Royal Society of Chemistry, Cambridge, UK.

Ahmed, A.-H., Moustafa, M.K. and El-Bassiont, T.A., 1986. Growth and survival of *Yersinia enterocolitica* in yoghurt. *Journal of Food Protection*, **49** (12), 983–985.

Anon, 1995. Food Safety (Miscellaneous Food Additives) Regulations, 1995. HMSO, London, UK.

Apella, M.C., Gonzalez, S.N., Nader de Macias, M.E., Romera, N. and Oliver, G., 1992. *In vitro* studies on the inhibition of the growth of *Shigella sonnei* by *Lactobacillus casei* and *Lactobacillus acidophilus*. *Journal of Applied Bacteriology*, **73**, 480–483.

Beard, P.J. and Cleary, J.P., 1932. The importance of temperature on the survival time of bacteria in acid foods. *Journal of Preventive Medicine*, **6**, 141–144.

Codex Alimentarius, 2000. Draft standard for fermented milks (A-11).

Critchley, E.M.R., Hayes, P.J. and Issacs, P.E.T., 1989. Outbreak of botulism in N.W. England and Wales, June 1989. *Lancet*, **ii**, 849–853.

Cuk, Z., Annan-Prah, A., Janc, M. and Zajc-Satier, J., 1987. Yoghurt: an unlikely source of *Camplyobacter jejuni/coli*. *Journal of Applied Bacteriology*, **63**, 201–205.

Elliason, D.J. and Tatini, S.R., 1999. Enhanced inactivation of *Salmonella typhimurium* and verotoxigenic *Escherichia coli* by nisin at 6.5 °C. *Food Microbiology*, **16** (3), 257–267.

Fleet, G.H., 1990. Yeasts in dairy products: a review. *Journal of Applied Bacteriology*, **68**, 199–211.

Frank, J.F. and Marth, E.H., 1977. Inhibition of enteropathogenic *Escherichia coli* by homofermentative lactic acid bacteria in skim milk. *Journal of Food Protection*, **40** (11), 749–753.

Gaenzle, M.G., Hertel, C. and Hammes, W.P., 1999. Resistance of *Escherichia coli* and *Salmonella* against nisin and curvacin A. *International Journal of Food Microbiology*, **48** (1), 37–50.

Goel, M.C., Kulshrestha, D.C., Marth, E.H., Francis, D.W., Bradshaw, J.G. and Read, R.B., 1971. Fate of coliforms in yoghurt, buttermilk, sour cream and cottage cheese during refrigerated storage. *Journal of Milk and Food Technology*, **34**, 54–58.

Hahn, S.S. and Appleman, M.D., 1952. Microbiology of frozen orange juice concentrate. 1. Survival of enteric pathogens in frozen orange concentrate. *Food Technology*, **6**, 156–158.

ICMSF, 1996. *Micro-organisms in foods*. Volume 5. *Characteristics of Microbial Pathogens*. Blackie Academic and Professional, London.

Leistner, L. and Gorris, L.G.M., 1994. Food preservation by combined processes. Final Report FLAIR Concerted Action no. 7, subgroup B. Directorate-General XII Science, research and Development, EUR 15776 EN.

Lund, B.M., Graham, A.F., George, S.M. and Brown, D., 1990. The combined effect of incubation temperature, pH and sorbic acid on the probability of growth of non-proteolytic type B *Clostridium botulinum*. *Journal of Applied Bacteriology*, **69**, 481–492.

Mihajlovic, V., Levi-Jovovic, E., Otasevic, M., Gasic, S., Stojicic, G. and Milosavljevic, R., 1980. Sopravivenza di enterobacteriaceae in differenti tipi di alimenti: sopravivenza di alcuni ceppi de *Shigella*, *Giornale di Malattie Infettive e Parassitarie*, Estratto del No. 8, **32**, 624–627. Quoted in ICMSF (1996).

Minor, T.E. and and Marth, E.H., 1972. Fate of *Staphylococcus aureus* in cultured buttermilk, sour cream and yoghurt during storage. *Journal of Milk and Food Technology*, **35** (5), 302–305.

Mocquet, G. and Hurel, C., 1970. The selection and use of some micro-organisms for the manufacture of fermented and acidified milk products. *Journal of the Society of Diary Technology*, **23** (3), 130–146.

Morgan, D., Newman, C.P., Hutchinson, D.N., Walker, A.M., Rowe, B. and Majid, F., 1993. Verotoxin producing *Escherichia coli* 0157 infections associated with the consumption of yoghurt. *Epidemiology and Infection*, **111**, 181–187.

Nogueira, C., Albano, H., Gibbs, P. and Teixeira, P.J., 1998. Microbiological quality of Portuguese yoghourts. *Journal of Industrial Microbiology and Biotechnology*, **21** (1/2), 19–21.

O'Mahony, M., Mitchell, E., Gilbert, R.J., Hutchinson, D.N., Begg, N.T., Rodhouse, J.C. and Morris, J.E., 1990. An outbreak of foodborne botulism associated with contaminated hazelnut yoghourt. *Epidemiology and Infection*, **104**, 389–395.

Park, H.S. and Marth, E.H., 1972. Survival of *Salmonella typhimurium* in refrigerated cultured milks. *Journal of Milk and Food Technology*, **35** (8), 489–495.

Park, H.S. and Marth, E.H., 1972. Behaviour of *Salmonella typhimurium* in skim-milk during fermentation by lactic acid bacteria. *Journal of Milk and Food Technology*, **35** (8), 482–488.

Parish, M.E. and Higgins, D.P., 1989. Behaviour of *Listeria monocytogenes* in low pH model broth systems. *Journal of Food Protection*, **52** (3), 144–147.

Parry, T.J. and Pawsey, R.K., 1984. *Principles of Microbiology for Students of Food Technology*. Hutchinson, London.

Rochaix, A. and Jacqueson, R., 1938. Pouvoir microbicide du jus de raisin frais, *Revue Hygiene*, **60**, 241–250.

Roering, A.M., Luchansky, J.B., Ihnot, A.M., Ansay, E., Kaspar, C.W. and Ingham, S.C., 1999. Comparative survival of *Salmonella typhimurium* DT104, *Listeria monocytogenes* and *Escherichia coli* O157:H7 in preservative-free apple cider and simula gastric juice. *International Journal of Food Microbiology*, **46** (3), 263–269.

Sahl, H.G. and Bierbaum, G., 1998. Lantibiotics: biosynthesis and biological properties of uniquely modified peptides from Gram positive bacteria. *Annual Reviews of Microbiology*, **52**, 41–71.

Shadbolt, C., Ross, T. and McMeekin, T.A., 2001. Differentiation of the effects of lethal pH and water activity: food safety implications. Letters in Applied Microbiology, **32**, 99–102.

Zhao, T., Doyle, P. and Besser, R.E., 1993. Fate of enterohaemorrhagic *Escherichia coli* O157:H7 in apple cider with and without preservatives. *Applied and Environmental Microbiology*, **59**, 2526–2530.

C. Risk

CHAPTER 10

Views of Risk

Key issues
- Semi-soft cheeses, and raw milk
- *Escherichia coli* O157:H7
- Cross contamination
- Control of *Escherichia coli* O157:H7 in cheesemaking
- HACCP
- Risk perception

Challenge

The case study presents you with the opportunity to consider how the application of HACCP assists the production of safe cheese, and whether raw milk cheese should be made. It further asks you to think about whether sampling of made cheeses can demonstrate the safety of a batch, and requires you to consider the perception of risk from different viewpoints – enforcer and producer.

10.1 THE CASE STUDY: RAW MILK CHEESE AND ITS SAFETY

10.1.1 The Outbreak, and Public Health Protection Measures

In England on 19th April 1998 a child was taken ill with food poisoning caused by a toxin producing strain of *E. coli* O157:H7 but after three weeks of illness, hospitalisation, and varied treatments including kidney dialysis fortunately the child recovered.

The food associated with the infection was semi-soft farm made cheese produced within the dates 4–6th April 1998. As a result the authorities (the Department of Health) ordered all this cheese to be withdrawn from retail outlets and prohibited from sale. The quantity was more than 3000 2 kg cheeses whose date codes extended from the first week in February 1998 to the first week in May 1998 and represented the entire stock of the cheese both in maturation and on retail sale.

Government scientists then tested 200 cheese samples recalled from sale from a number of outlets from around the country, and of these 200 ten were found to be positive for the presence of *E. coli* O157:H7. The ten found positive were coded with the dates of production 4–6th April 1998. The other 190 tested were not shown to contain *E. coli* O157:H7.

Nevertheless, on the basis of these 200 tests, the authorities, under new emergency powers made under the Food Safety Act, 1990, then wished to condemn and destroy all of the cheese considering that in the interest of public health protection it was too great a risk to re-release them onto the retail market.

The producer and the affineur contested this in High Court through Judicial Review and won their case, the Judge feeling that public health could have been protected by other measures. However the Government authorities – the Department of Health – appealed against the ruling and the outcome of the case, taken in the High Court in September 1998, was then reversed. This meant that the High Court agreed that the cheese was too high a risk for acceptable public health protection and must all be destroyed. Although the affineur did not have to pay the costs of the appeal, neither were they paid any compensation for the destruction of the product. The case and appeal had disastrous effect, severely adversely affecting the confidence of customers in all the cheeses and thus threatening the viability of the small cheese businesses.

10.1.2   The Appeal – Evidence from the Department of Health

Consider the following extract from the deposition of the Government expert to the High Court, dated 29 September 1998:

"Results of studies carried out on the *E. coli* O157 isolates

The cheese isolates
21. All the cheese isolates (10) have the same characteristics on molecular typing and are thus considered to be the same.

The human isolates
22. Both human isolates (from the boy who was ill and from [the cheesemaker] are virtually the same as the cheese isolates. They are the same phage type and produce the same toxins. When subjected to a technique known as pulsed field gel electrophoresis (PFGE) which separates the proteins of the organism, they have all the bands seen in the cheese isolates.
23. However, both human isolates have an additional band and these additional bands are different in the two human isolates. Further investigation of the two human isolates reveals the presence of an extra plasmid (a piece of genetic material which is not part of the fixed chromosomal genetic material and which can be gained or lost with relative ease). In each case this is the same size as the additional band on PFGE. Thus the additional band in each human isolate is considered to be due to the presence of an additional plasmid which is different in each case. Plasmids may be acquired during passage through the human intestine and these additional plasmids in the human isolates may therefore reflect their derivation from human faeces.
24. Thus while it would not be true to say that the isolates are exactly the same, they are so closely related as to make it highly probable that they are epidemiologically related.

The beef isolates

25. It is clear that the isolates from the beef cattle are completely different from the cheese and human isolates and, indeed, from each other. According to my information, there were five isolates from the beef herd. Of these, four were a different phage type from that found in both cheese and humans. The remaining isolate was of the same phage type but produced a different toxin.

26. The typing of the beef cattle isolates was not known on 19 May. However, it was clearly known that the positive results were from beef cattle not from dairy cattle supplying milk for the cheese. The positive results were of interest as they suggested a possibility that contamination of the cheese could have occurred as a result of contamination from the farmyard (for example, on footwear).

27. Once known, the fact that the strains isolated from the beef cattle were different from the human isolates made it less likely that the organism contaminating the cheese had come from elsewhere on the farm. Further evidence supporting this conclusion was the fact that environmental sampling in the dairy did not reveal the presence of *E. coli* O157. However, when the various possible explanations for the contamination were reviewed, it was considered that, since at least two strains of *E. coli* O157 were known to be present in the beef herd, the possibility of other strains being present could not be ruled out entirely."

[End of quotation from the Government expert, printed with permission of the Food Standards Agency, UK]

Later in court the Department of Health, protecting public health, argued that since *E. coli* O157 had been found in ten of 200 cheese samples, they could not rule out that the other cheeses made by the same company using raw milk from the same herd of cows, the same premises and equipment, the same staff and the same process might be similarly affected. Furthermore they argued that both because the infective dose of *E. coli* O157 is believed to be very low – perhaps as low as 100 cells, and because the initial *E. coli* O157 contamination could be very minute and would not necessarily affect every cheese in a batch, even intensive sampling could miss contamination and could create the false impression that a batch was acceptable. This view was accepted by the High Court judge.

10.1.3 The Producer and Affineur's Perspective

Each cheese weighed 2 kg (2000 g). Each was cylindrical in shape, about 20 cm high and about 15 cm diameter. Sampling such cheeses (ICMSF, 1986) involves taking portions of the cheese using a tool (a 'trier') which takes cylindrical paired samples from the diametrically opposite sides into the core of the cheese. These should be taken with care because such holes damage the cheese and cause loss of its value. The paired core samples together represent, when analysed, about 1/80th of the total weight and volume of the cheese. There would be no indication as to the best point to sample to find the pathogen, because *E. coli* O157 would not produce visible change to the outside surfaces of the cheese.

Nevertheless the ten samples did show the presence of the pathogen *E. coli* O157, possibly indicating that the pathogen must either have been present in very high numbers and/or more or less uniformly distributed in the cheese since the sampling points were random. The ten positive cheeses all were date coded 4/5/6 April which was also the production date of the cheese believed to have caused the illness in the child.

None of the other 190 cheese samples tested similarly were shown positive for the pathogen (see Figures 10.1 and 10.2 for the sampling pattern).

The daily farm cheese-making process used the evening and morning milk from the dairy herd. The milk was let into an open tank, warmed, renneted, and the starter added within 30 minutes and the warm milk was stirred round initially to achieve an even distribution of the starter organisms, then left for the curd to form, to be subsequently cut, drained, and then packed into the moulds, and further left to drain in the moulds. Such a mixing process would also, in the affineur's view, have evenly mixed in any contaminating organisms.

The producer and affineur accepted that *E. coli* O157 had been found in the samples of the date codes 4/5/6 April, but argued that the destruction of all the cheeses dating back to 1 February and forward to 3 May, when only samples of cheeses of the dates 4/5/6 April were shown to be positive while all other samples analysed did not demonstrate the presence of the pathogen, was not just.

10.1.4   The Cheese

The cheeses were made from raw cows' milk into 2 kg cheeses at the dairy farm. The on-farm stages produced the formed cheeses, and some of each week's production was matured off farm.

For those cheeses the maturation process took place in a cheese maturation cellar with a temperature of 16 °C and relative humidity of 85%. The surface bacterial culture *Bacillus linens* was added by immersion into a culture and allowed to grow on the surface. Each day each cheese was turned and smeared by hand to allow the surface to remain damp permitting the desired surface bacterial growth (which concurrently prohibited mould growth), while flavour developed within the cheese.

Cheeses were then sold to retailers at six weeks maturity. The 'best before' date was calculated from the day of production and given a period of three months.

10.2   BACKGROUND

In the UK there is a debate, often quite impassioned, about whether cheese made from untreated milk should be prohibited because of the risk that zoonotic pathogens in the raw milk will be carried over into the cheese (IFST, 1998; Campbell *et al.*, 1996, Rampling, 1996). The argument is countered by the view that safe raw milk cheese can be produced under good hygienic conditions, and that failures of hygiene cause problems not only in unpasteurised milk cheese but also in pasteurised milk cheeses (Kerr and Lacey, 1996; Rose, 1996; Cunynghame, 1996).

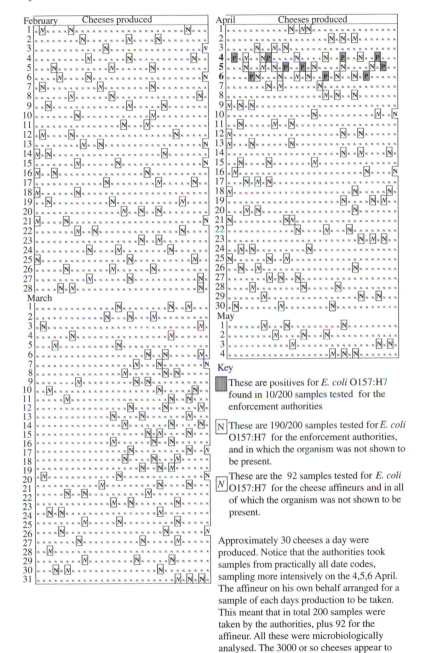

Figure 10.1 *Pattern of sampling of the cheeses*

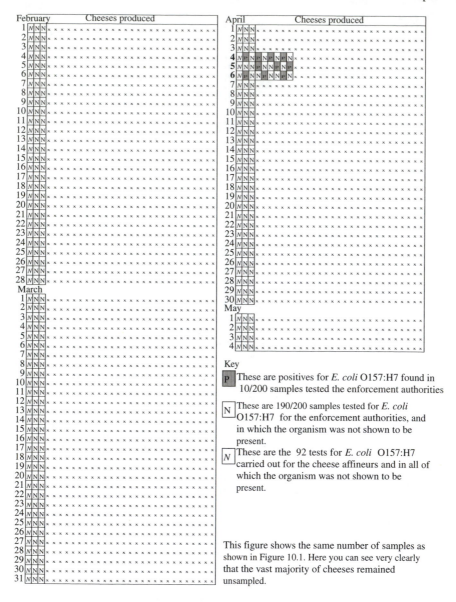

Key

P These are positives for *E. coli* O157:H7 found in
   10/200 samples tested the enforcement authorities

N These are 190/200 samples tested for *E. coli*
   O157:H7 for the enforcement authorities, and
   in which the organism was not shown to be
   present.

N These are the 92 tests for *E. coli* O157:H7
   carried out for the cheese affineurs and in all of
   which the organism was not shown to be
   present.

This figure shows the same number of samples as
shown in Figure 10.1. Here you can see very clearly
that the vast majority of cheeses remained
unsampled.

Figure 10.2 *Pattern of sampling of the cheese – a different view*

10.2.1 Raw Milk Quality

A one year survey (Department of Health, 1998) of 1674 raw milk samples, collected to assess the microbiological status of raw cows' drinking milk in England and Wales, showed that approximately 16% had a total viable count, and 25% had a coliform count which failed the standards required under the relevant legislation – the Dairy Products (Hygiene) Regulations 1995 (Table 10.1). The range and numbers of organisms found in this survey are shown in Table 10.2.

Table 10.1 *Microbiological standards for raw cows' milk for drinking[a] defined by the Dairy Products (Hygiene) Regulations, 1995[b]*

Plate count at 30 °C	Equal to or less than 2.0×10^4 cfu ml^{-1}
Coliforms	Less than 100 cfu ml^{-1}

[a]Direct sale to the ultimate consumer by a producer.
[b]The regulations relate to England and Wales.

An earlier survey (de Louvois and Rampling, 1998) conducted in 1996–97 of 1097 samples of raw milks evaluated the presence of pathogens and found:

1097 samples: free of pathogens	1056	
containing pathogens	41:	
	Campylobacter	19
	Staphylococcus aureus	12
	Salmonella	5
	E. coli O157:H7	3
	Streptococci, Group C or G	2

Thus it is certainly true that raw milk can contain a range of pathogens which, given the opportunity, cause illness in consumers – see for example Chapters 3, 4 and 11; that milk and milk products can permit the survival of pathogens (Desenclos *et al.*, 1996); and that pathogens associated with animals can cause illness in human beings (see Chapters 5 and 12).

10.2.2 Cheese Making

Cheese making is a complex process whose details are specific to individual cheese types. But the essence of cheese making is that liquid milk is coagulated by added rennet and by acid developed by starter lactic acid bacteria under warm conditions (perhaps 28–30 °C). The warmed milk has been pumped from where it was stored into a large vat, where the milk surface may be exposed to the air of the dairy. Carefully selected starter cultures, which are mixed populations of organisms selected and balanced for the contribution they make to the overall development of the cheese, are added while the bulk is mixed to ensure an even distribution of organisms of about 10^6 cfu ml^{-1}. Their growth produces organic

Table 10.2 Summary of results[a] from a survey of raw milk samples in England and Wales, May 1995 to May 1996

Test	Detected (%)	Detected per 25 ml (%)	Number of samples (%) in each range of cfu ml^{-1}							
			<1	≥1 to <10	≥10^1 to <10^2	≥10^2 to <10^3	≥10^3 to <10^4	≥10^4 to <10^5	≥10^5 to <10^6	>10^6
TVC	1591 (100)		0	0	5 (0.3)	69 (4)	1039 (65)	414 (26)	58 (3.6)	6 (0.4)
Coliforms	1469 (92)		122 (8)	442 (28)	637 (40)	329 (21)	53 (3)	8 (0.5)	0	0
E. coli	978 (62)		613 (39)	592 (37)	328 (21)	52 (3)	6 (0.4)	0	0	0
S. aureus	107 (7)		1484 (93)	18 (1)	36 (2)	39 (3)	13 (0.8)	1 (0.1)	0	0
Listeria spp. (plating method)	0			1591 (100)[b]	0	0	0			
Listeria spp. (enrichment method)		91 (5.7)								
Listeria monocytogenes		32 (2.0)								
Salmonella spp.		1(0.06)								
Campylobacter spp.		0								
E. coli O157:H7		0								

Total samples taken = 1674. Total samples tested 1674. Number phosphatase positive = 1591 (95%).
[a] Data based on phosphatase positive results.
[b] Detection limit for Listeria spp. 10 cfu ml^{-1}.
Source: Department of Health, October 1998.
Crown copyright, reproduced with permission.

acids, principally lactic acid, which causes the pH of the milk to drop from that of fresh milk (around pH 6.8) to values which may ultimately reach 3.5–4.5. The coagulation process takes 30–60 minutes, during which short period the native flora both survives and possibly begins to multiply. The curd is cut, shrinks a little and the whey is released from the coagulum, and in some cheeses heat is applied at this stage to aid the process. The warm pieces of curd are then removed, often by hand, and packed into moulds from which further whey drains. The moulds are turned either mechanically or by hand several times on the first day to ensure that the curds are packed evenly and form the texture required. All this while the curds are warm and moist, the starter organisms are growing and reaching populations higher than their initial population.

In their growth the starter organisms convert the milk sugars into lactic and other organic acids whose specific nature and the lower pH associated with them, together with other anti-microbial factors (Table 10.3), create an environment in which the growth of other, contaminating organisms is retarded or even prohibited.

If the starter growth is 'slow' for any reason (due perhaps to traces of antibiotics or other chemicals in the milk, or to the presence of bacteriophage multiplying and slowing the overall starter population increase) then contaminants may have a concomitant opportunity to increase in numbers.

Later the moulded cheeses are moved to maturing rooms where the temperature (between 6 and 15 °C) and relative humidity (normally above 85% RH) are controlled to allow development of texture and flavours in the cheese brought about by enzyme activity associated with the starter organisms. In the maturation process, which depending on the cheese type being made can take from a few days to several months, the environment within the cheese changes constantly as enzymes associated with microbial growth break down the milk proteins, fats and carbohydrates, and water is lost from the outer surfaces of the cheese. Both in spatial and temporal terms the environment in a cheese varies, and may support growth, may only permit microbial survival, or is hostile and causes microbial death.

Thus when for example a mature cheddar cheese is cut across it is possible to

Table 10.3 *Anti-microbial factors associated with lactic acid bacteria*

low pH
organic acids
bacteriocins
carbon dioxide
hydrogen peroxide
ethanol
low redox potential
nutrient depletion
crowding

Source: Reprinted from Adams, M.R. and Nicolaides, L., 1997. Review of the sensitivity of different foodborne pathogens to fermentation. *Food Control*, **8** (5/6), 227–229. With permission of Elsevier Science.

see the drier outer rind and the softer inner regions between which there may be an evident colour gradient. Conversely a cheese like Camembert matures by becoming more liquid, a process progressing from the outside towards the centre. But these visible gradients are also paralleled by other compositional differences. In certain cheeses there will be a distribution of bubbles in the curd where gas was formed, and there may be the development of mould – as in blue veined cheeses. In mould ripened cheeses the pH rises as the moulds break down the proteins and alkaline products are formed. In these complex environments the microbial population changes too.

The original starter organism population initially rises, but may eventually diminish in numbers. Furthermore some members of the starter population, initially in a minority, may grow later and contribute to flavour development in the maturing cheese when the environment changes to suit their needs. Non-starter lactic acid organisms, which enter the cheese from the cheese-making environment, may also contribute to the flavour characteristics of the cheese (Fitzsimons *et al.*, 2001), or may produce bacteriocins – an ability which could be exploited to control pathogens in the cheese (FLAIR-FLOW, 1993). In this dynamic microbial population undesirable organisms too, which enter with the milk, originating from the cow, and sheep or goat, or beef or farmyard cattle (Cizek *et al.*, 1999), or enter later in the dairy and its equipment or from the dairy workers, may multiply, survive or die (Feresu and Nyati, 1990).

10.2.3   *E. coli* O157:H7 in Cheeses

Massa *et al.* (1999) showed that some strains of *E. coli* O157:H7 can survive and grow in raw milk stored at as cool a temperature as 8 °C, and Maher and co-workers (2001) showed that *E. coli* O157:H7 could survive through the manufacture and ripening of a smear-ripened cheese made from raw milk. Although the work indicated above was experimental, in which the pathogen was purposefully inoculated into the milk or milk for cheese, accidental 'inocula-tion' of milk through cross-contamination in the cheese-making environment would have similar effect. Furthermore sometimes *E. coli* O157:H7 can enter milk directly from the [infected] udder, or from the teat duct in the forestream of milk (Wright *et al.*, 1994). Equally non-pathogenic strains of *E. coli* could be expected to gain entry to the raw milk, entering from traces of faecal matter from the cows. Clearly it is in the interest of safe milk that the cows' udders are sanitised before milking, and that all equipment used in handling the raw milk and in the cheese making processes is extremely clean.

In a survey (MAFF, 2000) of 801 raw milk cheeses on retail sale in England and Wales testing for the presence of *E. coli* as indicator of faecal contamination, and for *E. coli* O157:H7, although 1.4% of samples contained more than 100 000 *E. coli* g^{-1} (and the highest of those containing 1.5×10^8 *E. coli* g^{-1}) no sample was shown to contain *E. coli* O157:H7 (Table 10.4). In the UK legally enforceable standards for the microbiological quality of cheeses at the point of removal from the processing establishment exist and these are shown in Table 10.5.

The report of the survey said "As there were no isolations of *E. coli* O157:H7, no association could be drawn between the number of *E. coli* present and the likelihood of *E. coli* O157:H7".

Table 10.4 *Summary of E. coli results in 801 retail samples of raw milk cheeses on sale in England and Wales, January–February 1997*

E. coli $\log_{10}$ count g^{-1}	Number of samples				
	Cow	Sheep	Goat	Buffalo	Total
<1.00	520	116	22	6	664
1.00–1.99	32	13	1	3	49
2.00–2.99	42	6	1	1	50
3.00–3.99	13	2	1	0	16
4.00–4.99	8	2	0	0	10
5.00–5.99	5	1	0	0	6
6.00–6.99	3	1	0	0	4
7.00–7.99	0	0	0	0	0
>8.00	1	0	0	0	1
Total	624	141	25	10	800*

*One sample of unknown milk species ($\log_{10}$ count = 2.23 *E. coli* g^{-1}).
Source: MAFF, 2000. Crown copyright, reproduced with permission.

Table 10.5 *Microbiological criteria for cheese – The Dairy Products (Hygiene) Regulations (1995)*[a]
The following criteria are applicable to the manufactured product on removal from the processing establishment:

A	Product	Organism	Standard (ml, g)
	Cheese other than hard cheese	*Listeria monocytogenes*	Absent in 25 g[b], n = 5, c = 0
	Hard cheese	*Listeria monocytogenes*	Absent in 1 g, n = 5, c = 0
	Milk based products (other than milk powder)	*Salmonella* spp.	Absent in 25 g, n = 5, c = 0
	Milk powder	*Salmonella* spp.	Absent in 25 g, n = 10, c = 0

(*continued overleaf*)

Table 10.5 *(cont.)*

B	Product	Organism	Standard (ml, g)
	Cheese made from raw or thermised^c milk	*Staphylococcus aureus*	$m = 10^3$, $M = 10^4$, $n = 5$, $c = 2$
	Soft cheese (made from heat-treated milk)	*Staphylococcus aureus*	$m = 10^2$, $M = 10^3$, $n = 5$, $c = 2$
	Fresh cheese	*Staphylococcus aureus*	$m = 10$, $M = 10^2$, $n = 5$, $c = 2$
	Cheese made from raw or thermised milk	*Escherichia coli*	$m = 10^4$, $M = 10^5$, $n = 5$, $c = 2$
	Soft cheese made from heat treated milk	*Escherichia coli*	$m = 10^2$, $M = 10^3$, $n = 5$, $c = 2$

[a] The Regulations apply in England and Wales.
[b] The 25 g sample to consist of five specimens of 5 g taken from different parts of the same product.
[c] Thermised = the heating of raw milk for at least 15 s at a temperature between 57 °C and 68 °C such that after completion of the process the milk shows a positive reaction to the phosphatase test.
Source: The Dairy Products (Hygiene) Regulations, 1995.

10.3   EXERCISES

Exercise 1. Bear in mind that unlike many other food products cheese is the outcome of the controlled growth of micro-organisms. Determine how the drawing up and use of a HACCP system would reduce food safety risk in cheeses made in a small creamery in a farm environment.

Exercise 2. After weighing the arguments, determine whether you support the use of raw milk for the production of cheese.

Exercise 3. If you can, undertake the following exercise with a group of colleagues or fellow students. Divide yourselves into three groups:
    Group 1. Consider the detail of the information you would need, and then argue the case that the authorities were right to condemn all the cheeses in the interest of public health protection.
    Group 2. From the perspective of the cheese producers and affineurs determine how you would support the argument that some of the cheeses could be released for sale, thus allowing the opportunity for the small businesses to survive.
    Group 3. Listen to Groups 1 and 2 and determine what you would have decided had you been asked to make a judgement.

10.4 COMMENTARY

> **Exercise 1.** Bear in mind that unlike many other food products cheese is the outcome of the controlled growth of micro-organisms. Determine how the drawing up and use of a HACCP system should reduce food safety risk in cheeses made in a small creamery in a farm environment.

The skill in producing cheese is to provide conditions in which the starter organisms grow and produce the desired end product while, at the same time, preventing the growth of unwanted organisms, which would cause the cheese to spoil and/or be unsafe to eat.

Thus, before thinking about the role of HACCP in controlling the safety (and quality) of cheese it would be valuable for you to first consider how cheese is made. Try to visit a cheese-making dairy, or at least to view a video of the process. This will give you an appreciation of the type of equipment used and the opportunity to appreciate some points of contamination into the product.

In small creameries there may be a very small number of staff (one, two, three ...) who undertake all the tasks. The cheese rooms may be situated in older buildings attached to a working dairy farm. The farm may have animals other than dairy cows, a muddy farmyard, and staff who work with the animals. Cups of tea and coffee may perhaps be taken 'across the yard' which therefore involves staff leaving the cheese rooms, and later re-entering them. Other people may wish to enter the cheese rooms and have a quick chat with someone – about their pay, about an order, some social message and so on. You should think about whether these things are good practice and how such human wishes can be managed while guarding the safety of the cheese. For when staff and others enter the cheese room they could be bringing with them pathogens originating from the animals and the farm environment. So strict hygiene protocols determining who may enter the dairy and how they enter have to be observed.

In creating the HACCP system (Table 8.10) a logical sequence of steps must be followed (van Schothorst and Kleiss, 1994; Codex Alimentarius, 1993), but for the HACCP plan to be effective it must be sufficiently comprehensive. It must include consideration of the source of the milk for the cheese and consider how the pathogens potentially present in the milk will be controlled. In many cheese-making processes the critical control point for heat sensitive pathogens introduced in the milk will be pasteurisation, which will only be effective if the correct time and temperature of exposure are defined and achieved.

However, for raw milk cheeses, where pasteurisation of the milk is not an option, control of zoonotic organisms lies in other strategies some of which are common with production of pasteurised milk cheeses. Those strategies include

- the husbandry of the animals to manage mastitis, and to ensure that milk is not taken from cows with mastitis;
- that the raw milk is produced from animals managed to significantly reduce the risk of infection by *Listeria* or *Mycobacteria*. Regular testing of the dairy

cows will establish whether *E. coli* O157 strains are present;
- that the animals are milked in very hygienic conditions;
- that the milk is stored in well cleaned equipment;
- that milk is used for cheese immediately after milking – which exploits the natural anti-microbial systems of the milk, or is cooled to below 6 °C, and used within 36 hours;
- that the milk temperature is raised to the cheese-making temperature and correct quantities of rennet and starter are added to ensure that starter grows at the right rate developing the expected levels of acid. The times here are critical for if acidity develops at a slower rate than expected not only is the cheese quality less good, but also growth of pathogens could occur;
- that, where used, salt is added in the correct amounts;
- that cheeses are matured under the correct conditions of humidity and temperature, and that the equipment and environment are as clean as possible.

Such issues must not remain vague: 'as clean as possible', 'correct amounts' and so on. Methods and protocols must be developed and written down so that the procedures can be followed by staff, monitored and reviewed regularly.

In the dairy the HACCP plan must accord with the reality of the actual environment. Table 3.2 (Telemea salted cheese), and Figure 11.1 (Lanark blue cheese) give examples of the flow of steps taken one after another in making two different cheeses. But just as important are the spatial relationships of people and equipment. Could, for example, raw milk (maybe contaminated with pathogens) splash onto the cheese moulds stored nearby? Could the cheese-maker's hands pick up contamination from the door handle to the yard? Would the cheese-maker lift the curds from the cheese vat, and wipe their hands on their apron, or elsewhere? Must the cheese-maker move back and forth from cheese room, to refrigerator, to door, to test equipment, to mould store and so on? Are these routes of cross-contamination? What is to be done about them?

If cheese becomes infected – as in this case study – the route of infection may never be accurately identified. And even if infection has not entered with the milk, could it enter another way?

HACCP should identify and put in place controls for those potential infection routes taking account of the complex movements people make in the cheese-making process; ensuring that they are not themselves infected, nor that they pick up contamination from the farm and bring it into the cheese rooms. Water, air, insects, vermin and pet animals, as well as under-cleaned equipment, are all potential methods of introduction of contaminants. The HACCP plan must define quantities, temperatures, pH values and times in managing the milk, and in the cheese making, to ensure the potential for growth of pathogens is minimised, while cheese making is optimised.

Many of the controls in making raw milk cheese rely on the time periods being defined, for example:

- the time period in which the milk must be cooled to below 6 °C; the time for warming the milk for cheese making;
- the time to acid development;

- the maximum time intervals between cleaning routines, the time of exposure to sanitising materials;
- the time for cheese maturation.

In the latter case cheeses that require more than 60 days of maturation are considered to be safer than those requiring less than 60 days, because experience has shown that pathogens tend to diminish in numbers, often gradually becoming undetectable when cheeses are older than 60 days.

Thus in developing a HACCP scheme for a small creamery in a farm environment, whether or not pasteurisation is used as a critical control point, and remembering the potential for post-pasteurisation contamination, risk reduction should be achieved by the very process of analysing the whole process – thinking about it, noticing how people behave, evaluating the design of the equipment, evaluating the points where contamination can enter, and considering whether current practices manage the risk. Of course this is not enough – the HACCP plan has then to be drawn up, and practices changed where necessary, implemented and managed. The greatest point of difficulty for small companies is to have sufficient knowledge to ensure that their HACCP plan is adequate for the hazards likely to be present. Do they, for example, appreciate what is needed in the control of *E. coli* O157:H7? In this professional advice may be needed (Anon, 1997) and be related to the specific cheese production unit.

Exercise 2. After weighing the arguments, determine whether you support the use of raw milk for the production of cheese.

You may wish to refer to Chapters 3, 4 and 11 while thinking about Exercise 2 in this case study in Chapter 10.

This is an issue which is as much cultural as scientific (Desenclos *et al.*, 1996). In the UK most cheese is made from pasteurised milk, but as indicated in other chapters the argument continues. For pasteurisation of milk to be a possibility for small cheese producers specialised equipment of a size appropriate to their needs is required. For many years only batch pasteurisation equipment (LTH method) was available in the UK, and only in the last decade or so has small-scale continuous pasteurisation equipment facilitating HTST become available. Either represents costs unwillingly paid by small producers, yet it is undeniable that pasteurisation of milk destroys heat sensitive pathogens. But an equally potent counter-argument relates to the need to retain the quality and flavour of cheese made from raw milk. In France much cheese is made from raw milk and, in recognition of the superior flavour achieved and much valued by the French public, research into ways of managing the risk of zoonotic infection in raw milk, while retaining the flavour characteristics in cheese, have resulted in the development of new technologies for treating drinking milk – see Chapter 4.

But you are considering a slightly different argument from that put to you in Chapter 4. Here the milk is processed into cheese. So while pasteurisation can destroy pathogens in the milk, can it be argued that the cheese making and maturation processes manage the risk of pathogen survival or growth in the cheese?

You need to amass your data, perhaps take Chapter 3 and 4 into account, and weigh the evidence.

Where do you stand?

Exercise 3. If you can, undertake this exercise with a group of colleagues or fellow students. Divide yourselves into three groups:

Group 1. Consider the detail of the information you would need and then argue the case that the authorities were right to condemn all the cheeses in the interest of public health protection.

Group 2. From the perspective of the cheese producers and affineurs determine how you would support the argument that some of the cheeses could be released for sale – thus allowing the opportunity for the small businesses to survive.

Group 3. Listen to *Groups 1 and 2* and determine what you would have decided had you been asked to make a judgement.

Considering Group 1 first:

Group 1. In addition to the information supplied in the case consider what information you would need to argue that all the cheeses should be destroyed in the interest of public health.

Bear in mind the following:

- Connection of the cheese to the infection in the child was achieved through the epidemiological investigation.
- *E. coli* O157:H7 was shown to be present in the cheeses of 4/5/6 April, which were the same dates as the suspected production date of cheese eaten by the infected child.

You need to consider when these tests were carried out. The child became ill on 19th April. How quickly did the case control team assemble? How soon was cheese pinned down as suspect food? And thus, when were the tests done, and how old was the cheese when tested? The child ate very young cheese – about two weeks old, but some cheeses would have been over ten weeks (70 days old) when tested. Would the age of the cheese make any difference to the test, or to the survival and growth of *E. coli* O157:H7 if originally present?

Were actual counts of *E. coli* O157:H7 per gram obtained by the detection method, or simply 'presence' in 25 g? The reason for raising this point is thinking about the detectability of the organisms, and their distribution in a single cheese, or in a batch of cheese. You can see from Figure 10.1 that a great number of samples of cheese were taken and only some of those from date codes 4/5/6 April were found positive for *E. coli* O157:H7. But if you look at Figure 10.2 – which presents the same information – your perception is different and you can see that the vast majority of cheeses were not tested.

So how were the samples selected? What was their size in grams? Were they aggregated samples? Which method was used for analysis? It is known that the testing methodology, which must consider the type of food tested, affects the reliability of the results found (Scotter *et al.*, 2000).

High numbers of *E. coli* O157:H7 per gram would be likely to cause infection after consumption of even a very small piece of cheese, because the infective dose of the organism can be as low as 100 cells. High numbers per gram might indicate that from accidental infection of the milk with infected faecal material (a possible contamination route) the organism had multiplied in the cheese vat when the systems were liquid, and had led to all cheeses being infected. Several samples dated 4/5/6 April were shown to be infected. But several other cheeses of the same date codes were tested and not shown to be positive (Figures 10.1 and 10.2). Does that mean they were not infected, or that the organism was not present at those locations sampled, or that the testing method was at fault?

The fact that several cheeses were positive could indicate that the cheeses became infected at the bulk stages of production – in the milk itself, or in the cheese vat. Of 24 samples of date codes 4/5/6 April 10 samples were positive and the other samples were not shown to be positive. Clearly the organisms were distributed unevenly across those cheeses and it was easy to take a sample which missed a part of the cheese containing the organism. Very great care is taken of the curd at the cutting stage, so that although the curds express the whey it is not a time when general mixing of the bulk milk occurs. Thus one explanation might be that, if infection was somehow introduced at this stage, not all the curds would be infected, and hence distribution of the organisms would be irregular, and trapped in the curds at particular locations.

If, in the infected batches of 4/5/6 April, the *E. coli* O157:H7 were irregularly dispersed then the cheeses from which the samples were not shown to be positive for *E. coli* O157:H7 could not be assumed to be free of the organism. So it was clear that the whole batch of cheeses made on 4/5/6 April were at risk of containing *E. coli* O157:H7. No one argued that those cheeses not shown positive could be eaten. There was no dispute about these 90 cheeses.

So where did that leave the other batches which were sampled, not so intensively as 4/5/6 April, but which did not show the organism to be present (Figure 10.1). Was *E. coli* O157:H7 present in the daily batches or not? Were those cheeses safe to eat?

These results illustrate the difficulty of obtaining representative samples of the cheese on which to base judgements. Assuming the reliability of the testing method, the ability to detect pathogens present in a food depends on three major factors:

- the concentration of the pathogen [the greater the numbers of pathogen the easier they are to detect];
- the numbers of samples (n) taken [the greater the number of samples taken the more chance there is that one will demonstrate the presence of the pathogen];

- the size of the individual samples [the larger the size of the sample the greater the chance that enrichment methods will allow organisms at low original concentration to be detected].

Because of the nature of E. coli O157:H7 any batch of cheese is unacceptable if any sample fails, i.e. is shown to be positive. In sampling terminology when no sample may be allowed to fail, this is expressed as $c = 0$. But if samples are taken and the organism is not shown to be present then the risk of accepting the batch of cheese as safe is dependent on the number of samples taken and analysed. Greater numbers of samples not showing the organism provide greater assurance of the safety of the cheese. But 100% confidence that the food is safe could only be achieved if 100% of the product itself (not just samples from 100% of the items of the product) were to be tested and not shown positive – a process which is self-defeating because apart from practical and financial considerations, there would then be no product to sell.

Tables showing the probability of accepting batches of food, cheese in this case study, which are infected in fact, but in which the samples analysed do not detect the pathogen, are shown and explained in ICMSF (1986). Thus some risk has to be accepted and the size of the acceptable risk is what has to be decided, and this then determines the number of samples to be taken and tested.

In this case study cheese making was a daily process, producing cheeses each day for five or six days of the week – meaning more than 250 cheeses a week were produced, and the cheeses in the dispute representing the stock of cheese for maturation and could be broken down into 94 date codes.

In addressing the issue put to you here you would need to evaluate the company's HACCP plan and see how robust it was, and to examine its monitoring and verification records. You would need to ask whether the route by which the infected cheeses had become infected could have existed earlier and later than the dates on which cheeses were shown to contain E. coli O157:H7. This would guide your thinking.

The legally required microbiological standards for cheeses as they leave the production premises are shown in Table 10.5. Food law current in the UK (Anon, 1995) states that "where the standards in part A of that table are exceeded then the products should be excluded from human consumption, and excluded from the market"; and "in all cases where standards [in part B of that Table] are exceeded there should be a review of the implementation of the methods for monitoring and checking critical points applied in the processing establishment".

Although E. coli O157:H7 had only been found in one set of date codes (4/5/6 April), the authorities had to decide whether the risk that others might be infected was too great to accept. In fact the authorities considered that the ten samples shown positive for E. coli O157:H7 out of 200 tests taken across the production dates 1 February to early May was ample evidence on which to base a judgement that release of any of the cheeses was too high a risk to take.

Turning now to the producer's point of view:

> *Group 2.* From the perspective of the cheese producers and affineurs determine how you would support the argument that some of the cheeses could be released for sale – thus allowing the opportunity for the small businesses to survive.

Central to the problem put to you is the issue of sampling, and what is considered to be a batch.

The production of 30 cheeses a day represents a lot of hard work for the staff of one or two people at the small farm creamery using maybe up to 1000 litres of milk, but is not many when compared to the daily production of an industrial creamery processing thousands of litres of milk a day. Thus what may be proposed for one scale of industry may not be possible in another.

The small creamery would be unlikely to have laboratory facilities on the premises. Analysis of samples costs money. A well-implemented HACCP scheme should avoid the need for many microbiological analyses, which in any case may not demonstrate a problem even if one exists. Monitoring of controls whose nature should be capable of immediate interpretation – pH value, total acid development, temperature value – also facilitate immediate remedial action if required. The taking of samples for microbiological analysis do not provide data for immediate control of the cheese-making system, but information, available later, relating to microbiological status at those times of the cheese in production. Additionally microbiological tests are valuable for checking that cleaning regimes are achieving the reduction in microbial load required. It is the design of the HACCP system which should ensure that the microbiology of the products is within the limits required.

So how should a HACCP system for raw milk cheese deal with a dangerous pathogen like *E. coli* O157:H7?

At the outset it should be recognised that this is an exceptional organism.

- It is capable of causing illness through a very low dose, particularly in the more vulnerable sections of society – children, the elderly, the immuno-compromised for example.
- Secondly it is strongly associated with animals – beef cattle in particular.
- Thirdly it can be carried by beef cattle and dairy cattle without them showing any particular symptoms, yet can be present in their faeces, and sometimes in their milk.
- Fourthly, human beings can carry it without necessarily showing symptoms, yet can excrete it in their faeces.
- On farms faecal material may accidentally get into the water supply, may infect rodents and pet animals, and may be carried on the feet and bodies of flies and other insects, and on human beings.

Collectively these factors mean that the hygiene of the production of raw milk cheeses must be of the highest standards.

Furthermore *E. coli* O157:H7 is an acid tolerant organism – it can survive in low pH conditions (see Table 9.2 and Table 9.4) and could be expected to survive in the acid environment in young cheeses (ICMSF, 1996). It might also be able to multiply in them depending on the specific environmental conditions. Thus the strictest observance of on-farm hygiene and separation of farm and cheese making should be observed to avoid contamination of the milk and the cheese.

Another consideration is whether microbiological sampling would help in the control of the microbiological quality and safety of the cheese.

An outbreak of *Salmonella paratyphi* type B arose in France in 1993 and was found to be associated with goats' cheese made from raw milk. Details of this outbreak are shown in Box 10.1. The scale of production was bigger than the scale of production being considered in this case study but it illustrates that infection of raw milk can occur but may only be spasmodic, and as such not easy to detect. In Case Study 10 one question which arises is: would regular microbiological sampling have been appropriate?

BOX 10.1 *A Salmonella outbreak associated with goats' cheeses*

From Desenclos *et al.*, 1996, reproduced with permission of British Medical Journal Publishing Group: "Every other day two batches of cheese A are made (11 000 to 15 000 cheeses (200 g each) per batch) each corresponding to a pool of 40 farms supplying goats' milk. Cheeses are stored for 11 days at the plant for maturation before distribution. The 'use by' date is 45 days after the cheeses leave the plant.

Before October 1993 internal control for *Salmonella* at the plant consisted of a weekly culture on five cheeses picked from a single batch. Then, on 6 October, one brand A cheese grew *Salmonella*, later typed as *S. paratyphi* B. Subsequently, from 7 October, all batches of cheese stored at the plant, milk pools, and milk from all the farms that supplied each pool were sampled daily for *Salmonella*. The district public health authorities remained unaware of these matters until 8 November, when the district public health physician contacted the district veterinarian's office about a possible link with cheese A. The milk pool corresponding to the batch that grew *Salmonella paratyphi* B on 6 October was also positive for *Salmonella paratyphi* B on 9 October, but negative on the 7th, 11th and 13th.

Salmonella paratyphi B was recovered from the milk of only one of the 40 suppliers. No *Salmonella* was found in stool specimens from workers, cows, goats and pets at the farm. Cheese A and goats' milk isolates belonged to the epidemic phage type (1 var 3). IS 200 genotypic typing was done on three human and four cheese A 1 var 3 isolates. All seven strains exhibited a common IS 200 pattern (profile 2.7).

Around 30 tonnes of cheese, corresponding to the batches stored at the plant between 21 September and 6 October, were destroyed, after the isolation of *S. paratyphi* B from cheese A, and cheese production from the relevant milk pool was pasteurised until daily *Salmonella* control of each batch was implemented. Subsequently all batches produced have been tested for *Salmonella* on day 1 (milk pool), on days 2 and 6 of the maturation process, and on days 9 and 12 (packaging and distribution, respectively)." [. . .]

"The outbreak was detected during the third week of October, when contamination had gone unnoticed for almost three months."

After the outbreak, microbiological testing of the beef and dairy cattle did demonstrate the presence of the strains of *E. coli* O157 in the beef cattle, indicating that the farm environment was very likely to be contaminated, and that the dairy cattle were at risk in spite of not being shown to be positive. It would thus be appropriate to set up regular monitoring of the dairy cows for their carriage of *E. coli* O157:H7.

Unlike Salmonella, in which the infective dose is normally considered to be high, *E. coli* O157:H7 is a highly dangerous, low infective dose organism. Regular microbiological testing of the pooled milk could show the microbiological quality as used (through aerobic plate counts at 21 and 30 °C) and the general level of contamination (through the coliform test at 30 °C), or of faecal contamination through estimating the *E. coli* count. For routine purposes these tests provide good quality information about the general hygiene of production, and it would be desirable to implement a regular testing regime. Furthermore since coliforms and *E. coli* can grow in the early stages of cheese making, monitoring them and setting standards for their levels in the milk and in young formed cheese would be valuable.

But none of those tests detect *E. coli* O157:H7 and normally it would not be appropriate to test for *E. coli* O157:H7 routinely because negatives might create a false sense of security. With raw milk the possibility is always present and the best protection is in the strength of the HACCP plan. This would not only control as many infection routes into the cheeses as possible, and control microbiological hazards, which might enter, but also manage the cheese-making process to minimise the risk of the pathogens' growth. Furthermore it would identify whether *E. coli* O157:H7 would be likely to survive in the cheese beyond the normal maturation period to the time when the cheese would be eaten. Thus soft wet cheeses of short shelf life are the more vulnerable and very serious consideration would have to be given to how the cheese-making process could (or could not) control pathogens including *E. coli* O157:H7. It might have to be recognised that, for that particular cheese-making method and that type of cheese, adequate control of certain pathogens was not possible.

Assuming that the cheese making method and the HACCP plan implemented was capable of controlling the relevant pathogens then microbiological sampling would verify that the microbiological standards aimed for were being achieved.

The best argument that the small cheese maker could have made in favour of the release of some of the cheeses from 1 February to 19 May, and not found positive for *E. coli* O157:H7, would have been to show that the HACCP plan was robust for the production method and cheese type, had been strictly implemented, and that something unusual had happened on the days 4/5/6 April. This would be treating each day's production as a batch and as such microbiologically controlled by the HACCP system, recognising that each day many things change: the cows' health, the weather, the mud or dust in the farm environment, the staff, the milk quality and composition, and so on. It would not be treating all the cheeses as one batch to which the positive results related, as the authorities apparently did.

[Note also that under current UK law (Anon, 1995), unless a derogation relating to the production of traditional cheese has been obtained, every batch of raw milk cheese must meet the legal criteria set, shown in Table 10.4.]

The small cheese-maker's defence would also demonstrate what actions had been taken later in April after it was known that the cheese was the incriminated food.

These arguments would have been supported by information about the type of cheese which would indicate whether cheese ready for consumption would be older than 60 days, and whether it could be classified as a hard cheese, and it would explain whether it was normal that cheese as immature as 14 days old would be eaten.

Group 3. Listen to *Groups* 1 and 2 and determine what you would have decided had you been asked to make a judgement.

It is now up to you to consider the two sides of the argument.

10.5   SUMMARY

HACCP offers a system of control for raw and pasteurised milk cheeses. The level of risk associated with the products – the formed and mature cheeses – relates to the rigour with which the HACCP system is designed and implemented. Sampling and microbiological analysis can demonstrate the presence and concentrations of named categories of organisms but cannot totally assure their absence. The design of the sampling programme can only determine the probability of the acceptance of a batch of cheese which would be unacceptable if the predetermined concentrations of organisms were present. Furthermore, the view of the perceived risk depends on the perspective of the viewer – in this case study that of the enforcer or the producer.

10.6   REFERENCES

Adams, M.R. and Nicolaides, L., 1997. Review of the sensitivity of different foodborne pathogens to fermentation. *Food Control*, **8** (5/6), 227–239.

Anon, 1997. *The Specialist Cheesemakers' Code of Practice.* Published in the UK and obtainable through the Food Standards Agency, London.

Campbell, D.M., Cowden, J.M., Morris, G., Reilly, W.J. and O'Brien, S.J., 1996. All milk products should be heat treated. *British Medical Journal*, **312** (27 April), 1099.

Cizek, A., Alexa, P., Literak, I., Hamrik, J., Novak, P. and Smola, J., 1999. Shiga toxin-producing *Escherichia coli* O157 in feedlot cattle and Norwegian rats from a large scale farm. *Letters in Applied Microbiology*, **28**, 435–439.

Codex Alimentarius Commission: Codex Guidelines for the application of HACCP – adopted by the 20th session of the Joint FAO/WHO Codex Alimentarius Commission, 1993.

Cunynghame, A., 1996. Exceptional cheeses can be made only from raw milk. *British Medical Journal*, **312** (27 April), 1099–1200.

Dairy Products (Hygiene) Regulations, 1995. SI 1995/1086 as amended. HMSO, London.

de Louvois, J. and Rampling, A., 1998. One fifth of samples of unpasteurised milk are contaminated with bacteria. *British Medical Journal*, **316**, 625.

Department of Health, 1998. *Surveillance of the microbiological status of raw cows' milk on retail sale*. Number 7 in the series by the Joint Food Safety and Standards Group. Produced by the Department of Health, London, UK.

Desenclos, J.-C., Bouver, P., Benz-Lemoine, E., Grimont, F., Desqueyroux, H., Rebiere, I. and Grimont, P., 1996. Large outbreak of *Salmonella enterica* serotype *paratyphi B* infection caused by a goats' milk cheese, France, 1993: a case finding and epidemiological study. *British Medical Journal*, **312** (13 January), 91–94.

Dineen, S.S., Takenchi, K., Soudah, J.E. and Boor, K.J., 1998. Persistence of *E. coli* O157:H7 in dairy fermentation systems. *Journal of Food Protection*, **61** (12), 1602–1608.

Feresu, S. and Nyati, H., 1990. Fate of pathogenic and non-pathogenic *Escherichia coli* strains in two fermented milk products. *Journal of Applied Bacteriology*, **69**, 814–821.

Fitzsimons, N.A., Cogan, T.M., Condon, S. and Beresford, T., 2001. Spatial and temporal distribution of non-starter lactic acid bacteria in Cheddar cheese. *Journal of Applied Microbiology*, **90**, 600–608.

FLAIR-FLOW, 1993. Natural antimicrobial systems (F-FE 101/93), accessible at http://www.flair-flow.com/industry-docs/ffe 10193.htm

ICMSF, 1986. *Sampling for Microbiological Analysis: Principles and Specific Applications*. Second Edition. Blackwell Scientific Publications, Oxford, UK.

ICMSF, 1996. *Micro-organisms in Foods*, Volume 5. *Characteristics of Microbial Pathogens*, Blackie Academic and Professional, London.

IFST, 1998. Food safety and cheese. *Food Science and Technology Today*, **12** (2), 117–122.

Kerr, K.G. and Lacey, R.W., 1996. Cheese and salmonella infection. *British Medical Journal*, **312** (27 April), 1099.

MAFF, 2000. *Report on a study of E. coli in unpasteurised milk cheeses on retail sale*. Number 8 in the series by the Joint Food Safety and Standards Group. Produced by MAFF, London, UK.

Maher, M.M., Jordan, K.N., Upton, M.E. and Coffey, A., 2001. Growth and survival of *E. coli* O157:H7 during the manufacture and ripening of a smear ripened cheese produced from raw milk, *Journal of Applied Microbiology*, **90**, 201–207.

Massa, S., Goffredo, E., Altieri, A. and Nastola, K., 1999. Fate of *Escherichia coli* O157:H7 in unpasteurized milk stored at 8°C. *Letters in Applied Microbiology*, **28**, 89–92.

Rampling, A., 1996. Raw milk cheeses and *Salmonella*. *British Medical Journal*, **312** (13 January), 67, 68.

Rose, E., 1996. Cheese lovers should not be condemned to a pasteurised and tasteless product. *British Medical Journal*, **312** (27 April), 1099.

Scotter, S., Aldridge, M. and Capps, K., 2000. Validation of a method for the detection of *E. coli* O157:H7 in foods. *Food Control*, **11**, 85–95.

van Schothorst, M. and Kleiss, T., 1994. HACCP in the dairy industry. *Food Control*, **5** (3), 162–166.

Wright, D.J., Chapman, P. and Siddons, C., 1994. Immuno-magnetic separation as a sensitive method for isolating *E. coli* O157 from food samples. *Epidemiology and Infection*, **113**, 31–39.

Useful contact point:
www.food.gov.uk/news/newsarchive/44511
which details an initiative to help specialist cheese makers.

Hazards and Risks

Key issues
- Cheese – blue cheese
- *Listeria monocytogenes*
- Virulence and pathogenicity: when is an organism a food hazard?
- Risk to public health and food safety policy
- Microbiological tests and their meaning
- Risk management

Challenge

The case study relates to a consignment of raw sheep's milk cheese in which *Listeria monocytogenes* was reportedly found "at high level", and the authorities required its removal from the market. The small family cheese-making business, put at risk of bankruptcy, challenged this decision through the courts.

The case study allows you to think about when the identification of *Listeria* strains in foods represent a risk to public health.

11.1 THE CASE STUDY: *LISTERIA MONOCYTOGENES* AND LANARK BLUE CHEESE, SCOTLAND 1995

11.1.1 The Newspaper Case

In 1994–1995 a semi-soft, raw sheep's milk cheese – Lanark Blue Cheese, made in Lanarkshire, Scotland – received a lot of publicity due to the alleged presence in it of unacceptable levels of *Listeria monocytogenes*. Examination of the London *Times* index covering December 1994 to December 1995 indicates the entire coverage *The Times* dedicated to the case:

March 2, 1995 (The Times, page 16, column a) – article on cheese firm that will go bankrupt if ruling that Lanark Blue is unfit to eat because of listeria contamination; photo.

March 7, 1995 (page 2, column a) – court orders destruction of Lanark Blue cheese worth £27 000 because of presence of listeria bacteria.

March 9, 1995 *(page 16, column a)* – *medical briefing note on Lanark Blue cheese withdrawn because of listeria contamination.*

March 9, 1995 *(page 17, column g)* – *article takes example of Lanark Blue cheese to reinforce argument that too many controls on what we eat or do are detrimental to mankind.*

April 25, 1995 *(page 5, column d)* – *maker of Lanark Blue to learn soon whether court will order him to destroy listeria contaminated stocks.*

April 29, 1995 *(page 8, column g)* – *farmer who was ordered to destroy Lanark Blue cheese wins reprieve.*

June 2, 1995 *(page 36, column a)* – *law report. Errington v Wilson.*

December 3, 1995 *(pages 2 and 11, column a)* – *article on how Whitehall regulations threaten Scottish cheese producer's business.*

December 7, 1995 *(page 3, column a)* – *Scottish cheese maker celebrates victory in long legal battle to save gourmet Lanark Blue cheese from being branded unfit to eat because of listeria contamination.*

December 10, 1995 *(pages 2 and 10, column a)* – *Scottish cheesemaker wins court ruling preventing Clydesdale District Council from destroying stocks of Lanark Blue cheese.*

January 16, 1996 *(page 6, column e)* – *Prince of Wales congratulates Lanark Blue cheese-maker on his victory over council officials who tried to ban his product because it allegedly contained high levels of listeria bacteria.*

Thus it can be seen from these extracts that on the one hand a small cheese producer fought (and won) to save his business. But on the other hand, not mentioned in these abstracts, the food law enforcement officers sought, in the light of findings on the microbiological quality of the cheese and on the best advice they received, to protect public health.

11.1.2 The Chronology

Although the UK national newspapers picked up the case in March 1995, in fact the case really started in **early December 1994**. At that time environmental health officers in Edinburgh had undertaken a routine food surveillance sampling exercise examining the microbiological quality of cheeses randomly selected from Edinburgh retail shops. Among them were samples of Lanark Blue cheese and these tests, undertaken at a public health laboratory (PHL1), indicated the presence of "unacceptable" levels of *Listeria monocytogenes* (*i.e.* over $1000 \, g^{-1}$). (Table 11.1 Part A).

The Edinburgh environmental health officers (EHOs) did two things: they contacted their counterparts in Clydesdale (the district in which the Lanark Blue cheese was made) and informed them of their findings, and additionally took new samples from whole wrapped Lanark Blue cheese on retail sale for microbiologi-

Table 11.1 *A comparison of the test results for Clydesdale Local Authority and Mr Errington's December 1994 test results*

Sampling date	Sample taken by	Date received and analysed by laboratory (where recorded)	Analytical laboratory	Cheese sample month of production and date codes	Analytical laboratory's comments	Analytical laboratory's method
Table 1/Part A: Edinburgh tests November/December 1994						
18/11/94	Edin DC			not given	Listeria monocytogenes 1.3 × 10^4 g^{-1}	
2/12/94	Edin DC			F/23/24	Listeria monocytogenes 3.85 × 10^5 g^{-1}	
Table 1/Part B: Clydesdale Local Authority tests December 1994						
13/12/94	SDC		PHL3	MAY 94 E15	Listeria monocytogenes 1.5 × 10^5 g	
				JUNE 94		
9/12/94	CDC		PHL2	F21	Listeria monocytogenes 6.8 × 10^5 g^{-1}	
8/12/94	Edin DC		PHL1	F22	Listeria monocytogenes 5.5 × 10^5 g^{-1}	
7/12/94	Edin DC		PHL1	F23	Listeria monocytogenes 9.0 × 10^5 g^{-1}	
2/12/94	Edin DC		PHL1	F23/24	Listeria monocytogenes 3.85 × 10^5 g^{-1}	
7/12/94	Edin DC		PHL1	F24	Listeria monocytogenes 7.0 × 10^5 g^{-1}	
7/12/94	Edin DC		PHL1	F27	Listeria monocytogenes 1.35 × 10^5 g^{-1}	
8/12/94	Edin DC		PHL1	F30	Listeria monocytogenes 5.0 × 10^4 g^{-1}	
				JULY 94		
7/12/94	Edin DC		PHL1	G1	Listeria monocytogenes 9.5 × 10^5 g^{-1}	
9/12/94	CDC		PHL2	G1	Listeria monocytogenes 3.7 × 10^3 g^{-1}	
8/12/94	Edin DC		PHL1	G4	Listeria monocytogenes 1.25 × 10^6 g^{-1}	
7/12/94	Edin DC		PHL1	G5	Listeria monocytogenes 9.0 × 10^5 g^{-1}	
9/12/94	CDC		PHL2	G6	Listeria monocytogenes 5.3 × 10^5 g^{-1}	
9/12/94	CDC		PHL2	G8	Listeria monocytogenes 1.0 × 10^7 g^{-1}	

Date	Lab	Sample	Result	Protocol
9/12/94	CDC	G16	*Listeria monocytogenes* 1.2×10^6 g^{-1}	
9/12/94	CDC	G16	*Listeria monocytogenes* 6.1×10^4 g^{-1}	
9/12/94	CDC	G28	*Listeria monocytogenes* 1.7×10^5 g^{-1}	
		AUGUST 94		
7/12/94	CDC	H1	*Listeria monocytogenes* 6.0×10^2 g^{-1}	
9/12/94	CDC	H1	*negative*	
8/12/94	Edin DC	H9	*Listeria monocytogenes* 3.5×10^5 g^{-1}	
13/12/94	SDC	H30	*Listeria monocytogenes* 2.8×10^4 g^{-1}	

Table 1/Part C: Errington tests December 1994

Date	Lab	Sample	Result	Protocol
		JUNE 94		
10/12/94	AL1	F24	*Listeria* spp. not detected	AL1-M14
		JULY 94		
13/12/94	AL1	G16	*Listeria* spp. not detected	AL1-M14
1/12/94	AL1	G28	*Listeria* spp. not detected	AL1-M14
13/12/94	AL1	G28	*Listeria* spp. not detected	AL1-M14
		AUGUST 94		
8/12/94	AL1	H1	*Listeria* spp. present and identified as *Listeria monocytogenes*	AL1-M14
13/12/94	AL1	H4	*Listeria* spp. not detected	AL1-M14
13/12/94	AL1	H9 and others	*Listeria* spp. not detected	AL1-M14
9/12/94	AL4	HI	Absent from 25 g	Method not referenced
		SEPTEMBER 94		
11/10/94	AL1	I3	*Listeria* spp. not detected	AL1-M14
13/12/94	AL1		*Listeria* spp. not detected	AL1-M14

Key: Public Health Laboratories 1, 2, 3 : PHL1, PHL2, PHL3.
Private Analytical Laboratory 1: AL1; AL1-M14 = *Listeria monocytogenes* detection protocol used by AL1 – see Table 11.7.
Private Analytical Laboratory 4: AL4.
SDC = Suffolk Coastal District Council; CDC = Clydesdale District Council; Edin DC = Edinburgh District Council.

Source: Court Hearing papers (see Section 11.1.3) viewed with the permission of the Sheriff, Sheriff Court Lanark. Permission to publish granted by Mr H. Errington, and by South Lanarkshire Council on behalf of the former Clydesdale District Council.

cal testing, tests which subsequently also demonstrated the presence of *L. monocytogenes* (Table 11.1, part B).

Clydesdale officers, wanting to check the mature cheese at source, contacted the owner of the company making Lanark Blue, Mr Errington, and at his production unit both discussed the Edinburgh findings with him and took nine samples for analysis on **9 December**. These too were analysed, but at different laboratories from the original samples. They were two other public health laboratories: (Public Health Laboratory 2 [PHL2] and Public Health Laboratory 3 [PHL3]), and the majority of samples were shown to be positive for *L. monocytogenes*, and some had *L. monocytogenes* counts well over $1000 \, \text{cfu} \, \text{g}^{-1}$ (Table 1, part B). When isolates were serotyped at the Central Public Health Laboratory in Colindale, London, they were found to be *L. monocytogenes* serotype 3a.

On **13 December 1994** the Clydesdale EHOs gained Mr Errington's agreement that he would voluntarily withdraw all Lanark Blue from the market, and Mr Errington signed an undertaking to this effect. Additionally Mr Errington and the EHOs agreed in principle certain further testing steps. On **15 December 1994**, because of the seriousness of listeriosis to some vulnerable groups, the Scottish Office issued a public warning to health concerning eating Lanark Blue – relating of course to cheese already purchased by the public. The withdrawal meant that neither Mr Errington's existing stocks of Lanark Blue held at his production unit nor stock already at wholesalers and retailers could be sold, and the latter would probably be returned by them to him. This was a significant commercial blow in the period coming up to Christmas when sales were likely to be substantial. From the EHOs point of view there were difficulties too for at this time of year their department was liable to be short staffed with various team members wanting to take holiday leave and the management of this problem probably fell at the worst time of year for them.

Lanark Blue, a sheep's milk cheese, was only produced during the months of the year when the sheep were lactating – namely between April and October, so from November 1994 no further Lanark Blue cheese had been made, and production would not have restarted until April 1995.

The role of the EHOs in a situation like this was not only to protect public health, but also to co-operate with the producer and help him find a solution to the problem, although the final responsibility under Scottish food law (as in the food law relevant to other parts of the UK) rested with the producer.

The EHOs adopted a risk assessment approach, prioritising the actions they could take. First they had controlled a problem identified in the Lanark Blue cheese by its withdrawal from the market; second they planned, by working with the producer, to evaluate the risk presented by the stock of cheese through a systematic sampling and analysis programme of it so that some cheese could be released onto the market as soon as possible.

From mid-December, over the Christmas and New Year holiday period, wholesalers and retailers were returning cheese to Mr Errington, and as owner of the cheese he was entitled to send samples for analysis – which he did. The results (Table 11.1, part C) he obtained were worrying to him because they did not seem to indicate any problem with Listeria, or *L. monocytogenes*. By the **end of January 1995** Mr Errington had decided that he would withdraw from the voluntary agreement he had made, and on the 26th wrote to Clydesdale to inform them of this, his voluntary agreement ending at 12 noon on 31 January 1995. There is no available record of the activities of the EHOs in the period 15 December 1994 to 31 January 1995, and it is not clear whether they did return to the farm to give advice or take samples.

Thus on **27 January 1995** Clydesdale issued a formal 'Notice of Seizure' under the Food Safety Act, 1990, which covered all Lanark Blue cheese of date codes F, G, H, I – that is to say 956 cheeses, each 3 kg, the entire stock of Lanark Blue which had been produced between June and September 1994. Additionally the EHOs issued to Mr Errington a 'Food Condemnation Warning Notice' the legal Hearing for which was to be before the local lay magistrate on **30 January 1995**, at 3.0 p.m. [The magistrate has the power to, and is the legal authority who may condemn the food as unfit for human consumption. After condemnation the food is taken away and destroyed under a legally defined supervisory process.] The EHOs' technical justification for this were the results of the nine samples taken by them from the production premises in December 1994, and for which the PHL2 had demonstrated "unacceptable" levels of *L. monocytogenes* (see Table 11.1, part B).

However, at the Hearing a legal representative [a barrister, and Queen's Counsel (QC)] for Mr Errington indicated that Mr Errington had test data which did not agree with that of the EHOs (Table 11.1 parts B and C), and the magistrate decided it was more appropriate for the cheeses to remain withdrawn from the market, while further tests on cheeses from date code G16 onwards (July 16) were undertaken. She instructed that the enforcement officers and the producer agreed all details of the process – namely the sampling procedures, the laboratory for analysis and the analytical methods. The intention was that through this agreement process, results jointly paid for by the producer and Clydesdale Local Authority would also be accepted by both sides.

On **1 February 1995**, the day after the Hearing, the EHOs thus issued the formal Notice of Detention – which under the FSA (1990) had a validity of 21 days, and was the period during which the agreed sampling and testing took place. The cheeses remained at Mr Errington's premises, and sampling of 50 batches of cheese took place on 6th February, and analysis at the agreed laboratory 'AL2' on **7 February 1995**. It was also agreed that only cheeses whose date codes were after G16 would be analysed, the earlier codes representing cheeses whose 'best before' date had by then been exceeded.

The preliminary results were sent by the laboratory by fax to both parties and received on **13 February 1995**, and these were confirmed by letter on **20 February**. The results (Table 11.2) were very alarming. Forty-four of the 50 cheese batches tested were reported as having levels both of Listeria species and of *L. monocytogenes* above $1000 \, \mathrm{cfu \, g^{-1}}$, some samples having over $100\,000 \, \mathrm{g^{-1}}$.

After consultation with the regional consultant in Public Health Medicine, and with the Central Public Health Laboratory Service, Colindale, London, the EHOs felt justified in their original view, and on **22 February 1995** issued a new

Table 11.2 *Results of analyses of cheese samples undertaken by Private Analytical Laboratory Number 2 (AL2), February 1995*

Cheese code number	Dates sampled	Dates received	Date tested	Listeria cfu g^{-1}	Listeria monocytogenes cfu g^{-1}	Listeria spp. in 25 g	Listeria monocytogenes in 25 g
Code F not tested							
G16	06 Feb 95	06 Feb 95	07 Feb 95	4.5×10^4	4.5×10^4		
G18	06 Feb 95	06 Feb 95	07 Feb 95	2.3×10^5	2.3×10^5		
G19	06 Feb 95	06 Feb 95	07 Feb 95	2.7×10^5	2.7×10^5		
G20	06 Feb 95	06 Feb 95	07 Feb 95	6.0×10^3	6.0×10^3		
G21	06 Feb 95	06 Feb 95	07 Feb 95	1.1×10^5	1.1×10^5		
G22	06 Feb 95	06 Feb 95	07 Feb 95	1.0×10^4	1.0×10^4		
G23	06 Feb 95	06 Feb 95	07 Feb 95	2.0×10^3	2.0×10^3		
G25	06 Feb 95	06 Feb 95	07 Feb 95	2.3×10^5	2.3×10^5		
G26	06 Feb 95	06 Feb 95	07 Feb 95	8.1×10^4	8.1×10^4		
G27	06 Feb 95	06 Feb 95	07 Feb 95	5.5×10^4	5.5×10^4		
G28	06 Feb 95	06 Feb 95	07 Feb 95	1.9×10^5	1.9×10^5		
G29	06 Feb 95	06 Feb 95	07 Feb 95	1.5×10^3	1.5×10^3		
G30	06 Feb 95	06 Feb 95	07 Feb 95	1.6×10^4	1.6×10^4		
I1	06 Feb 95	06 Feb 95	07 Feb 95	1.4×10^4	1.4×10^4		
I3	06 Feb 95	06 Feb 95	07 Feb 95	7.5×10^4	7.5×10^4		
I5	06 Feb 95	06 Feb 95	07 Feb 95	5.0×10^5	5.0×10^5		
I8	06 Feb 95	06 Feb 95	07 Feb 95	1.4×10^5	1.4×10^5		
I9	06 Feb 95	06 Feb 95	07 Feb 95	5.0×10^2	5.0×10^2		
I12	06 Feb 95	06 Feb 95	07 Feb 95	1.2×10^5	1.2×10^5		
I14	06 Feb 95	06 Feb 95	07 Feb 95	1.2×10^4	1.2×10^4		
I16	06 Feb 95	06 Feb 95	07 Feb 95	1.4×10^4	1.4×10^4		
I19	06 Feb 95	06 Feb 95	07 Feb 95	1.9×10^3	1.9×10^3		
I21	06 Feb 95	06 Feb 95	07 Feb 95	7.3×10^3	7.3×10^3		
I23	06 Feb 95	06 Feb 95	07 Feb 95	1.3×10^3	1.3×10^3		
H1	06 Feb 95	06 Feb 95	07 Feb 95	3.8×10^4	3.8×10^4		
H2	06 Feb 95	06 Feb 95	07 Feb 95	6.0×10^2	6.0×10^2		
H3	06 Feb 95	06 Feb 95	07 Feb 95	1.0×10^6	1.0×10^6		

					present	**present
H4	06 Feb 95	07 Feb 95	2.1×10^5	2.1×10^5		
H5	06 Feb 95	07 Feb 95	2.8×10^3	2.8×10^3		
H6	06 Feb 95	07 Feb 95	<10	<10	present	**present
H8	06 Feb 95	07 Feb 95	3.4×10^3	3.4×10^3		
H9	06 Feb 95	07 Feb 95	8.2×10^4	8.2×10^4		
H10	06 Feb 95	07 Feb 95	1.9×10^4	1.9×10^4		
H11	06 Feb 95	07 Feb 95	1.3×10^5	1.3×10^5		
H12	06 Feb 95	07 Feb 95	3.7×10^4	3.7×10^4		
H13	06 Feb 95	07 Feb 95	1.3×10^4	1.3×10^4		
H15	06 Feb 95	07 Feb 95	<10	<10	present	**present
H16	06 Feb 95	07 Feb 95	8.0×10^2	8.0×10^2		
H17	06 Feb 95	07 Feb 95	2.0×10^5	2.0×10^5		
H18	06 Feb 95	07 Feb 95	8.5×10^3	8.5×10^3		
H20	06 Feb 95	07 Feb 95	*300	*300		
H22	06 Feb 95	07 Feb 95	2.8×10^3	2.8×10^3		
H23	06 Feb 95	07 Feb 95	5.1×10^6	5.1×10^6		
H24	06 Feb 95	07 Feb 95	7.2×10^3	7.2×10^3		
H25	06 Feb 95	07 Feb 95	8.2×10^3	8.2×10^3		
H26	06 Feb 95	07 Feb 95	5.4×10^4	5.4×10^4		
H27	06 Feb 95	07 Feb 95	5.9×10^3	5.9×10^3		
H29	06 Feb 95	07 Feb 95	7.8×10^4	7.8×10^4		
H30	06 Feb 95	07 Feb 95	2.7×10^5	2.7×10^5		
H31	06 Feb 95	07 Feb 95	4.2×10^4	4.2×10^4		

*Only 3 isolates recovered from direct plating.
**Only 2 isolates recovered from enrichment.
Sampling method: each numbered sample normally comprised 5 core samples from 3 cheeses of the same date code except in the cases of G18, G19 and G30 where the bulk sample comprised cores from 4 cheeses.
Analytical method: see Figure 11.2.
Plating method: use of spiral plater: see Figure 11.2.
Criteria for believing isolates to be *Listeria monocytogenes*: colony appearance and morphology on Oxford Agar; colony appearance on TSA; catalase reaction; haemolysis on horse agar; CAMP reaction on sheep blood agar.

Source: Court Hearing papers (see Section 11.1.3) viewed with the permission of the Sheriff, Sheriff Court Lanark. Permission to publish granted by Mr H. Errington, and by South Lanarkshire Council on behalf of the former Clydesdale District Council.

'Notice of Seizure', and a new 'Food Condemnation Warning Notice', this time, however, listing very precisely to which cheeses it referred. The EHOs expected the same procedure as before to take place – namely before a magistrate; but the producer, Mr Errington, requested that the Hearing was before a Sheriff (a judge) at which evidence could then have been presented and cross examination of witnesses undertaken. However, this was not agreed and the second Hearing on **24 February 1995** was again heard by a magistrate. The magistrate reserved judgement – the Council not offering any evidence that the cheese could cause harm – but later that day took a decision that the cheese should be condemned. This meant that the cheese would be condemned the following day – **25 February 1995**.

However, Mr Errington and his QC acted very quickly that same day (**24 February**) and succeeded in being granted another legal process – a Judicial Review – on the basis that because he (the producer) had not been allowed the opportunity to cross-question the local authority's officers (the EHOs) regarding the basis of their decision and which was resulting in his stock of cheese, and most probably his business, being destroyed, he was being denied natural justice. This was granted and a date in March was set for a Hearing before Lord Weir.

In **March** Lord Weir held the Judicial Review, and ruled in favour of Mr Errington: he had been denied natural justice, and the decision to condemn the cheese, taken in the magistrates court on 24 February, was not valid. During this Hearing the QC representing Mr Errington asked for permission, which was granted, for the stock of cheese to be frozen, because while these legal processes proceeded the quality of the cheese was liable to pass beyond ripeness and begin to deteriorate. Up until that time the entire stock of Lanark Blue cheese had been held in chill store at $-2\,°C$ at Mr Errington's production unit.

However, the EHOs were still faced with a difficult situation. The February tests had shown the cheese in question (the 44 batches) to be contaminated with *L. monocytogenes* at levels greater than $1000\,\mathrm{cfu\,g^{-1}}$ thus categorising it as "unsatisfactory/potentially hazardous". The guidelines from the Central Public Health Laboratory Service, Colindale, London (PHLS, 1992) – (see Table 11.9) were those used and were officially recognised by the Scottish Office to be the guidelines for EHOs to follow.

Furthermore the European Community Directive 92/46 (EC, 1992), and the draft Dairy Products (Hygiene) Regulations designed to comply with it and by which the EHOs were also being guided were more stringent, requiring an absence of *L. monocytogenes* in a 25 g sample of ready-to-eat food. The EHOs had to decide in the face of this body of knowledge (which did not at that time have statutory force) whether they could allow the cheeses on the market and risk fatal illness in vulnerable consumers, or accept the advice of the nation's top experts and attempt to prevent the release of the cheese on the market.

They took the view that the cheese was still unfit to eat, and a few days after Lord Weir made his ruling they announced they would appeal against the decision of the Judicial Review. This had the effect of still preventing any Lanark Blue being released for sale.

For Mr Errington, although he had obtained Lord Weir's permission to freeze the cheese, effecting this proved difficult because a number of food stores, although approached by him, would not accept this cheese of suspect quality. A solution was found in his purchase of seven deep freezers, which he installed at his production unit, and which he filled during **April 1995**. This arrangement was initially satisfactory to both him and the EHOs, but later (on **1 May**) the EHOs wrote to indicate that the cheese was to be taken away to a neutral place of storage. However, this action was strongly contended by Mr Errington and the outcome was that the cheese remained in freezers at his premises, and on **2 May** they were officially locked and sealed by the EHOs, who were then satisfied with this arrangement.

In **early June 1995** the Council's appeal against Lord Weir's ruling in his Judicial Review was turned down. But they still held to their view regarding the risk they perceived associated with the cheese, and immediately applied for a Hearing before the Sheriff. This was granted, and a date – **14 August 1995** – was fixed.

11.1.3   The Hearing before the Sheriff in the Sheriff Court in Lanark

Clydesdale District Council
v
Humphrey Errington of H.J. Errington and Co.
14 August 1995, and the following days

Mr Errington and his legal and microbiological advisors were pleased about this. It gave them the chance they had not had before to produce evidence, and to cross examine the Council officers (the EHOs) and their witnesses.

Clydesdale Council's Case

Within the legal framework of the Food Safety Act (1990), Part II, Section 9(6), and Section 8(2) a and b, the Council submitted that the food in question – the cheese – was "unfit for human consumption" meaning "not suitable for, or proper for human consumption". They submitted that the Court could order the 44 batches of cheese to be destroyed or disposed of if, on the evidence, they were found in any respect unsuitable for human consumption.

They contended that the February tests (Table 11.2) for *Listeria* spp. and *L. monocytogenes* and on which the seizure had been based, were reliable, and that the best scientific opinion available considered that regardless of strain all *L. monocytogenes*, if found in food, must be considered potentially pathogenic.

Mr Errington's defence

His defence rested on interpretation of the term 'unfit for human consumption' and his submission indicated prior agreement between the two parties, Clydesdale and Errington, that food is 'unfit for human consumption' if it is, or is likely to be, injurious to health.

He contended that while listeriosis is a serious, if rare, disease caused by strains of *L. monocytogenes*, the specific implementation of the testing methodology for the presence of that organism in 44 batches of his cheeses left the results, by which his cheese was liable to be condemned, open to doubt. Secondly, although he recognised that *L. monocytogenes* had been found on occasion in his cheese and in the sheep's milk from which it was made, he argued that there was no proven connection between the serotype of the strain found (3a) and either virulence or pathogenicity thus both refuting and challenging the then normally accepted practice of regarding the presence in ready-to-eat food samples of levels of $> 1000\,\mathrm{cfu\,g}^{-1}$ as unacceptable, and all strains of *L. monocytogenes* as potentially pathogenic, and the appropriate trigger for the removal of such food from sale in the interest of protection of public health.

11.1.4   The Judgement

After a Hearing which extended from mid-August to late October and occupying 19 days of court time, and involving a number of witnesses and experts for both sides, the Sheriff ruled in favour of Mr Errington – namely that the cheese was not unfit for human consumption and that the cheese should not be condemned.

In his 'findings' (Ref.: Court record, Lanark Court) he found that the reliability of the results of the tests implemented on behalf of both sides on 7 February by AL2 (Table 11.2), and which appeared to show unacceptable levels of *L. monocytogenes* in 44 samples remained unresolved. He further found that the tests commissioned by Mr Errington during May 1995 (Table 11.3), and which showed no, or low counts of *L. monocytogenes* were the more reliable.

Furthermore he formed the view

- that the evidence did not support the claim that all strains of *L. monocytogenes* are potentially pathogenic;
- that all *L. monocytogenes* should not be regarded as potentially pathogenic;
- no evidence was presented to him that convinced him that the strain of *L. monocytogenes* found in the cheese was pathogenic to human beings and thus the cheese was not a risk to public health.

Finally

- the epidemiological evidence relating to the estimated 63 000 portions of the suspect cheese eaten and from which no known cases of listeriosis had arisen supported his view.

Thus he could not accept Clydesdale's argument that the cheese was 'unfit' there having been no convincing case made that it was injurious to health.

11.2   BACKGROUND TO THE TECHNICAL ARGUMENTS PUT IN THE HEARING, AUGUST 1995

Note: The text presented in this section relies on references quoted in evidence in the Hearing and thus none are more recent than 1995. This is to permit you to view the case from the perspective of the time.

Table 11.3 *Test results on a range of cheese samples (February and May 1995), and referred to by the Sheriff in his judgement*

	Private Analytical Laboratory Number 2 (AL2) results February 1995		Private Analytical Laboratory Number 1 (AL1) results May 1995			
	Date of analysis	Result recorded	Sample (as described by AL1)	Date received	Date of analysis	Result recorded (Method used: AL1-M14)
G16	7/2/1995	*Listeria monocytogenes* 4.5×10^4 g^{-1}	Cheese sample (G16)	23/05/95	23/05/95	*Listeria* count <100; *Listeria* species present and identified as *Listeria monocytogenes*
G18	7/2/1995	*Listeria monocytogenes* 2.3×10^5 g^{-1}	Lanark Blue 1/2 moon G18 98 kg **BB** 7/2/95 Fat in dry matter 52%; moisture 40%	10/5/1995	10/5/1995	Not detected
G28	7/2/1995	*Listeria monocytogenes* 1.9×10^5 g^{-1}	Lanark Blue Wedge G28 **BB** 6/1/95 Fat in dry matter 52%; moisture 40%	10/5/1995	10/5/1995	*Listeria* count <100; *Listeria* species present and identified as *Listeria monocytogenes*
G4	7/2/1995	not tested	Cheese sample (14) (Lanark G4)	23/05/95	23/05/95	*Listeria* count <100; *Listeria* species present and identified as *Listeria monocytogenes*
H23	7/2/1995	*Listeria monocytogenes* 5.1×10^6 g^{-1}	Cheese sample (20) (Lanark H23)	23/05/95	23/05/95	*Listeria* not detected
H3	7/2/1995	*Listeria monocytogenes* 1.0×10^6 g^{-1}	Cheese sample (17) (Lanark H3)	23/05/95	23/05/95	*Listeria* count <100; *Listeria* species present and identified as *Listeria monocytogenes*
H4	7/2/1995	*Listeria monocytogenes* 2.1×10^5 g^{-1}	Cheese sample (18) (Lanark H4)	23/05/95	23/05/95	*Listeria* count <100; *Listeria* species present and identified as *Listeria monocytogenes*
I3	7/2/1995	*Listeria monocytogenes* 7.5×10^4 g^{-1}	Cheese sample (21) (Lanark I3)	23/05/95	23/05/95	*Listeria* count <100; *Listeria* species present and identified as *Listeria monocytogenes*

Methods: AL1 – see Table 11.7; AL2 – see Figure 11.2.

You will see more recent references in Section 11.4.

11.2.1   Lanark Blue Cheese

The cheese-making process in operation in 1994–95 is described below.

Lanark Blue cheese was made (as now) in the Lowlands of Scotland from unpasteurised sheep's milk, salted and mould ripened using a method similar to that of the famous French Roquefort cheese. It still is only made in one location, and from the milk of the sheep on that farm.

The sheep are normally in lactation in the late spring, the summer and early autumn months – and in that period in 1994 they were out in the field grazing on grass. They may have been fed silage in the winter months, but never during the milking season. Thus the cheese could only be made during the period April to October. However, cow's milk was purchased all year and brought from another local farm and made into another blue cheese, 'Dunsyre', by a similar process. By using both the sheep's milk in the summer, and the cow's milk all the year round the cheese production unit was also in use all the year round.

The sheep were milked in a rotary milking parlour, which experience had shown facilitates the milking of each sheep in a matter of a minute or so, and induced very much less stress in them than do shearing or dipping. At the times of the year when the latter processes are of necessity implemented it had also been found that the viable bacterial counts in the raw milk increased. Making the cheese is a skilled hand process, which is outlined in Figure 11.1.

The skilled cheese maker would adjust the process to produce consistent quality cheeses in spite of the inevitable variability of the milk which occurs throughout the season. The sheep were milked twice a day, evening and morning, so the two quantities of milk were used for the daily cheese making which was started at around 8 a.m. The same staff made the cheese, turned it, salted it and wrapped it. The processes involved in each day's make were recorded – particularly the timings, the temperatures and the acidity developed in the whey. On average the sheep yielded sufficient milk between the pooled evening and morning milkings to make 12–14 3 kg Lanark Blue cheeses. Samples were regularly sent off to a local microbiological testing laboratory.

11.2.2   Listeria and *Listeria monocytogenes*

This section covers the following

- Identifying characteristics and taxonomy
- Listeriosis – the illness
- Epidemiology of listeriosis
- Pathogenicity and virulence factors
- Causative organisms and their subtypes
- Growth and survival
- Chilled foods of extended shelf life
- Testing foods for the presence of *Listeria* spp. and *Listeria monocytogenes*.

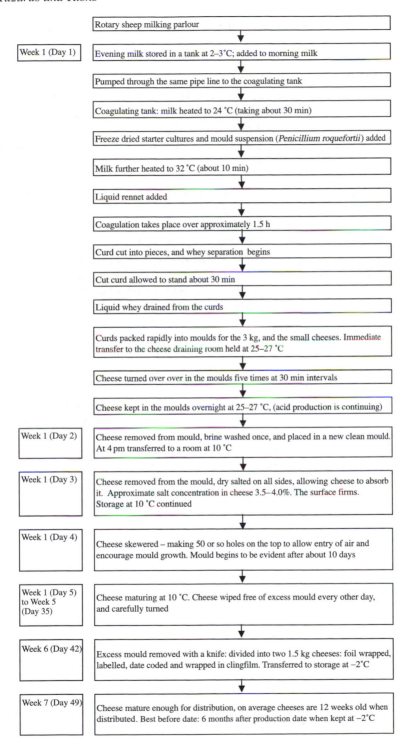

Figure 11.1 *Production of Lanark Blue cheese*

- Standards for Listeria and *Listeria monocytogenes* in foods current in 1994 and 1995.

Identifying characteristics and taxonomy

Listeria are Gram positive, short rods, motile at 22 °C but not 37 °C, catalase positive and facultatively anaerobic (ICMSF, 1996). The named species *Listeria monocytogenes, L. ivanovii, L. welshimerii, L. seeligeri, L. innocua, L. murrayi* (also named *L. grayi*) are differentiated by their various abilities to ferment a range of carbohydrates, by the CAMP test, and by their ability to produce β-haemolysin (Table 11.4). Some members of the group cause listeriosis.

Listeria are organisms which are very widely distributed in nature having been isolated from many wild and domestic animals. In milking animals the organisms can be carried in the udder and excreted in the milk with or without symptoms of mastitis (Gitter *et al.*, 1980). Additionally *Listeria* spp. have been isolated from water, soils, muds, sewage, plants, vegetables, animal faeces, food processing surfaces, refrigerators, and in many foods (Bilney *et al.*, 1991). From food surveys the only foods which appear to be exceptions are those which have received a severe heat treatment after their final packaging stage (Roberts, 1994) the packaging thus protecting them from post-processing contamination. Some strains of *Listeria* are capable of entering certain human and animal cells – particularly the subgroup of white blood cells, the monocytes, and growing and multiplying within them. It is from there that they gain entry to other tissues and cause localised or generalised pathogenic effects. Thus while *Listeria* spp. appear to be harmless for most people, some strains are opportunistic pathogens largely affecting the vulnerable groups of the old, young children, pregnant women and others who are immuno-compromised. The majority of cases occur in this vulnerable group in whom case mortality rates can be as much as 50% (Farber and Peterkin, 1991). The organism most commonly causative of listeriosis is *Listeria monocytogenes*.

Listeria monocytogenes, although found in the environment and in animals, also occurs in the human gastro-intestinal tract, and has been shown to be present in the faeces of about 5–10% of symptomless healthy human carriers (MacGowan *et al.*, 1994) while additionally many healthy people can be shown to carry antibodies to *Listeria* species (Farber and Peterkin, 1991) indicating prior infection.

Listeriosis – the illness

Listeriosis infection occurs principally through eating infected food, but can also occur through skin infection, and from mother to child in pregnancy and/or in childbirth. The symptoms can range from a mild flu-like condition to very serious, even fatal, illness particularly affecting the heart and brain. Pregnant woman who are infected may only experience a mild 'flu', but the baby they carry may be very seriously affected and possibly abort, be born before term, and/or suffer life threatening septicaemia and possibly meningitis with case mortality ranging between 20 and 40% (McLauchlin, 1993). In 1994/5 the infective dose

Table 11.4 *Listeria and the strain differentiating characteristics*

Genus Listeria:

Gram positive short rods
Catalase positive
Facultatively anaerobic
Tumbling motility in culture at 25 °C
Non motile at 37 °C
Colonies greyish blue, changing to blue green in oblique light
Methyl red reaction positive
Voges–Proskauer reaction positive
H_2S production negative
Produce acid but not gas from glucose
Indole production negative
Citrate utilisation negative
Urease activity negative
Mannitol fermentation negative
Nitrate reduction negative
Gelatin hydrolysis negative

Listeria species	Normally considered potential to cause human infection	Sugars fermentation			CAMP test[a]		
		Mannitol	L-rhamnose	D-xylose	SA	RE	β-haemolysis
L. monocytogenes	pathogenic	−	+	−	+	−	+
L. seeligeri	rarely pathogenic	−	−	+	+	−	+
L. ivanovii	rarely pathogenic	−	−	+	−	+	+
L. welshimerii	rarely pathogenic	−	±	+	−	−	−
L. innocua	non-pathogenic	−	±	−	−	−	−
L. murrayi (L.grayi)	non-pathogenic	+	+	−	−	−	−

[a]CAMP test (Christie, Atkins, Munch–Peterson phenomenon), *i.e. L. monocytogenes* shows typical haemolytic zone on blood plates when streaked together with haemolytic *Staphylococcus aureus* (SA) and/or *Rhodococcus equi* (RE).
+ = sugar fermented and acid formed.
Source: ICMSF, 1996.

was not known, likely to vary between individuals, and speculatively believed to range between 10^2 and 10^9 organisms. For example in an immuno-competent patient an estimated dose of $2.5–4.3 \times 10^9$ cells of *Listeria monocytogenes* caused meningitis (Azadian, 1989), but in the immuno-suppressed the dose would be lower. Estimating the infective dose is not easy because the symptoms of the illness may occur many weeks after infection meaning that relating a food (and its load of organisms) to listeriosis is uncommon unless surveillance has caused a product recall, as for example in 'the Mexican-style cheese outbreak' (Linnan *et al.*, 1988). But even in the latter estimating the infective dose proved impossible; nor did it prove possible in the Vacherin d'Or outbreak (Bille, 1990; Bula *et al.*, 1995) (Table 11.5).

Epidemiology of listeriosis

Listeriosis is a rare disease. For example, the figures for England and Wales, population close to 60 million people, showed approximately 100–120 cases occurring in each of the five years 1990–1994 inclusive. Data reviewed by McLauchlin (1993) indicated that most cases of listeriosis are sporadic, yet occasionally large outbreaks have occurred in which the source has been traced back to infected foods such as the Mexican-style cheese, Vacherin Mont d'Or cheese (see above) and pâté (McLaughlin *et al.*, 1991). These large outbreaks rapidly led to categorisation of ready-to-eat foods as 'high risk' and to *Listeria* control programmes for these foods comprising a number of elements:

* surveillance for the occurrence of the organism in the environment, in food processing production units, and in retail foods;
* implementation of HACCP systems in production of foods, but particularly in the high risk ready-to-eat chilled foods sector;
* adoption (in the USA) of a zero tolerance policy for the organism and development of maximum acceptable limits (in the UK);
* public education on the correct handling of high risk foods; and
* public education schemes to warn vulnerable groups not to eat the high risk foods.

Food surveys demonstrated that many foods including raw vegetables, shellfish, raw fish, raw and pasteurised milk, raw and pasteurised milk cheeses, ice-cream, salami, pâté are liable to contain *Listeria* spp. and *Listeria monocytogenes* (MacGowan *et al.*, 1994). A survey of milk in the UK showed that, out of 4172 samples taken, about 2.5% were positive for *Listeria* species (and some of these were *L. monocytogenes*), which were found in both raw and pasteurised samples (Greenwood *et al.*, 1991). *Listeria* spp. were also found in cheeses (McLauchlin *et al.*, 1990) with the World Health Organisation report (WHO, 1988) particularly commenting on their occurrence in white mould ripened cheese and red smeared-surface cheeses.

Investigations of the linkage between organisms isolated from those suffering the symptoms of illness and the source of infection most commonly used serotyping and phagetyping alone or combined (Bannister, 1987; Linnan *et al.*, 1988;

Bula *et al.*, 1995) although molecular methods were also coming into use.

Pathogenicity and virulence factors

Investigations of outbreaks of listeriosis have shown many *Listeria monocytogenes* strains to be pathogenic, *i.e.* capable of causing disease, but those strains are not necessarily equally pathogenic. Actual disease arises as a result of a combination of the capabilities of the infecting organism and the susceptibility of the host.

The ease with which the *L. monocytogenes* overcomes the hosts defences and establishes itself is a measure of its virulence; the amount of damage the organism does in this process and in its growth in its host is a measure of its pathogenicity. Often the two are closely linked.

For pathogenesis to occur the strain of organism has to be capable of initial penetration of the host and for example, resisting the acidity of the stomach before penetrating gut wall cells. It also has to survive and multiply in the host and invade its target cells. At cellular level the organism it has to be capable of completing a number of steps

- *infection*: needing entry protein (known as 'p60', size 60 000 Da) to get into the host cell;
- *withstanding the defences of the lymphocytes it invades*: producing for example superoxide dismutase – an enzyme which breaks down the toxic free radical superoxide which is produced as part of the cell defences of activated phagocytes;
- *be able to break down the vacuole cell membrane inside the host cell in which it initially resides*: haemolysin (listeriolysin O) an enzyme which breaks down the membrane of the phagosome – *i.e.* the membrane of the vacuole surrounding the cell;
- *to grow and metabolise inside the host cell*: haemolysin, for example, helps this process because it also breaks the cell membrane of red blood cells thereby releasing iron containing compound which the listeria needs for growth;
- *then to become motile and migrate in order to invade other cells*: possession of actin assembly chemical – which produces polar tails of organised molecules of actin on the *Listeria* cells and facilitates *Listeria* cell movement, and *Listeria* spread from host cell to host cell.

(Farber and Peterkin, 1991; Ryser and Marth, 1991).

The individual factors which facilitate the infection, growth, multiplication and spread of the organisms are called **virulence factors** and in their production, if it occurs within a host cell or tissues, various degrees of pathogenesis occur. These factors are separately encoded on the genome, meaning individual strains of *L. monocytogenes* are variously invasive and variously potentially pathogenic. Strains which have been shown to be pathogenic produce β-haemolysin and it appears to be essential for pathogenicity. Thus its possession is a strong indicator of potential pathogenicity (Tabouret *et al.*, 1991). However, a few strains have

Table 11.5 *Human listeriosis*

A Examples of some large outbreaks			
Canada – Nova Scotia (1981)	41 cases, 11 deaths	Contaminated coleslaw containing raw cabbage fertilised with raw sheep manure		Klima and Montville, 1995
USA – Boston, Massachusetts (1983)	49 cases, 11 deaths	Pasteurised milk – but no *L. monocytogenes* isolated		Klima and Montville, 1995
California (1985)	142 cases, 48 deaths	Soft Mexican-style cheese made from pasteurised milk subsequently contaminated with raw milk. Cheese factory closed	Serotype 4b, all of one phage type	Linnan *et al.*, 1988
Switzerland (1983–1987)	122 cases, 34 deaths	Vacherin Mont d'Or cheese – a locally made soft cheese	Serotype 4b with a unique 'Swiss epidemic' phage type	Bille, 1990
UK (1989–1990)		Pâté	Surveillance showed that samples of retail pâté were frequently contaminated with *Listeria monocytogenes*, and the fall in cases of listeriosis coincided with Government warning (directed to vulnerable groups) against eating it	McLaughlin *et al.*, 1991
USA – Illinois (1994)	Non-invasive listeriosis	Pasteurised chocolate milk containing 10^9 *L. monocytogenes* ml^{-1} due to post-pasteurisation contamination followed by temperature abuse in storage		Klima and Montville, 1995

B Human listeriosis cases, England and Wales, 1983–1999

	Non-pregnancy associated cases	Pregnancy associated cases	Not known	Total
1983	67	44	–	111
1984	77	35	–	112
1985	77	59	–	136
1986	87	42	–	129
1987	136	102	–	238
1988	164	114	–	278
1989	123	114	–	237
1990	91	25	–	116
1991	94	32	–	126
1992	80	26	–	106
1993	86	17	–	103
1994	89	26	–	115
1995	77	10	–	87
1996	103	17	–	120
1997	100	24	–	124
1998	87	21	–	108
1999[a]	89	17	–	106

Note: Mothers and babies are counted as one case.
[a]Provisional data.

Source: Laboratory reports to CDSC: PHLS Food Hygiene Laboratory, updated to 16 May 2000, from PHLS Website accessed January 20, 2001.

been shown to be haemolytic without being pathogenic, presumably because they lack some or all of the other virulence factors (Ryser and Marth, 1991). Other strains, possessing other virulence factors but not producing haemolysin, are non-pathogenic (Farber *et al.*, 1991; Tabouret *et al.*, 1991).

Thus it is clear that "not all strains of *Listeria monocytogenes* are pathogenic; rough varients possess only reduced virulence; [and] non-haemolytic mutants have completely lost their pathogenic potency" (Hof and Rocourt, 1992).

Listeriosis: causative organisms and their subtypes

At the time of the Lanark Blue cheese problem (1994/5) evidence showed that most human listeriosis was caused by three serotypes of *Listeria monocytogenes*: 1/2a, 1/2b and, predominantly, 4b, although 13 serovars had been identified (McLauchlin, 1990). Serotype 3 strains had rarely been associated with human listeriosis, and type 3a only once, and in that case there was no known food connection. Phage typing of those serotypes has indicated a number of subtypes for each showing that organisms which possess the same serotype are not necessarily identical. Yet some serotypes have been shown not to be phage typable. Hof and Rocourt (1992) pointed out the value of molecular methods (multilocus enzyme analysis, isoenzyme electrophoresis, ribotyping, chromosome restriction patterns and so on) for identifying relationships between strains of *L. monocytogenes* and differences within strains (Baxter *et al.*, 1993). For example, genomic divisions in the biotype *L. monocytogenes* demonstrate virulent clones within serotypes, thus showing variability within a serotype (Brosch *et al.*, 1993, 1994).

Although clinical investigations have shown that *L. monocytogenes* is the *Listeria* biotype most commonly associated with human listeriosis there is evidence that not all strains of *L. monocytogenes* are pathogenic when tested in mice (Conner *et al.*, 1989; Farber *et al.*, 1991). Yet research indicates that there may be an association between serotype and virulence (McLauchlin, 1990), although Menudier *et al.* (1991) showed that four strains of the same serovar (1/2a) (originating from four different food sources) were differently virulent for mice, with two being avirulent. Interestingly Conner *et al.* (1989) also demonstrated that a 'non-pathogenic' strain of *L. monocytogenes* killed test non-immunocompromised mice inoculated with a high dose ($LD_{50} = 10^8$ cells) of the organism, as compared to $LD_{50} = 10^5 - 10^7$ cells of 'pathogenic' strains.

Growth and Survival

Listeria monocytogenes is capable not only of growing within lymphocytes and causing disease but also of growing in a wide range of foods and other environments. Its known growth tolerances (where other conditions are optimal) are shown in Table 11.6.

In cheeses the environment changes during its maturation. In mould ripening cheeses after an initial period when the pH is low (pH 4.0–4.5) the later growth of the mould causes protein breakdown and the pH in that vicinity then rises. So although the pH change is not uniform across the cheese, in areas where mould is

Table 11.6 *Growth and survival characteristics of Listeria monocytogenes*

	Temperature °C	pH	Water activity (a_w)	Sodium chloride tolerance
Range	−0.4 to 45	4.4–9.4	>0.92	At least up to 10% w/v; tolerance is dependent on other constraints
Optimum	37	7		

Note: some growth rates for *L. monocytogenes* in cheeses can be found in McLauchlin, Greenwood and Pini (1990).
Source: ICMSF, 1996.

growing well – particularly at the outer edges – and within air pockets in the cheese the pH rises, progressing upwards from approximately pH 4.0 towards an eventual pH 6 or 7. So that, whereas it would be unlikely that *Listeria* strains could grow in freshly made cheese, the cells (if they survived and were not destroyed by either bacteriocins or organic acids present – see Kinderlerer and Lund, 1992) might later grow in the maturing cheese (McLauchlin *et al.*, 1990). But Papageorgiou and Marth (1989), studying the ripening of blue cheese, observed that the 1.0 log increase in population of *L. monocytogenes* seen in the first 24 h ceased when the pH dropped below 5.0. The population dropped an average 2.68 $\log_{10}$ cfu g^{-1} during 50 days storage at 4 °C, after which as pH began to rise improved rates of survival, but no growth, were observed.

Growth conditions can also affect virulence. For example virulence can be increased by passaging strains (which have been stored as cultures in the laboratory) through mice, or conversely virulence decreases when rough varients of the strain emerge through exposure to unfavourable growth conditions on laboratory media (Hof and Rocourt, 1992). Finally, *L. monocytogenes* is a heat sensitive organism – destroyed by normal regimes of pasteurisation (Bryner *et al.*, 1989).

Chilled foods of extended shelf life

Aspects of the food safety problem with *Listeria monocytogenes* are associated with two of its qualities. Firstly its ability to grow under refrigerated conditions means that if present in chilled foods with expected shelf life of weeks or even months its population may slowly increase to critical levels. Secondly it is tolerant of low concentrations of sodium chloride which in food could prohibit the growth of many organisms which would otherwise compete with it. Thus mild salting, as for example practised in making Lanark Blue or other similar blue veined cheeses can give *Listeria* a selective advantage.

While *Listeria* have been shown to be present in some cheeses, to grow in others (see earlier text) it is also possible for the growth of *Listeria* in cheese to be counteracted by the presence of bacteriocins produced by the starter organisms which are normally present in very high numbers, although the phenomenon seems to be very dependent on the cheese type and the specific starter cultures used and their specific bacteriocins (if any) (Genigeorgis *et al.*, 1991).

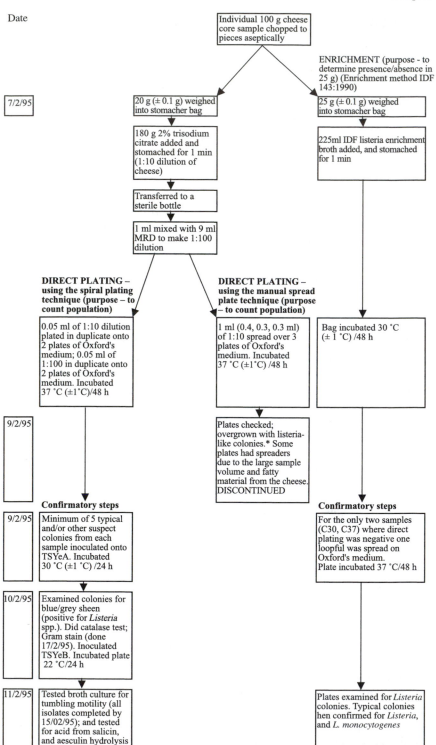

Figure 11.2 *Analysis for Listeria monocytogenes – methods used by AL2, February 1995*

Testing foods for the presence of Listeria and *Listeria monocytogenes*

In laboratory culture *Listeria* can be shown to grow well in a range of non-selective laboratory media. However for the isolation of *Listeria* from foods, particularly where there is likely to be a mixed flora, or high numbers of other organisms, practice has developed which uses carefully designed selective media which inhibit many of the non-*Listeria* organisms (Golden *et al.*, 1988). There was no internationally recognised method for the isolation of *Listeria* from foods in 1994/5 but well respected protocols such as those of the USDA had gained wide acceptance.

The test methods used to isolate and quantify *Listeria monocytogenes* found in Lanark Blue (results recorded in Tables 11.2 and 11.3) are given in Figure 11.2 and Table 11.7, with details about the media used in Table 11.8.

Standards for Listeria and *Listeria monocytogenes* in foods current in 1994 and 1995

In 1994, in the UK, there were no legally enforceable microbiological standards for *Listeria* or *Listeria monocytogenes*. There were guidelines (Table 11.9), draft regulations (the draft Dairy Products (Hygiene) Regulations) and the EC Directive 92/46. In the USA a policy of 'zero tolerance' was being implemented – that is to say where *L. monocytogenes* was found present in 25 g of tested food, the food had to be withdrawn from the market. All these policies, and laws, treated all *L. monocytogenes* as potentially pathogenic.

Figure 11.2 *(cont.)*

Key	Criteria used by AL2 to define	
TSYeA = trypticase soy yeast extract agar	*Listeria*	*Listeria like colonies*
TSYeB = trypticase soy yeast extract broth	Gram positive rods	colonies surrounded by dark
MRD = maximum recovery diluent	Non pigmented colonies	brown haloes. If found streak
Oxford's medium = Oxford listeria selective agar	Catalase positive	five of these onto TSYeA at
IDF = International Dairy Federation	Oxidase negative	30 ˚C/1 – 2 days. Colonies with
PHLS = Central Public Health Laboratory	Tumbling motility at 22 ˚C	blue-green sheen, and
Service, Colindale, London.	Aesculin hydrolysis	catalase positive interpreted
CAMP test – for details see Figure 11.4	Acid (not gas) from glucose	as *Listeria*
C30,C37 = AL2 laboratory code for H6,H15	Acid (not gas) from salicin	
AL2 = Analytical Laboratory 2		
	L. monocytogenes	
	β-haemolysis on horse blood agar	
	acid from rhamnose	
	no acid from xylose	
	CAMP test	
	L. monocytogenes confirmation	
	Serogrouping using Difco antisera	
	At the PHLS: all 61 isolates were	
	serotype 3; 21 of which were 3a.	

Source: Court papers – third inventory of productions Document 2.

Table 11.7 *Analyses for Listeria monocytogenes – methods used by Analytical Laboratory 1 (AL1), and compared to the USDA method*

Method M14 for enrichment of samples in order to detect presence of *Listeria* species as detailed in AL1 protocols	**USDA method**
Test portion 25 ± 0.2 g	25 g sample
Add 225 ± 0.2 g buffered listeria enrichment broth. (This is 1/10 dilution) Mix by stomaching 30 s	Add 225 ml Listeria enrichment broth
Incubate at 30 ± 1 °C for 3–6 h	Incubate 24 h at 30 °C
Add *Listeria* selective enrichment supplement at ratio 2 ml reconstituted in 500 ml broth	Inoculate 0.1 ml into 10 ml Fraser broth
Incubate at 30 °C for 48 h (± 4 h)	Incubate 24–48 h at 35 °C
Streak 0.1 ml onto two media: 1. Oxford agar – incubate 30 °C (± 1 °C) and 2. Palcam agar – incubate 37 °C (± 1 °C) – both for 48 h (± 4 h)	Streak onto Oxford agar after 24 and 48 h regardless of aesculin hydrolysis. Incubate 48 h at 30 °C
If typical *Listeria* colonies present streak two from each plate onto TSA/blood agar. Incubate 37 °C (± 1 °C)/24 h anaerobically	Suspect *L. monocytogenes* colonies typically appear surrounded by dark brown or black haloes. Inoculate 5 suspect colonies onto TSYeA. Incubate 37 °C 24–48 h (*i.e.* until adequate growth is obtained)
Record *β*-haemolysis on sheep's blood agar	
Confirmation tests for *Listeria* spp.	**Confirmation tests for *L. monocytogenes***
Gram stain (+ ve); rod shaped cells Oxidase test (+ ve) Catalase test (− ve) Wet slide for tumbling motility API *Listeria* test kit	Gram stain (+ ve); rod shaped cells Wet slide for tumbling motility Haemolysis on sheeps blood agar (+ ve) Carbohydrate fermentation (xylose − ve; rhamnose + ve) CAMP test (+ ve)
Result: *Listeria* present/absent in 0.1 ml enriched broth. If present specify *Listeria* species.	

Notes:
1. For composition of media see Table 11.7.
2. 'Palcam' = Polymixin B, Acriflavin, Lithium chloride, Ceftazidime, Aesculin, Mannitol.
3. TSA = tryptone soy agar.
4. Other abbreviations are as used in Figure 11.2.
5. The AL1 method 'M14' was the method accredited by NAMAS and was based on the USDA method. The descriptions came from the evidence presented in the case.

Source: Court Hearing papers (see Section 11.1.3) viewed with the permission of the Sheriff, Sheriff Court Lanark. Permission to publish granted by Mr H. Errington, and by South Lanarkshire Council on behalf of the former Clydesdale District Council.

Table 11.8 *Selective media for listeria (as used by AL1)*

Listeria enrichment broth

Base	(g/l)
proteose peptone	5
tryptone	5
'Lab-Lemco' powder	5
yeast extract	5
sodium chloride	20
disodium hydrogen phosphate	12
potassium dihydrogen phosphate	1.35
aesculin	1
pH after sterilisation	7.2 (±0.2)
Before use add 1 and 2 below	
1. Supplement per 500 ml base:	
	(mg/500 ml)
magnesium sulfate (anhydrous)	2.5
ferrous sulfate.7H$_2$O	0.15 g
water	10 ml
2. Primary enrichment supplement	
acriflavin HCl	6 mg
nalidixic acid	10 mg

Fraser broth (FB)

Base	(g/l)
proteose peptone	5
tryptone	5
'Lab-Lemco' powder	5
yeast extract	5
sodium chloride	20
disodium hydrogen phosphate	12
potassium dihydrogen phosphate	1.35
aesculin	1
lithium chloride	3
pH after sterilisation	7.2 (±0.2)
Before use add 1 below	
1. Supplement per 500 ml base:	
ferric ammonium citrate	0.25
acriflavin HCl	12.5 mg
nalidixic acid	10 mg

Oxford agar

	(g/l)
Columbia blood agar base (see below)	39
aesculin	1
lithium chloride	15
ferric ammonium citrate	0.5
pH after sterilisation	7.0 (±0.2)
[Columbia blood agar base:	
special peptone	23
starch	1
sodium chloride	5
agar	10]
Before use add 1 below	
1. Supplement per 500 ml Oxford agar base:	
acriflavin	2.5 mg
cycloheximide	200 mg
colistin sulphate	10 mg
cefotetan	1 mg
fosfomycin	5 mg
ethanol	2.5 ml
water	2.5 ml

See Kornacki, J.L., Evason, D.J., Reid, W., Rowe, K. and Flowers, R.S., 1993. Evaluation of the USDA protocol for detection of *L. monocytogenes*. *J. Food Protection*, **56**, 441–443.

Source: Court Hearing papers (see Section 11.1.3) viewed with the permission of the Sheriff, Sheriff Court Lanark. Permission to publish granted by Mr H. Errington, and by South Lanarkshire Council on behalf of the former Clydesdale District Council.

Table 11.9 *Provisional guidelines for microbiological acceptability of some ready-to-eat foods (PHLS 1992)*

	Microbiological quality ($cfu\ g^{-1}$) (unless otherwise stated)			Unacceptable/ potentially hazardous (Note 3)
	Satisfactory	Fairly satisfactory	Unsatisfactory (Note 2)	
Aerobic plate count (35–37 °C/36–48 h)[a]				
Cooked pies, pasties, quiches *etc.*	$<10^3$	10^3–10^5	$>10^5$	Note 1
Confectionery products – without dairy cream	$<10^3$	10^3–10^5	$>10^5$	Note 1
Cooked meats	$<10^4$	10^4–10^6	$>10^6$	Note 1
Sandwiches – without salad	$<10^4$	10^4–10^7	$>10^7$	Note 1
Sandwiches – with salad; cooked seafoods; confectionery products with dairy cream prepared mixed salads	$<10^5$	10^5–10^8	$>10^8$	Note 1
Salmonella	not detected in 25 g			present in 25 g
V. parahaemolyticus (seafoods)	not detected in 25 g			present in 25 g
Campylobacter (thermotolerant)	not detected in 25 g			present in 25 g
L. monocytogenes	not detected in 25 g	present in 25 g to $<10^2$	10^2–10^3	$>10^3$
E. coli	<20	20–$<10^2$	10^2–10^4	$>10^4$
S. aureus	<20	20–$<10^2$	10^2–10^4	$>10^4$
C. perfringens	<200	200–$<10^3$	10^3–10^4	$>10^4$
B. cereus and other *Bacillus* spp.	<200	200–$<10^4$	10^4–10^5	$>10^5$

For 1996 guidelines see Table 5.4; for 2000 guidelines see Table 11.10/11.11.

[a] Guidelines for aerobic plate counts may not apply to certain fermented foods: *e.g.* salami, soft cheese and unpasteurised yoghurt.

Note 1: Prosecution based solely on high aerobic plate counts in the absence of other criteria of unacceptability is unlikely to be successful.

Note 2: The guidelines have no formal standing or status but samples falling in this category indicate further sampling may be necessary and that EHOs may wish to carry out a detailed inspection of the premises, food production and handling processes *etc.*

Note 3: Samples falling into this category might form the basis for prosecution by Environmental Health Departments.

Source: PHLS, 1992. Reproduced with permission of the PHLS Communicable Diseases Surveillance Centre. © PHLS.

11.3 EXERCISES

Exercise 1. On the information given in Sections 11.1 and 11.2 what possible explanations could there have been for

- the seriously conflicting counts of *Listeria monocytogenes* detected in the cheese samples and shown in Table 11.1;
- the Sheriff's disquiet over the reliability of the results shown in Table 11.2; and
- the differences shown between samples of the same date code shown in both Table 11.2 and Table 11.3?

Exercise 2. Bearing in mind that the *L. monocytogenes* strain found may not have been pathogenic, do you think the risk management policy implemented (which resulted in the removal of Lanarkshire Blue cheese from the market in the interest of public health, with its potential consequences for the producer) was justified?

Exercise 3. Following the Appeal against the outcome of the Judicial Review – [the Hearing] (Section 11.1.3) what options did the producer have to manage the risks associated with the farm production of raw milk blue cheese?

Exercise 4. How is the discovery of the presence of *L. monocytogenes* in food managed today?

11.4   COMMENTARY

> **Exercise 1.** On the information given in Sections 11.1 and 11.2 what possible explanations could there have been for
> - the seriously conflicting counts of *Listeria monocytogenes* detected in the cheese samples and shown in Table 11.1; for
> - the Sheriff's disquiet over the reliability of the results shown in Table 11.2 and
> - the differences shown between samples of the same date code shown both in Table 11.2 and Table 11.3?

Environmental Health Officers in the UK (England and Wales, Northern Ireland, and Scotland) have the statutory right and duty to take samples of foods in order to ascertain their quality. In this case study samples were first obtained from retail shops in Edinburgh. Later samples were taken at the Lanark Blue production unit by the 'home authority' environmental health officers, and additionally samples were taken for the producer. According to the tests done for the EHOs *Listeria monocytogenes* was found at concentrations which official guidance indicated should be withdrawn from sale in order to protect public health. In his defence in the Hearing in August 1995 the producer challenged the reliability of the tests.

In developing your view you need to consider the influence of a number of factors:

- the distribution of the organism – *Listeria monocytogenes* – in each cheese;
- the distribution of the organism in a batch of cheeses of the same date code;
- the sampling methodology;
- the potential for survival and/or growth of the organism in the cheese;
- the analytical methods;
- the potential sources of error in the implementation of the methods;
- the interpretation of the results achieved.

The distribution of the organism *Listeria monocytogenes* in each cheese

It is possible for *L. monocytogenes* to be present in raw milk cheese for two reasons – it has entered with the milk, or alternatively has contaminated the cheese in the production process. In the former case the organisms would be more likely to occur in every cheese of a batch; in the latter the organisms might also be distributed more or less evenly through the milk and thus through the cheeses of that batch. But if they contaminated the product after the moulding process it is possible that a batch of cheeses could contain some which were contaminated, others not so. Once in or on the cheese the organisms might die, survive or grow depending on the environmental parameters in the particular cheese (Genigeorgis *et al.*, 1991) but in some, such as some soft cheeses, *Listeria* are likely to grow well on the surfaces (Ryser and Marth, 1987).

For detectability by the methods used the organisms would have needed to be viable at the time of testing and present at a cell concentration of at least one per 25 g of sample (enrichment method); or 200 per g (spiral plate count method) (calculation: 1 per plate, inoculated with 0.05 ml of 1 : 10 dilution); or 10 per g (spread plate method) (calculation: 1 per 3 plates, inoculated with 1.0 ml of 1 : 10 dilution).

The sampling methodology

The cheese sampling method used to obtain the retail samples in Edinburgh evidently detected *L. monocytogenes* beneath the packaging. Thus the organisms must have been present in those samples when supplied to the retail shop, but not necessarily at the levels detected. Growth might have happened in the time between distribution from the production unit and the moment of sampling if storage conditions were permissive (Table 11.1, part A).

Thus the first question in relation to Table 11.1, parts B and C is: were the tests done on comparable samples?

Then – where samples were taken at the production unit – how was sampling done? Remember that each day's production comprised perhaps 10–12 3 kg cheeses, as well as a number of smaller size cheeses (see Figure 11.1). Would the sampling method detect low numbers of *Listeria*? Would high numbers (if they occurred) be uniform in distribution?

Remember too that wedge and core sampling of cheese damages its appearance and it is not the role of the sample-taker to reduce the value of the product, but to check whether the product is contaminated. This consideration might constrain the sampler in how they approached the sampling method. Were standard sampling methods, which require samples penetrating from the outside to deep within the cheese to be taken from a number of points in the whole cheese, used? *L. monocytogenes* would be more likely to occur and grow at locations in the cheese where the pH was rising due to ripening (closer to the surface and near air pockets) and conversely would be less detectable where the ripening mould was growing less rapidly. It is improbable that within the mass of a cheese the concentrations of *L. monocytogenes* would be evenly distributed. Would each cheese in a day's production carry the same load? Could the sampling methods used which resulted in the data in Table 11.1 have themselves have been responsible for the different results?

No data are available in the case papers which describe how sampling for results in Table 11.1 was undertaken. But when samples were later taken in February 1995 each sample, of total weight of approximately 100 g, comprised five core samples from each of three cheeses. Thus the data recorded in Table 11.2 part B provide a composite result for a batch of cheeses of the same date code. The source of those organisms is assumed to be the cheese. In the process of sampling the 50 date codes and taking material from three cheeses, the two EHOs working together (following the protocols of correct practice) took an average of 4 minutes per code, demonstrating that great care is needed in the sampling to avoid contamination of the samples in the process.

The irregularity of distribution of the organism (if present at all) in the cheese means that low numbers could be missed; and if it occurred in pockets at high level the data generated might not be representative of the whole cheese.

[Sampling methods which were current at the time are in FSA (1990), Code of Practice No. 7; The Creamery Proprietors' Association, 1988; International Dairy Federation, 1990.]

The potential for survival and/or growth of the organism in the cheese

Many of the counts shown in Tables 11.1, 11.2 and 11.3 show counts over 1000 cfu g^{-1}, indeed many counts are in the 10^5–10^6 cfu g^{-1} range. Such high counts indicate that either those levels were present at the initial production stage and had survived in the cheese throughout its storage and distribution history, or that growth of the detected organisms had occurred in or on the cheese during or after production. In the Hearing before the Sheriff (Section 11.1.3) these points – whether *L. monocytogenes* can survive and/or grow in blue cheese – provided material for argument. It was not clear. Evidence was quoted to support the view that it is possible, while other evidence based on experience of the production of the French raw ewe's milk cheese – Roquefort – indicated that it was unlikely (see Section 11.2). Of course at issue was whether it could grow specifically in Lanark Blue – for which no systematic research was available and only inferences could be drawn from the disputed results. Raw milk cheese contains a wide flora derived from the raw milk and the cheese starters and there is a possibility of some members of this range of organisms producing bacteriocins which might inhibit *Listeria*, as might the added salt, the falling pH or organic acids concentrations within the cheese. Conversely these restraining factors might not be sufficient to kill the *Listeria* but merely to prevent their growth until the environment changed, for example during cheese maturation the rising pH creating a more permissive environment. If *Listeria* entered the cheese at low level and *L. monocytogenes* could not grow in Lanark Blue cheese, then the high counts shown would have been more likely to have been due to testing methodology error. From the evidence presented relating to the health of the sheep the Sheriff, in his judgement, ruled out the raw milk as the source of high and variable levels of *Listeria*. He also ruled out cross-contamination in the production process since no sample of the cow's milk cheese – made alongside the sheep's milk cheese – had been shown to contain *Listeria*. The doubt further emphasised the need for absolute reliability of the analytical procedures.

The analytical methods

A number of different testing laboratories produced the results shown in the three Tables 11.1, 11.2 and 11.3. The laboratories themselves were a mixture of accredited public health laboratories and private laboratories staffed by qualified 'food examiners' as defined under the FSA (1990), together with trained technicians. In the court case in August 1995 both the written procedures for the methods used and their implementation in practice were scrutinised in detail. All the laboratories used widely accepted methods and while at that time (1994/5) there was no reference method available the use of the USDA method came close to it (Kornacki *et al.*, 1993). However, the Milk Marketing Board guidelines for good hygienic practice for small cheese makers (MMB, 1989) recommended (and detailed) the FDA method (FDA, 1984).

Examination of Table 11.8 shows that the media used for the detection and speciation of *L. monocytogenes* are selective media using a range of antibiotics to suppress the other organisms in the microbial flora (bacteria, yeasts and moulds) of the sample, raw milk cheese in this case. The media are also indicative – exploiting the fact that *L. monocytogenes* hydrolyses aesculin resulting in black zones around the colonies due to the formation of black iron phenolic compounds. Thus initial presumptive identification of the presence of *L. monocytogenes* depends on the unique appearance of these colonies in the presence of any other organisms which may have grown on the plates. Oxford agar suppresses Gram negative organisms and most unwanted Gram positive species but some strains of enterococci grow poorly (suppressed by the inclusion in the medium of lithium chloride) and exhibit a weak aesculin reaction, usually after 40 hours incubation (Oxoid, 1992).

Colonies presumptively identified as *L. monocytogenes* on their colonial morphology must then be confirmed through biochemical and serological tests.

In evaluating of these media Golden *et al.* (1988) indicated that they are effective in isolating *Listeria* strains from the background flora, that the colonies have a black appearance together with a black halo in the medium, that the development of the black colour requires at least 24 hours incubation at 35 °C to achieve it, that 48 hours incubation may be necessary to ensure the growth of slow growing strains of *Listeria*, and that on occasion other organisms can grow. Furthermore there were different protocols developed depending on the nature of the material being examined – faecal, food or environmental samples. However, in the Hearing (Section 11.1.3) no evidence was put forward to suggest that the basic protocols themselves were inappropriate.

Modification of the media – to reduce the suppression of *Listeria* by the ranges of antibiotics used – has resulted in a range of *Listeria* selective media (USDA method, Fraser's medium, Palcam, and so on – (Tables 11.7 and 11.8). Considering the results reported in the Hearing, unwanted suppression of *Listeria* would have been important where detection of the presence of even low numbers per gram was significant. Detection of low numbers of *Listeria* usually relies on enrichment techniques in which the *Listeria* population, resisting the selective antibiotics, multiplies and reaches cell concentrations which can then be cultured and detected on plates. In the results in the Lanark Blue cheese Hearing the issue was of the conflict between high counts resulting from the plate counts for the local authority and interpreted as *L. monocytogenes*, against the 'Listeria not detected' reported for many samples analysed for the cheese producer. Whereas the latter may have suffered from over suppression due to media composition or other causes, if the disputed high counts were at fault, how may that have come about? There are of course a number of ways – contamination of samples, inaccurate dilutions, population growth in diluents prior to plating, counting colonies which did not have typical morphology, inaccurate counting, incorrect calculation, use of wet plates leading to population increase during plate incubation, contamination, to list some main areas of error.

In the Hearing (started in August 1995, Section 11.1.3) the use of the spiral plating technique, which produced the results recorded in Table 11.2, was closely examined. The technique relies on the mechanical inoculation of a rotating plate, starting at the centre and moving towards the perimeter, dispensing a progressively smaller volume over increasing length of track. The volume dispensed over any area of the plate is known. After plate incubation colonies are crowded in the central region, dispersed towards the edges of the plate, and any area (defined by the use of a template) where there are clear discrete colonies can be used for calculation of organism concentration in the original inoculum. If you calculate back from the reported *Listeria* count of 4.5×10^4 cfu g^{-1} (Sample G16, Table 11.2) you will see that the mean colony population on the 1:10 plates must have been 225 colonies. In the event of a count of 2.3×10^5 g^{-1} (Sample G18) it must have been 1150 (and 115 at 1:100); and at 5.0×10^6 cfu g^{-1} (Sample H23) it must have been 25000 at 1:10 (and 2500 at 1:100), and so on. The latter two plates would have been very crowded and it is probable that it could have been appropriate for further dilution to be plated. Thus how the operator counted the plates is important to the achievement of accurate data. It is standard practice to count spread plates with between 30 and 300 colonies evident; values lower and greater than this are subject to unacceptable error. But it is not necessary to count the whole spiral plate – duplicate subsections containing a known inoculum are enough – reducing both the use of operator time and operator counting error. The technique allows, through dispensing 0.05 ml onto a single plate, a count range of three orders of magnitude – such as 6×10^2 to 6×10^5 cfu ml^{-1} (Jarvis *et al.*, 1977). Although potentially a big saver of labour its main disadvantage is that the fine dispensing tube can become blocked with food particles, or contaminated with oily material from the food which is not easily flushed out of the tube. Prior to use the machine should be regulary serviced and calibrated to ensure that the volume dispensed is indeed the expected volume (IFST 1995a; 1995b). The proper technique in use requires the sterilisation of the tube between samples (by hypochlorite for example), followed by its rinsing in sterile water. In use, control samples should be run to check the sterility of the machine both prior to use and intermittently between samples. Equally some controls using a range of dilutions of a pure culture of the test organism should be run both through the plater and using, for example, conventional spread plates, to check on the consistency of counts being achieved under the testing and incubation conditions used.

It can be seen from Figure 11.2 that a second technique – direct plating of 0.3 or 0.4 ml of the 1:10 dilution of cheese – was also undertaken but encountered problems of spreading colonies. Looking at the results in Table 11.2 it can be seen that Sample G16 for example would have produced 4500 colonies over the three plates – so crowding on the spread plates is not surprising. It is not normally the case that a volume as small as 0.4 ml fails to be absorbed into an agar plate – unless it is very newly poured and wet itself. These plates, in addition to the control of checking that a reference strain of *L. monocytogenes* could grow on the media under the same incubation conditions, could have acted as a positive control for the spiral plater results – and the lack of use of controls was one of the points raised by the Sheriff in his judgement, and one of the reasons which led to his lack of confidence in the results in Table 11.2. For the spread plates to have acted as an effective positive control for the spiral plates they would, of course, have had to have been of the same thickness, and dryness – in other words made from the same batch of medium, and dried under the same conditions for the same periods of time so that the organisms within the inocula grew under the same conditions.

The results which the Sheriff found more reliable are those performed by the private analytical laboratory [AL1] and reported in Table 11.3. This accredited company did not use a spiral plater but used an enrichment technique – M14 (see Table 11.7) – followed by conventional spread plating which even after enrichment did not detect *Listeria* species for the majority of samples. It was argued from the Clydesdale District Council side that the flora of the frozen samples analysed in May 1995 (Table 11.3) would have been injured and not detected, but on the evidence the Sheriff did not accept this point.

The interpretation of the results achieved

This is problematic. When *Listeria monocytogenes* is cultivated in pure culture it might be expected that all the colonies would be of the same form and appearance. But even colonies of a pure culture do not look exactly alike because they achieve different colony sizes depending on how crowded the colonies are in the different regions of the plate. But when the sample is food with its diverse flora, and the analysis is undertaken to explore whether or not *Listeria* species, or *L. monocytogenes*, is present and at what counts per gram, expertise in colony recognition in the operator is needed. Organisms which clearly have the expected morphology may be scored as 'typical'; but are colonies with slight variations in size, colour, profile and so on the sought after organisms or not? Should they be ignored or counted? Should they be subcultured or not? It is known that, for example, *Listeria innocua* can occur with *L. monocytogenes* (Pini and Gilbert, 1988) and when together in culture *L. innocua* may mask *L. monocytogenes* (Mpamugo, 1997).

L. monocytogenes can also easily be overgrown by other organisms in culture – which is one of the reasons for the inhibitors in the selective media. Enterococci, however, can survive the inhibitors and can have similar appearance to *Listeria* colonies on selective agars. A point contended in the Hearing in August 1995 was the fact that all *Listeria* found in the tests in February 1995 (see Table 11.2) were recorded as *L. monocytogenes*. It was asked: "surely, if *Listeria* spp. are present in this raw milk cheese, isn't it unlikely that they would be present in pure culture?" and in the exploration of this point the process of examining the plates, how colonies are picked off for further testing, how colony choices are made, were all examined. It became clear that one of the (many) weaknesses of plate counts is the subjectivity of the operator.

Clearly if the confirmatory tests (biochemical and serological) used five colonies of both 'typical' form and 'suspect' form (*i.e.* more or less typical but through observation differing in some way), then a conclusion might even so have erroneously been drawn that all the colonies seen on the original plate were of the confirmed type. But if only identical 'typical' colonies were selected and biochemical tests confirmed the diagnosis '*Listeria monocytogenes*' would that have led to accurate calculation of the counts of the organism, and what would the significance of the other organisms present have been? There was conflict, to which the Sherriff drew attention, between the reported findings of pure culture of *L. monocytogenes* (by private analytical laboratory AL2–Table 11.2) and mixed *Listeria* populations found by private analytical laboratory AL1 (data not shown).

If you examine the tables included in this chapter, including undertaking a detailed appraisal of Table 11.3 there will be other issues about the whole sampling, testing, and results interpretation as well as issues about when *Listeria* can or cannot grow in cheeses (particularly blue cheeses) which should occur to you. But in determining your own view about the reliability of the results on which both public health risk and the well-being of the small business hung, you should now see that the quality of implementation of the methodology is vitally important.

Exercise 2. Bearing in mind that the *Listeria monocytogenes* strain found may not have been pathogenic do you think the risk management policy implemented (which for a period in 1994/95 resulted in the removal of Lanarkshire Blue cheese from the market in the interest of public health, but with its potential consequences for the producer) was justified?

In addressing this question you are really being asked to evaluate the scientific basis of risk management policy concerning a food-borne pathogen and a serious public health issue – listeriosis, and relate that to the outcomes of that policy. In thinking about what was done in 1995, you may then be in a position to form a view of whether you think the policy was justified or not.

In the late 1980s and early 1990s based on the recent experience that in the current food manufacturing and distribution systems large outbreaks of listeriosis could occur (Table 11.5) and the knowledge of the widespread incidence of *Listeria* in foods (Bilney, 1991), the presence of *L. monocytogenes* was considered an unacceptable risk, and so microbiological guidelines were set accordingly. In the USA a policy of 'zero-tolerance' operated in which the presence of *L. monocytogenes* in 25 g of retail food was unacceptable. At the time guidelines (Creamery Proprietors' Association, 1988) for UK farm based producers of soft and fresh cheeses made similar recommendations that if testing demonstrated the presence of pathogenic *Listeria* (defined as *L. monocytogenes*, *L. ivanovii* and *L. seegleri*) in any 25 g sample derived from a 125 g composite sample (5 × 25 g samples per lot of cheese) withdrawal of the affected lot of cheese from the market should be considered, while confirmation of pathogenic and/or non-pathogenic strains indicated that cleaning practices, management control systems and product segregation should be reviewed.

But the guidelines in use in the UK in 1994/1995, which gave food examiners* guidance on the interpretation of microbiological results for food samples taken at point of sale, were those of the PHLS (PHLS, 1992) (Table 11.9). Evidence in the Lanark Blue cheese Hearing indicated that these were formally backed by the Scottish Office [government authority] as appropriate to guide the enforcement officers.

At the time the infective dose was not known (see page 274), but MacGowan *et al.* (1994) commented "It is very difficult to relate the level or degree of food contamination with *Listeria monocytogenes* to health risks as the number of ingested bacteria required to establish infection in man is unknown although between 10^2–10^9 cfu g^{-1} has been suggested [by Farber and Peterkin, 1991]." The provisional PHLS guidelines (1992) set the 'unacceptable/potentially hazardous' threshold at $> 10^3$ cfu *L. monocytogenes* g^{-1} for product at the point of sale.

At the same time the EC Directive 92/46/EEC on Milk and Dairy Products (Anon, 1992), but not then implemented in UK legislation, required the absence of *L. monocytogenes* in 25 g of cheese, other than hard cheese, and from 1 g of other dairy products, and these standards were proposed in the draft Dairy Products (Hygiene) Regulations. The standard says "where this standard is exceeded the food must be excluded from consumption and the product withdrawn from the market:

Products	**Type of micro-organism**	**Standard**
cheese	*Listeria monocytogenes*	Absent in 25 g
(other than hard cheese)		where $n = 5, c = 0$

The sample of 25 g referred to above shall consist of 5 specimens of 5 g taken from different parts of the same product".

When the Lanark Blue strains of *L. monocytogenes* were serotyped by the most senior laboratory in the UK (the Central Public Health Laboratory, Colindale, London) all were found to be serotype 3a. In informing the actions of the EHOs the serotype was not vital to them, although in surveillance terms such information is useful. This was not an outbreak of listeriosis, rather it was a case of preventing an outbreak. It was known from where the organisms had originated – the cheese – and with the test results showing the presence of *L. monocytogenes* at levels in excess of the provisional PHLS guidelines (1992) (which did not specify serotypes) the enforcement officers felt fully justified in their actions.

However in court, counsel for Mr Errington drew attention to the fact that the actions of the EHOs were not, on their own admission, based on any form of assessment of the risk which the presence of *L. monocytogenes* serovar 3a presented to the consuming public for they were being guided by the provisional PHLS guidelines on the unacceptability of the biotype *L. monocytogenes*. It was further argued that there was no proof of the pathogenicity of this particular strain; that since some haemolytic strains are known to be non-pathogenic the

*Defined in the FSA, 1990.

haemolytic capability of this strain did not prove its pathogenicity; that pathogenicity itself in part depends on the vulnerability of the consumer groups. Since 1989 the known high risk groups had already been defined and targeted in official pamphlets advising against eating high risk foods such as raw milk and soft and blue veined cheeses. Later in 1989 both in England and Wales, and in Scotland, public notices were issued warning immuno-compromised people and pregnant women not to eat pâté, another high risk food. As Roberts (1994) observed following the latter notices, a sharp decline in listeriosis cases was seen, perhaps reflecting compliance with that advice.

The Sheriff considered these arguments and as detailed at the end of Section 11.1.4 accepted them not least because the doubt concerning the actual pathogenicity of the strain present did not support the case that the specific cheese – Lanark Blue – was a risk to public health.

What did not enter the Hearing was any debate concerning the acceptable and non-acceptable levels of *L. monocytogenes* – but had that question arisen the issues of infective dose, exposure, human susceptibility and the range of risk factors for listeriosis cases would have to have been explored.

At the time of the hearing the Government was still developing policy, as well as implementing legislation required as a member of the EC. Roberts (1994) outlined national approaches to the listeria problem currently being undertaken:

- development of guidelines for microbiological criteria through the PHLS (which recognised the difficulty of achieving 'absence' of *L. monocytogenes* from ready-to-eat foods);
- funding of research to establish facts;
- providing advice to vulnerable groups about which foods to avoid, and additionally seeking to educate the general public in how to safely handle ready-to-eat foods;
- development of Regulations, such as the [then] new Food Hygiene (Temperature) Regulations, 1995 (Anon, 1995b), which required high risk foods to be held chilled at less than 8 °C (recognising that good low temperature storage reduces the growth rates of cold tolerant organisms such as *L. monocytogenes*).
- encouraging industry to evaluate its own procedures and implement HACCP based systems (required anyway under the FSA, 1990) from production through to sale;
- encouraging industry to develop industry Codes of Practice. The Government was not introducing legislation for a zero tolerance policy, although some published criteria, recommended for the production of soft cheeses, did (Creamery Proprietors Association, 1988) [as did also one Government Department for 'cook chill' foods immediately before cooking (DoH, 1989)].
- through end product and environmental testing determine that good management practices were working.

Thus the policies operating in 1995 regarded all *L. monocytogenes* as potentially pathogenic, while in surveillance routine analytical techniques did not normally differentiate beyond serotype and phagetype, and did not normally use molecular techniques for finer strain definition. Guideline criteria for foods at the point of production or the 'ready-to-eat' point only tolerated the presence of low numbers of the organisms because of the difficulty of achieving effective absence of the organism, and also because of lack of information relating to the infective dose. But the zero tolerance policy seen in the USA was required by the EC directive 92/46/EEC (Anon, 1992) on the hygiene of dairy products. Available evidence showed that while most foods are liable to contain the organism, most people were evidently exposed without ill effect and the vulnerable group who might be affected had, apparently, responded to public warning of what to avoid eating. In the absence of both robust techniques to routinely determine genotype and in the absence of data necessary to really evaluate the risks to public health, policy in the USA and in the EC, in government and in the scientific community was based on the precautionary principle.

This policy made it possible for legislators and enforcers to protect public health on the premise that regardless of strain the presence of *L. monocytogenes* in ready-to-eat food beyond guideline levels represented an unacceptable risk.

But at a price – why should a food be regarded as unsafe if in fact it is not?

So, in the end it is for you to judge not only whether policy and its implementation was appropriate, but if you think it was not, what alternatives were there?

Beyond the Lanark Blue Hearing

This Hearing challenged the current thinking in the scientific and enforcement communities yet did not itself provide an answer to the management of the risk associated with *Listeria monocytogenes* in Lanark Blue cheese specifically, and in ready to-eat-foods generally.

In 1994 Roberts quoted the World Health Organisation (WHO, 1988) working group on listeriosis and said:

"the elimination of *Listeria monocytogenes* from all food is impractical and probably impossible . . . the critical issue is not how to prevent its presence but how to control its survival".

As required under the Food Safety (General Food Hygiene) Regulations, 1995 (Anon, 1995c) the implementation of HACCP in the manufacture of high risk foods – including raw milk cheeses – would reduce microbiological risk associated with them but the judgement forced the questions:

- how should the incidence of *L. monocytogenes* in foods now be managed?
- would a zero tolerance policy, at least for high risk foods which undergo no further cooking, be more appropriate?
- should routine microbiological analysis for *L. monocytogenes* be changed?
- should routine tests discriminate between strains of *L. monocytogenes*?

The Lanark Blue Hearing itself did not, of course, eliminate the presence of *L. monocytogenes* in foods! Daily the authorities still had to deal with incidences, when the organisms were found, on a case by case basis evaluating risk factors such as:

- the actual type of food in which the organisms were found;
- whether the organism could grow in or on the food and increase further;

- what could stop that growth;
- at what point in the shelf life of the product was the test undertaken;
- would 'low' levels now (at the time of the test) imply 'higher' levels later?
- would the food be cooked before being eaten?

At least the zero tolerance policy would eliminate the problem of variability in strain virulence and potential pathogenicity, but at the same time it would impose stringent, possibly unachievable, standards of hygiene on producers. The standards proposed in draft regulations and described earlier in Section 11.4 were included in The Dairy Products (Hygiene) Regulations 1995 (Anon, 1995a) and thus did impose the 'zero tolerance' standard for cheese (other than hard cheese) on its removal from the processing establishment in spite of this being a very hard target to meet. They also impose a slightly less stringent standard on milk based products of the absence of *L. monocytogenes* in 1 g. However, inserted in the Regulations, and not proposed in the draft, was the opportunity for producers of cheese made in traditional manner to apply for derogations from this standard. The derogation facility recognised that for small traditional producers it might take time for them to achieve the hygiene standards required and to implement effective HACCP systems. These derogations did not, however, apply to the microbiological standards at the point of sale and these foods also had to meet the general requirements under the FSA (1990) and be 'not injurious to health'.

UK policy was not to legislate, and has not legislated to provide official action level for *L. monocytogenes* contamination of food. If a microbiological criterion other than zero tolerance were to be required for foods it could pose the question of whether strain sub-typing should also be applied routinely (which could lead to clearer understanding of virulence and potential pathogenicity of strains found). In the UK there was clearly a need for standard methodology for isolation of *Listeria* and its biotypes, the methods in use at the time of the Lanark Blue Hearing varying between laboratories, although the method included in section 3.15 of the British Standard BS4285 (1993) 'Microbiological examination for dairy purposes' for *L. monocytogenes* was available.*

The PHLS produced new guidelines for ready-to-eat foods in 1996 (PHLS, 1996), and modified them further in 2000 (PHLS, 2000) giving the following criteria for *Listeria* in all ready-to-eat foods including cheese which solve the dilemma another way as shown in Table 11.10. Failure to meet these criteria is not itself a basis for prosecution, for the Guidelines indicate that it is up to the enforcer to decide whether to pursue a prosecution under the FSA (1990) (see Tables 11.10 and 11.11).

The PHLS has clearly taken an approach which does not involve the determination of subtype when samples are taken for routine surveillance purposes, and thus avoids the issue of the actual virulence and potential pathogenicity of the strain(s) present, and recognizes that where *Listeria* are found *L. monocytogenes* is likely to be present.

---

* The PHLS recommended methods (Roberts, Hooper and Greenwood, 1996) are based on the FDA method on which the method BS4285 was also based.

In the explanations within the guidelines it says:

"Quantitative levels in the counts of *Listeria* spp. in previous versions of the guidelines,* excluded *Listeria monocytogenes*. This has been changed to include *Listeria monocytogenes* and hence the term is fully inclusive of all *Listeria* spp. The reasons for this are because of the changes to the quantitative criteria for *Listeria monocytogenes* explained below and to represent what happens in practice when examining food samples by the standard method. Although *Listeria* spp. other than *Listeria monocytogenes* are rarely implicated in illness they are indicators of the likely presence of *Listeria monocytogenes* and so concerns described below about the presence of low levels of *Listeria monocytogenes* in certain products also apply to other species of *Listeria*.

Listeria monocytogenes

The quantitative microbiological quality levels for *Listeria monocytogenes* have been modified and the classification of 'unsatisfactory' is now not applicable in this case. Some quality standards require a zero level for *Listeria monocytogenes* at the production stage of the food (Dairy Products (Hygiene) Regulations, 1995) thus 10^2 cfu/g at point of sale/consumption represents a potential risk to health. Counts of this level may also indicate a significant failure of hygiene standards in the preparation and/or storage of such foods. None of the figures within the guidelines can be said to carry an absolute risk to health. Nevertheless, opinions and published reports support the contention that the numbers quoted in [the criteria above] under the heading 'unacceptable/potentially hazardous' represent unacceptable microbiological quality and are a potential hazard to those who eat such food.†

On the basis of current information it is the opinion of the ACFDP‡ that it is unacceptable that ready-to-eat foods contain any serogroup of *Listeria monocytogenes* at levels at or above 10^2 cfu/g. Some serotypes/phage types of *Listeria monocytogenes* may rarely be associated with human infection, but their presence represents an inadequate level of hygiene.

Listeria monocytogenes is widely distributed in the environment and is able to multiply slowly at 4 °C. The shelf life of foods [listed in Table 11.11] varies enormously. Certain foods – such as soft ripened cheese, vacuum packed pâté, and sliced meats – have long shelf life under refrigeration, and the presence of *Listeria monocytogenes* at any level may be of significance due to its potential for growth during storage. The use of enrichment procedure, in addition to enumeration, should therefore be considered to ensure that the organism is absent from the product."
[From Gilbert *et al.*, 2000. Printed with permission of the PHLS Communicable Disease Surveillance Centre. © PHLS.]

* See PHLS, 1992; PHLS, 1996.
† See European Commission, 1999.
‡ ACFDP = Advisory Committee for Food and Dairy Products (of the PHLS).

Table 11.10 *Guidelines for the microbiological quality of various ready-to-eat foods*

Food category (see Table 11.11)	Criterion	Microbiological quality (cfu g^{-1} unless stated)			
		Satisfactory	Acceptable	Unsatisfactory	Unacceptable/ potentially hazardous[a]
Aerobic colony count[b] 30 °C/48 h					
1		$<10^3$	$10^3-<10^4$	$\geq 10^4$	NA
2		$<10^4$	$10^4-<10^5$	$\geq 10^5$	NA
3		$<10^5$	$10^5-<10^6$	$\geq 10^6$	NA
4		$<10^6$	$10^6-<10^7$	$\geq 10^7$	NA
5		NA	NA	NA	NA
Indicator organisms[c]					
1–5	Enterobacteraceae[d]	<100	$100-<10^4$	$\geq 10^4$	NA
1–5	E. coli (total)	<20	$20-<100$	≥ 100	NA
1–5	Listeria spp. (total)	<20	$20-<100$	≥ 100	NA
Pathogens					
1–5	Salmonella spp.	not detected in 25 g			not detected in 25 g
1–5	Campylobacter spp.	not detected in 25 g			not detected in 25 g
1–5	E. coli O157 and other VTEC	not detected in 25 g			not detected in 25 g
1–5	V. cholerae	not detected in 25 g			not detected in 25 g
1–5	V. parahaemolyticus[e]	<20	$20-<100$	$100-<10^3$	$\geq 10^3$
1–5	L. monocytogenes	$<20^g$	$20-<100$	NA	≥ 100
1–5	S. aureus	<20	$20-<100$	$100-<10^4$	$\geq 10^4$
1–5	C. perfringens	<20	$20-<100$	$100-<10^4$	$\geq 10^4$

1–5	*B. cereus* and other pathogenic *Bacillus* spp.[f]	$<10^3$	$10^3-<10^4$	$10-<10^5$	$\geq 10^5$

For 1992 guidelines see Table 11.9; for 1996 guidelines see Table 5.4

Notes: [a]Prosecution based on high colony counts and/or indicator organisms in the absence of other criteria of unacceptability is unlikely to be successful.

[b]Guidelines for aerobic colony counts may not apply to certain fermented foods, *e.g.* salami, soft cheese and unpasteurised yoghurt. These foods fall into category 5. Acceptability is based on appearance, smell, texture, and the levels or absence of indicator organisms or pathogens.

[c]On occasions some strains may be pathogenic.

[d]Not applicable to fresh fruit, vegetables and salad vegetables.

[e]Relevant to seafood only.

[f]If the *Bacillus* counts exceed 10^4 cfu g^{-1} the organism should be identified.

[g]Not detected in 25 g for certain long shelf-life products under refrigeration.

NA Not applicable.

Source: Gilbert *et al.*, 2000. Reproduced with the permission of the PHLS Communicable Disease Surveillance Centre. © PHLS.

Table 11.11 Food categories used in the guidelines for the microbiological quality of various ready-to-eat foods (see Table 11.10)

Food	Category				
	1	2	3	4	5
Meat	Beefburgers; meat pies (steak and kidney, pasty); pork pies; scotch egg	Faggots; kebabs; meat meals (shepherds/cottage pie/casseroles); poultry (unsliced); sausages (British)	Meat, sliced (beef, haslet, pork, poultry)	Brawn; meat, sliced (cooked ham, tongue); tripe and other offal	Ham – raw (Parma/country style); salami and fermented meat products; sausage roll; sausages (smoked)
Seafood	Herring/roll mop and other raw pickled fish		Crustaceans (crab, lobster, prawns); other fish (cooked); seafood meals	Molluscs and other shellfish (cooked); smoked fish; taramasalata	
Dessert	Mousse/dessert	Cakes, pastries, slices and desserts – without dairy cream; tarts, flans and pies	Cakes, pastries, slices and desserts – with dairy cream; trifle		
Savoury	Bhaji (onion, spinach, vegetable)	Cheese-based bakery products; flan/quiche; mayonnaise/dressings; samosa	Pate (meat, seafood, or vegetable); satay; spring rolls	Homous, tzatziki, and other dips	Fermented foods
Vegetable		Vegetables and vegetable meals (cooked)	Coleslaw; fruit and vegetables (dried); rice	Prepared mixed salads and crudities	Fruit and vegetables (fresh)

		Without salad	With salad, with cheese
Dairy	Ice cream, milk shakes (non-dairy); ice lollies, slush, and sorbet		Cheese, yoghurt/frozen yoghurt (natural)
Ready-to-eat meals	Pasta/pizza; meals (other)		
Sandwiches and filled rolls			

Source: PHLS, 2000.

These guidelines have no statutory force, but are based on the collective experience of the PHLS which in the year 2000 examined 190 000 food samples. That evidence is being "collated to provide an evidence base for quantitative microbiological risk assessment and for the implementation of HACCPs for food safety" (PHLS, 2000).

Exercise 3. Following the Appeal against the outcomes of the Judicial Review [the Hearing] what options did the producer have to manage the risks associated with the farm production of raw milk blue cheese?

It is for you to explore how you think the farm production system could address the risk of the presence of *Listeria monocytogenes* in raw milk cheese.

Approaches could include the following:

- evaluation of the entire process to implement effective HACCP system including considering the health of the sheep, and continued avoidance of feeding the sheep silage;
- strict separation of the sheep from cows;
- cleanliness of milking;
- minimisation of the time between milking and use of the raw milk to ensure the natural antimicrobial factors of the milk are protective;
- consideration of the use of bacteriocin producing starter strains;
- strict monitoring of starter activity to ensure the rate of acid production remains optimal.

Exercise 4. How is the discovery of the presence of *Listeria monocytogenes* in food managed today?

This exercise should allow you to explore current outbreaks in your country and see what policy has been adopted and how it being implemented. You may find the following reference useful:

Lund, B.M., 2000. *Listeria monocytogenes* in ready to eat foods. *Food Safety Express*, **4**, (Oct./Dec.), 3–4.

and references quoted in the above paper:

Buchanan, R.L. *et al.*, 1997. Use of epidemic and food survey data to estimate a purposefully conservative dose-response relationship for *Listeria monocytogenes* levels and incidence of listeriosis. *Journal of Food Protection*, **60**, 918–922.

Buchanan, R.L. and Lindqvist, R., 2000. Hazard identification and hazard characterisation of *Listeria monocytogenes* in ready to eat foods. Preliminary report prepared for Joint FAO/WHO Expert Consultation on Risk Assessment for Microbiological Hazards in Foods, FAO Headquarters, Rome, Italy, 17–21 July 2000.

Centers for Disease Control and Prevention (CDC), 1999. Update: Multistate outbreak of listeriosis – United States, 1998–1999. *Morbidity and Mortality Weekly Report*, **47**, 1117–1118.

Farber, J.M. and Harwig, J., 1996. The Canadian position on *Listeria monocytogenes* in ready-to-eat foods. *Food Control*, **7** (4/5), 253–259.

Lindqvist, R. and Westoo, A., 2000. Quantitative risk assessment for *Listeria monocytogenes* in smoked gravid salmon and rainbow trout in Sweden. *International Journal of Food Microbiology*, **58**, 181–196.

Lyytikainen, O. *et al.*, 2000. An outbreak of listeriosis due to *Listeria monocytogenes* serotype 3a from butter in Finland. *Journal of Infectious Diseases*, **181**, 1838–1841.

[List quoted with permission of Dr B.M. Lund]

11.5 SUMMARY

This case study concerning the disputed role of *Listeria monocytogenes* strains found in raw milk mould ripened blue cheese – Lanark Blue demonstrates a number of issues:

- that microbiological sampling and testing methodologies should be meticulously implemented using recognised standard protocols;
- that all strains of *L. monocytogenes* cannot be assumed to be pathogenic;
- that the presence of *Listeria* spp. is indicative of the probable presence of *L. monocytogenes*;
- that the presence of *L. monocytogenes* in ready-to-eat foods represents a risk to public health in that the strain found present could be pathogenic, and so criteria defining maximum tolerable levels have to be identified to ensure that the level of risk is acceptable and the overall risk to public health is managed.

11.6 REFERENCES

Anon, 1990. Food Safety Act, HMSO, London. Food Safety Act, Code of Practice no. 7: Sampling for analysis and examination, HMSO, London, UK.

Anon, 1990. Food Safety Act, 1995, HMSO, London, UK.

Anon, 1992. Council Directive 92/46/EEC of the 16th June 1992. Laying down the health rules for the production and placing on the market of raw milk, heat treated milk and milk based products. *Official Journal*, **L268**, 14.9, 1992.

Anon, 1993. Microbiological examination for dairy purposes. Methods for detection and/or enumeration of specific groups of microorganisms. Detection of *Listeria monocytogenes*. British Standards Institution. BS 4285-3.15:1993 [ISO 10560:1993], London, UK.

Anon, 1994. *The Times Index*.

Anon, 1995. *The Times Index*.

Anon, 1995a. The Dairy Products (Hygiene) Regulations 1995. HMSO, London, UK.

Anon, 1995b. The Food Safety (Temperature Control) Regulations 1995. HMSO, London, UK.

Anon, 1995c. The Food Safety (General Food Hygiene) Regulations, 1995, HMSO, London, UK.

Azadian, B.S., Finnerty, G.T. and Pearson, A.D., 1989. Cheese-borne listeria meningitis in an immunocompetent patient. *Lancet*, February, 322–323.

Bannister, B., 1987. *Listeria monocytogenes* meningitis associated with eating soft cheese. *Journal of Infection*, **15**, 165–168.

Baxter, F., Wright, F., Chalmers, R.M., Low, J.C. and Donachie, W., 1993. Characterisation by Multilocus Enzyme Electrophoresis of *Listeria monocytogenes* isolates involved in ovine listeriosis outbreaks in Scotland from 1989 to 1991. *Applied and Environmental Microbiology*, **59** (9), 3126–3129.

Bille, J., 1990. *Epidemiology of human listeriosis in Europe, with special reference to the Swiss outbreak*. In Miller, A.J., Smith, A.J. and Somkuti, G.A. (eds). *Foodborne Listeriosis*. Society for Industrial Microbiology, Elsevier, Amsterdam.

Bilney, F., Armstrong, R., Vickerman, A. *et al.*, 1991. Listeria in food: report of the North and West Yorkshire joint working group on a two year survey of the presence of *Listeria* in food. *Environmental Health*, **99** (June), 132–137.

Brosch, R., Catimel, B., Milon, G., Buchreiser, C., Vindel, E. and Rocourt, J., 1993. Virulence heterogeneity of *Listeria monocytogenes* strains from various sources (food, human, animal) in immunocompetent mice and its association with typing characteristics. *Journal of Food Protection*, **56** (4), 297–301.

Brosch, R., Chen, J. and Luchansky, J.B., 1994. Pulsed field fingerprinting of Listeriae: identification of genomic divisions for *Listeria monocytogenes* and their correlation with serovar. *Applied and Environmental Microbiology*, **60** (7), 2584–2592.

Bryner, J., Wesley, I. and van der Maaten, M., 1989. Research on listeriosis in milk cows with intramammary inoculation of *Listeria monocytogenes*. *Acta Microbiologica Hungarica*, **36**, 137–140.

Bula, C.J., Bille, J. and Glauser, M.P., 1995. An epidemic of food borne listeriosis in Western Switzerland: description of 57 cases involving adults. *Clinical Infectious Diseases*, **20**, 66–72.

Conner, D.E., Scott, V.N., Sumner, S.S. and Bernard, D.T., 1989. Pathogenicity of foodborne, environmental and clinical isolates of *Listeria monocytogenes* in mice. *Journal of Food Science*, **54** (6), 1553–1556.

Creamery Proprietors Association, The, 1988. Guidelines for good hygienic practices in the manufacture of soft and fresh cheese. 1st edition. The CPA, 19 Cornwall Terrace, London, NW1 4QP, UK.

Department of Health, 1989. Chilled and frozen: Guidelines on cook-chill, and cook-freeze catering systems, HMSO, London, UK.

European Commission, 1999. Opinion of the Scientific Committee on Veterinary measures relating to public health on *Listeria monocytogenes*, 23 Sept., 1999 (www.europa.eu.int/comm/dg24/health/sc/scv/out25-en.html).

Farber, J.M. and Peterkin, P.I., 1991. *Listeria monocytogenes* – a food borne pathogen. *Microbiological Reviews*, **55** (3), 476–511.

Farber, J.M., Carter, A.O., Varughese, P.V., Ashton, F.E. and Ewan, E.P., 1990. Listeriosis traced to the consumption of alfalfa tablets and soft cheese. *New England Journal of Medicine*, **322**, 338.

Farber, J.M., Spiers, J.I., Pontefract, R. and Conner, D.E. 1991. Characteristics of non-pathogenic strains of *Listeria monocytogenes*. *Canadian Journal of Microbiology*, **37**, 647–650.

FDA, 1984. *Bacteriological Analytical Manual*, 6th edition, Chapter 29, Section 13.

Genigeorgis, C., Carniciu, M., Dutulescu, D. and Farver, T.B., 1991. Growth and survival of *Listeria monocytogenes* in market cheeses stored at 4–30°C. *Journal of Food Protection*, **54** (9), 662–668.

Gilbert, R.J., de Louvois, J., Donovan, T., Little, C., Nye, K., Ribeiro, C.D., Richards, J., Roberts, D. and Bolton, F.J., 2000. Guidelines for the microbiological quality of some ready-to-eat foods sampled at the point of sale. *Communicable Disease and Public Health*, **3** (3), 163–167.

Gitter, M., Bradley, R. and Bampied, P.H., 1980. *Listeria monocytogenes* infection in bovine mastitis. *Veterinary Record*, **107**, 390–393.

Golden, D.A., Beuchat, L.R. and Brackett, R.E., 1988. Evaluation of selective direct plating media for their suitability to recover uninjured, heat-injured and freeze-injured *Listeria monocytogenes* from foods. *Applied and Environmental Microbiology*, **54**, 1451–1456.

Greenwood, M.H., Roberts, D. and Burden, P., 1991. The occurrence of *Listeria* species in milk and dairy product: a national survey in England and Wales. *International Journal Food Microbiology*, **12**, 197–206.

Hof, H. and Rocourt, J., 1992. Review: Is any strain of *Listeria monocytogenes* detected in food a health risk? *International Journal of Food Microbiology*, **16**, 173–182.

ICMSF, 1996. *Micro-organisms in Foods*. Volume 5. *Characteristics of Microbial Pathogens*, Blackie Academic and Professional, London.

International Dairy Federation (IDF), 1990. Milk and milk products: detection of *Listeria monocytogenes*: Enrichment method. IDF 143:1990. Brussels.

IFST, 1995a. Professional Food Microbiology Group Accreditation Advisory Group Guidelines No.1 (12/95). Recommended checks for mechanically driven spiral platers. Institute of Food Science and Technology, 5 Cambridge Court, 210 Shepherds Bush Road, London, W6 7NJ, UK.

IFST, 1995b. Professional Food Microbiology Group Accreditation Advisory Group Guidelines. No. 1a (12/95). Recommended checks for automatic spiral platers. Institute of Food Science and Technology, 5 Cambridge Court, 210 Shepherds Bush Road, London, W6 7NJ, UK.

Jarvis, B., Lach, V.H. and Wood, J.M., 1977. Evaluation of the spiral plate maker for the enumeration of micro-organisms in foods. *Journal of Applied Bacteriology*, **43** (1), 149–157.

Kampelmacher, E.H. and van Noorle Jansen, L.M., 1972. Further studies on isolation of *Listeria monocytogenes* in clinically healthy individuals. *Zentraalblat Bakteriologie und Hygiene*, I.Abt.Orig. A221, 70–77.

Kinderlerer, J.L. and Lund, B., 1992. Inhibition of *Listeria monocytogenes* and *Listeria innocua* by hexanoic and octanoic acids. *Letters in Applied Microbiology*, **14**, 271–274.

Klima, R.A. and Montville, T.J., 1995. The regulatory and industrial responses to listeriosis in the USA: a paradigm for dealing with emerging foodborne pathogens. *Trends in Food Science and Technology*, **6**, 87–93.

Kornacki, J.L., Evason, D.J., Reid, W., Rowe, K. and Flowers, R.S., 1993. Evaluation of USDA protocol for detection of *Listeria monocytogenes*. *Journal of Food Protection*, **56**, 441–443.

Linnan, M.J., Mascola, L., Lou, X.D. *et al.*, 1988. Epidemic listeriosis associated with Mexican style cheese. *New England Medical Journal*, **319**, 823–828.

MacGowan, A.P., Bowker, K., McLauchlin, J., Bennett, P.M. and Reeves, D.S., 1994. The occurrence of *Listeria* spp. in human faeces, shop bought food stuffs, sewage and soil from urban source. *International Journal of Food Microbiology*, **21**, 325–334.

McLauchlin, J., 1990. Distribution of serovars of *Listeria monocytogenes* isolated from different categories of patients with listeriosis. *European Journal of Clinical Microbiology and Infectious Disease*, **9**, 201–203.

McLauchlin, J., 1993. Listeriosis and *Listeria monocytogenes. Environmental Policy and Practice*, **3**, 201–214.

McLauchlin, J., Greenwood, M.H. and Pini, P.N., 1990. The occurrence of *Listeria monocytogenes* in cheese from a manufacturer associated with a case of listeriosis. *International Journal of Food Microbiology*, **10**, 255–262.

McLauchlin, J., Hall, S.M., Velani, S.K. and Gilbert, R.J., 1991. Human listeriosis and pate: a possible association. *British Medical Journal*, **303**, 773–775.

Menudier, A., Bosiraud, C. and Nicolas, J.-A., 1991. Virulence of *Listeria monocytogenes* serovars and *Listeria* spp. in experimental infection in mice. *Journal of Food Protection*, **54**, 917–921.

Milk Marketing Board (MMB), 1989. Guidelines for good hygienic practice in the manufacture of soft and fresh cheese in small and farm based production units. MMB, Thames Ditton, UK.

Mpamugo, O., 1997. An investigation into the interactions of *Listeria monocytogenes* and *Listeria innocua* growing together in co-culture. Dissertation for MSc Food Safety and Control, South Bank University, London, October 1997.

Oxoid, 1992. *Oxoid Manual – media sheets*, March 92, p. 2–130b. [Note 'Oxoid' is now the Unipath Company.]

Papageorgiou, D.M. and Marth, E.H., 1989. Fate of *Listeria monocytogenes* during manufacture and ripening of blue cheese. *Journal of Food Protection*, **52** (7), 459–465.

PHLS, 1992. Provisional microbiological guidelines for some ready-to-eat foods sampled at point of sale. *PHLS Laboratory Digest*, **9**, 98–99.

PHLS, 1996. Microbiological guidelines for some ready-to-eat foods sampled at point of sale: an expert opinion from the PHLS. *PHLS Laboratory Digest*, **13**, 41–43.

PHLS, 2000: see Gilbert, R.J. *et al.*, 2000.

Pini, P.N, and Gilbert, R.J., 1988. The occurrence in the UK of *Listeria* species in raw chickens and soft cheeses. *International Journal of Food Microbiology*, **6**, 317–326.

Roberts, D., 1994. *Listeria monocytogenes* and food: the UK approach. *Dairy and Environmental Sanitation*, **14**, 198–204.

Roberts, D., Hooper, W. and Greenwood, M., 1996. *Practical Food Microbiology. Methods for the examination of food for micro-organisms of public health significance.* PHLS, 61 Colindale Avenue, London, NW9 5DF, UK.

Ryser, E.T. and Marth, E.H., 1987. Fate of *Listeria monocytogenes* during manufacture and ripening of camembert cheese. *Journal of Food Protection*, **50** (5), 373–378.

Ryser, E.T. and Marth, E., 1991. Chapter 1: *Characteristics and classification* in *Listeria, Listeriosis, and Food Safety.* Marcel Dekker, New York.

Tabouret, M., de Rycke, J., Audurier, A. and Poutrel, B., 1991. Pathogenicity of *Listeria monocytogenes* isolates in immuno-compromised mice in relation to listerioslysis production. *Journal of Medical Microbiology*, **34**, 13–18.

WHO working group, 1988. Foodborne listeriosis. *Bulletin of the World Health Organisation*, **66**, 421–428.

The Need for Food Hygiene

Key issues
- Cooked meat pies
- *Escherichia coli* O157:H7
- Infection and cross contamination in the food chain
- Food hygiene training and management

Challenge

The case study relates to an outbreak of food poisoning caused by *E. coli* O157:H7 in which of 496 known cases, 21 people died, with 17 as a direct result of the infection. A Government commissioned report by the Pennington Group made a number of recommendations for future strategies to minimise the risk to the public from this organism, and a Fatal Accident Enquiry identified the errors made and what safety measures could reasonably be expected in the production and retail sale of cooked products. Although the case study may enable you to consider a wide number of issues both in and outside the scope of this book, the focus of the exercises presented in Chapter 12 is to challenge you to consider how cross-contamination may arise, and the reasonable precautions, including food hygiene training, which may manage and reduce that risk.

12.1 THE CASE STUDY: COOKED MEAT PRODUCTS, CENTRAL SCOTLAND, NOVEMBER AND DECEMBER 1996

This case study is about the infection and illness of 496 people among whom 21 people died, 17 of whom died directly due to consumption of food infected with *E. coli* O157:H7. The outbreak occurred in November and December 1996. Investigations showed that the infection was associated with meat products which came from one butchers' shop in Wishaw in Lanarkshire, Scotland. The reasons for the outbreak were investigated very fully by the Pennington Group (see Section 12.1.1), and through criminal proceedings (Section 12.1.2), legal proceedings in relation to food hygiene offences (Section 12.1.3) and a Fatal Accident Enquiry (Section 12.1.4).

On Sunday 17 November 1996 Wishaw Parish Church entertained many of its elderly parishioners to Sunday lunch. Eighty-seven over-70s turned up and enjoyed a sociable meal which had been prepared by a number of volunteers in the kitchens of the church hall. On Wednesday 20 November a number of these elderly people began exhibiting symptoms of gastric illness and diarrhoea (in some cases bloody diarrhoea). More cases occurred on the following days and eventually eight of these elderly people died of the infection.

By Friday 22 November the public health department became aware of a number of other cases of infection showing similar symptoms both in the Wishaw area and further afield and an Outbreak Control Team (OCT) was set up. Some case histories had by that time been taken and it was becoming clear that a linking factor was that a number of the people affected had consumed food from one butchers' shop – J. Barr and Son, Butchers, Wishaw. On Saturday 23 November and Sunday 24 November the very elderly residents of the Bankview Nursing Home were given cooked meat sandwiches for afternoon tea – and very soon afterwards a number of them and also some staff of the nursing home were showing similar symptoms. Later in the Fatal Accident Enquiry, the Sherriff (Cox, 1998) determined a link to John Barr's products.

On Wednesday 27 November 1996 the voluntary closure of the entire Barr's business was announced.

On investigation it became clear that the butchers' shop was an extensive business selling raw meats, cooked meats and pies in the shop and also supplying about 85 other shops in the Lowlands of Scotland. The shop had a bakery attached and at the back of the two shops a wide range of cooked meats and meat pies were manufactured. Although the permanent staff was a small number investigation showed that at the time of the outbreak about 40 people were employed by the business in various capacities, full and part-time. These facts were not initially known to the OCT, and only became evident through patient investigation, questioning, analysis of invoices and other evidence. These facts took time to accumulate and delayed control of the outbreak.

12.1.1   The John Barr Case and the Pennington Group Report

The Pennington Group* was convened in early December 1996 – while the cases in the outbreak were still occurring – and it reported on 27 March 1997. The Group was required to investigate the reasons for the outbreak – not to attribute blame but to make recommendations to the Government for actions which would prevent further outbreaks occurring.

For the purposes of this case study, extracts from the Pennington Group Report†, which ran to 58 pages, have been selected to indicate both the progress of the outbreak and its investigation:

"Large quantities of meat and meat products were taken from Barr's, or from other premises supplied by Barr's, for microbiological testing. These included

---

*The Scottish Office set up an expert group to investigate the circumstances surrounding the outbreak under the chairmanship of Professor Hugh Pennington of Aberdeen Royal Hospital's Trust. This became known as 'The Pennington Group'.

† Extracts from the Pennington Group Report (1997) quoted with permission. Crown copyright.

cold cooked meats supplied to other butchers and the remains of gravy supplied with cooked steak to the Wishaw Parish Church lunch on 17 November. Microbiological swabbing of Barr's premises was undertaken and some items of equipment sent for detailed examination. The large numbers of food samples placed a heavy burden on the testing laboratories. The staff of Barr's were subsequently screened. Members of the Pennington Group visited Barr's premises during the course of investigations in order to familiarise themselves with the physical layout and to gain a first hand impression of the circumstances and nature of the operation of the business.

The premises were long established and had been extensively converted and extended. However the premises were similar to many other premises of this age. The layout and design constrained the measures that could be taken to ensure effective product flows and separation of cooked and raw products. However a significant capital investment had been made in the premises which had modern chillers, equipment and work surfaces."

"The number of cases of suspected or confirmed infection increased dramatically from the outset of the outbreak" [See Figure 12.1].

"By Sunday 24 November reports indicated that distribution of products from Barr's had extended beyond the local authority area into the central belt of Scotland. Cases of infection were subsequently reported in the Forth Valley, Lothian and Greater Glasgow. In the early days of the outbreak, before full

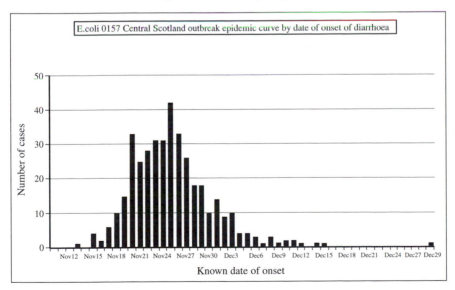

Figure 12.1 *The E. coli O157:H7 Central Scotland outbreak epidemic curve by date of onset of diarrhoea*
(From the Pennington Group Report (1997), with permission. Crown copyright)

exposure histories were obtained and in the absence of comprehensive information about supply and distribution, the OCT could not assume that there was
only one source of contaminated food."

"Epidemiological and subsequent microbiological evidence showed that the
outbreak comprised several separate but related incidents:

* relating to a lunch (attended by about 100 people) held in Wishaw Parish
  Church Hall [17 November],
* a birthday party held in the Cascade Public house on 23 November 1996,
* retail sales in Lanarkshire and the Forth Valley.

All isolates of *E. coli* O157 from individuals in the outbreak belonged to phage
type 2, and possessed verocytotoxin gene VT2."

[At the time the Pennington Group report was written of these isolates (which
totalled 272) 262 had been subjected to pulsed field electrophoresis (PFGE): all
had indistinguishable profiles.]

"As of 27 March 1997, the final date of onset of illness in a confirmed case of
illness associated with the outbreak was 15 December 1996. A possible case was,
however, reported with a date of onset of 29 December 1996. The outbreak was
declared over on Monday 20 January 1997 – although it was recognised that
further cases could occur as a result of secondary (person to person) spread, or
could be identified retrospectively as laboratory profiles became available. It was
also recognised that further deaths could occur following prolonged illness."

"After the outbreak was declared over, microbiological and serological results
continued to be accumulated and clinical and exposure histories reviewed. The
27 March 1997 figures were provisional but unlikely to change significantly.
They showed 496 cases of infection with *E. coli* O157 linked to the outbreak, of
which 272 were confirmed, 60 probable and 164 possible. It was the largest ever
outbreak of infection with the organism in the UK. A further breakdown of these
figures is as follows:

	All Scotland	Lanarkshire	Forth Valley	Lothian	Greater Glasgow
Confirmed*	272	195	73	4	0
Probable†	60	50	10	0	0
Possible‡	164	128	35	0	1
Total	496	373	118	4	1

* Confirmed case: someone with *E. coli* O157 identified in their stool, irrespective of
their history.
† Probable case: someone with bloody diarrhoea and positive serology.
‡ Possible case: someone who has non-bloody diarrhoea with positive serology; or
someone who has no symptoms with positive serology; or someone who has bloody
diarrhoea without positive serology.

"The outbreak placed substantial pressure on local health resources. In Lanarkshire, the Wishaw clinic carried out batches of tests on 969 people with diarrhoea (and there will have been, in addition, a substantial number of people who attended their GP). There were admitted to hospital 127 people, of whom 13 required dialysis (all transferred to Glasgow). Twenty seven people were diagnosed as having evidence of [. . .] haemolytic uraemic syndrome (HUS) or thrombotic thrombocytopaenia purpura (TTP)."

"There were 18 deaths in all (all adults) associated with the outbreak – the second highest number of deaths associated with an outbreak of *E. coli* O157 infection in the world. Of these, 8 people had attended the luncheon at Wishaw Old Parish Church on 17 November 1996, and 6 were residents of Bankview Nursing Home in Bonnybridge, Forth Valley. The age range of the 12 residents of Lanarkshire who died was 69 to 90 years, and in Forth Valley it was 70 to 93 years."

Chapter 3 [of the Pennington Group Report] sets out

"some of what is (and is not) known about *E. coli* O157, its characteristics and behaviour. The Central Scotland outbreak brought some of that very sharply into focus and has caused the Pennington Group to examine a number of general issues and questions. These include:

- how and why fresh meat becomes contaminated with *E. coli* O157 in the first place;
- the likely distribution in the food chain;
- the measures which can and should be taken to minimise contamination/cross contamination;
- how these measures are regulated; and
- once an outbreak has occurred, the steps that need to be taken to manage and control it – and the adequacy of the systems and arrangements for that."

"The potential for contamination/cross contamination with the organism, its virulence and the very severe effects it can have on particularly vulnerable groups of the community has been tragically underlined. Of particular significance is the issue of the asymptomatic excretion of the organism, which may have very significant implications in terms of the potential for spread of infection, and outbreak management and control."

[end of text extracts from the Pennington Group Report]

12.1.2 Criminal Proceedings against Mr J. Barr

On 10 January 1997 Mr John Barr of J. Barr and Sons was charged with culpable and reckless conduct arising from the alleged supply of cooked meats in relation to a function at the Cascade Public House in Wishaw. He was then bailed pending the hearing which took place later in October 1997.

OCTOBER 28, 1997 *The Times* **reported:**
"Butcher cleared of world's worst E. coli outbreak.
(From *The Times*, London, 28 October 1997, p. 8, column f. © Times Newspapers Limited, 1997)

The butcher whose shop was linked to the world's worst recorded *E. coli* O157 food poisoning outbreak, in which 20 people died and 400 were infected, walked free from court yesterday after the case against him collapsed. John Barr, 52, of North Lanarkshire, was found not guilty of culpably, wilfully and recklessly supplying cooked meat for an 18th birthday party on November 23 last year, the day after being told not to do so by environmental health officers.

The verdict, delivered by Sheriff Alexander MacPherson on the 6th day of the trial at the Hamilton Sheriff Court was cheered from the public benches by relatives of Mr Barr.

After a week of prosecution evidence, the Sheriff sustained the defence's submission that Mr Barr had no case to answer. George Moore, Mr Barr's solicitor said that the prosecution case hinged on a telephone conversation alleged to have taken place between the butcher and a customer who had collected meat for the party.

David Moon, 66, said that after being told of the health scare he rang Mr Barr. He said he was told by a man whom he believed to have been Mr Barr that the meat was fine. Mr Barr denied the conversation took place, and his solicitor argued that the evidence was not corroborated. After the verdict, a smiling Mr. Barr embraced members of his family before leaving the court by a back door. He refused to comment, but Mr Moore said: "I am pleased at the outcome and pleased for him and his family. I have nothing more to say. There are other matters that are outstanding and there is going to be a full enquiry, hopefully early next year, which will raise a number of important considerations."

Mr Barr, his wife Elaine, and son Martin, partners in the firm John Barr and Sons of Wishaw, face a further trial in January charged with contravening food safety laws."
[end of quote from *The Times*]

12.1.3   John Barr and Sons Fined for Food Hygiene Offences

On January 21, 1998 J. Barr and Sons was fined £2250 for breaches in the food hygiene laws, and Mr. Barr himself cleared of any personal blame in the outbreak.

12.1.4   The Fatal Accident Enquiry (FAI) Chaired by Sir Graham Cox, Reported 20 August 1998

The Crown Office (5 December 1996) announced that an FAI would be held into 21 deaths appearing to arise from the outbreak of infection with *E. coli* O157 in North Lanarkshire, but it could not take place until after any criminal proceedings were completed.

Sheriff Principal Graham Cox reported in August 1998 (Cox, 1998). From the evidence he determined in the case of each death:

- when and where it occurred;
- the cause of death;
- whether it was a result of an accident, *i.e.* accidental ingestion of *E. coli* O157;
- and if it was an accident where the accident took place;
- what the causes(s) of the accident were;
- whether there were reasonable precautions which could have been taken to avoid the accident;
- where *E. coli* was deemed to be the cause whether any defect could be identified in the systems of work in the food chain, and in the treatment of the patient;
- any other relevant facts.

Of the 21 deaths considered he determined that 17 were directly due to the ingestion of food infected with the outbreak strain of *E. coli* O157:H7. In the process of his determination he evaluated the whole food chain – abattoirs, meat wholesalers, retail, enforcement officers role and actions, Government guidance supporting the Meat Product (Hygiene) Regulations 1994, and the actions of individuals involved in the case. The FAI was conducted in three sections – the deaths of people attending the church lunch, the deaths of people in an old people's nursing home and individual cases.

The church lunch

Every year the church arranged a lunch for those members of the congregation who were over 70 years of age. In determining the cause of death of eight people who had attended the church lunch it was necessary to determine how the food they had eaten had become contaminated with *E. coli* O157:H7. This involved the detailed questioning of witnesses to determine precisely what food was delivered for the lunch and how it was handled and cooked. After thorough consideration of the evidence the Sheriff determined that all eight deaths he considered were due to the accidental consumption of *E. coli* O157 present in the stew or gravy supplied by Barr's.

The menu which was served was:

- home-made meat broth
- cooked stew and gravy (from Barr's)
  pastry tops (from Barr's)
  potatoes and vegetables
- ice-cream and fruit salad

On Saturday 16 November Barr's delivered two bags of cooked stew, the pastry tops, and raw meat for the soup. These lay overnight in the church premises, unrefrigerated.

On Sunday 17 November the stew was heated up in the oven of the church hall kitchen and served up to the 87 guests, 45 of whom became ill on or after Wednesday 20 November.

The procedures used to create the lunch are shown in Figure 12.2.

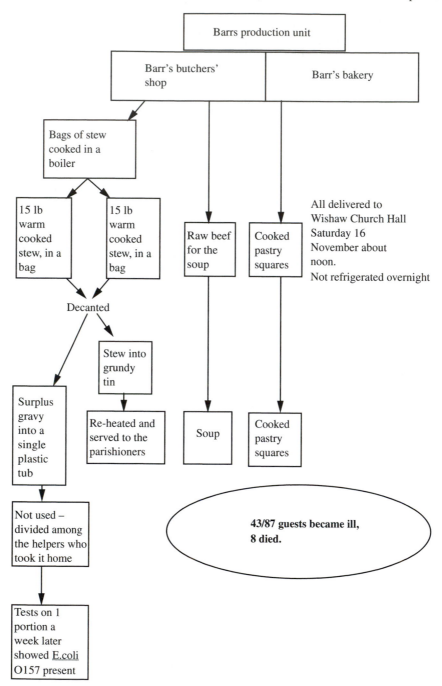

Figure 12.2 *The Wishaw Church lunch, 17 November 1996: production procedures*

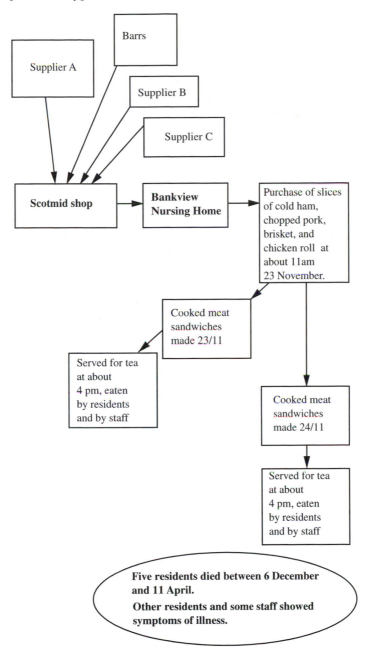

Figure 12.3 *The Bankview Nursing Home outbreak*

The Bankview Nursing Home

At the FAI the Sheriff found that of five deaths of very elderly people, three were directly due to the consumption of meat sandwiches infected with the outbreak strain of *E. coli* O157:H7. Figure 12.3 shows the sequence of purchase of meat which was incriminated in the outbreak, and sandwich making which occurred on 23 and 24 November 1996.

Individual cases

Although some people attending the birthday party at the Cascade Public House became ill, fortunately there were no deaths, so that group of cases did not fall within the consideration of the FAI. However, a quantity of cooked cold meat was supplied to the party and samples were found to be infected with the 'outbreak strain'. Cooked cold meats from Barr's were sold to a number of retail outlets in the Lowlands of Scotland and then either directly infected consumers, or cross-infected other meat products at those outlets leading to infection on consumption. Six deaths occurred in this way.

Having determined the cause of death of each of the affected persons the Sheriff turned to identifying reasonable precautions which might have been taken before and after 22 November – the date on which the authorities had sufficient evidence to associate the outbreak with Barr's meat products.

12.2   BACKGROUND

12.2.1   Recommendations from the Pennington Group Report (1997)*

The Pennington Group were studying the outbreak as it occurred – for it was imperative to identify as far as possible what had gone wrong, and thus to be in a position to make recommendations to the Government which, if implemented would greatly reduce the risk of another such outbreak occurring. The Pennington Group report was comprehensive and the full list of recommendations it made is quoted below. One of the main findings was that appropriate training to give understanding of the causes of food-borne disease, particularly *E. coli* O157, and of food hygiene practices generally needed to be implemented throughout the meat/food chain.

Summary of Recommendations from the Pennington Group Report (1997)*

Chapter 5 – Farms and Livestock

1. There should be an education/awareness programme for farm workers, repeated and updated periodically as appropriate, to ensure that they are aware:
a. of the existence, potential prevalence and nature of *E. coli* O157;
b. of the potential for the spread of infection on farms in a number of ways, including notably from faecal material, and of the consequent need for scrupulous personal hygiene;

*Extracts from the Pennington Group Report (1997) quoted with permission. Crown copyright.

c. of the need for care in the use of untreated slurry or manure; and
d. of the absolute requirement for the presentation of animals in an appropriate, clean condition for slaughter.

2. All of this must be backed up by rigorous enforcement by the Meat Hygiene Service at abattoirs.

Chapter 6 – Slaughterhouses

3. The Meat Hygiene Service should urgently implement its scoring system for clean/dirty animals, should ensure that official veterinary surgeons and the trade are educated and trained in its use, and should pursue consistent and rigorous enforcement.

4. The Meat Hygiene Service must take forward urgently, with the help and support of Government Departments and the industry, the identification and promotion of good practice in slaughterhouses – including specifically in the areas of hide and intestine removal.

5. Abattoir workers should be trained in good hygiene practice during slaughter and the Meat Hygiene Service should concentrate enforcement on slaughter and subsequent handling of carcasses.

6.The Hazard Analysis and Critical Control Point System should be enshrined in the legislation governing slaughterhouses and the transportation of carcasses and meat. Meanwhile, enforcers and the trade should ensure that HACCP principles are observed.

7. The Meat Hygiene Service should be given additional powers to enforce at the abattoir standards for the transportation of meat and carcasses between licensed and non-licensed premises.

8. Further consideration should be given, involving the industry and consumer interests, to the potential use and benefits of end-process treatments such as steam pasteurisation.

9. In line with the approach recommended for more general enforcement, the efforts and resources of the Meat Hygiene Service should be targeted at higher risk premises – especially those abattoirs with hygiene assessment scores of under 65.

Chapter 7 – Meat Production Premises and Butchers' shops

10. HACCP (*i.e.* the approach and all 7 principles) should be adopted by all food businesses to ensure food safety. While this is being negotiated into European Union and domestic legislation, implementation and enforcement of the HACCP principles contained in existing legislation should be accelerated.

11. The Government should seek to have HACCP elements enshrined in the review and consolidation of the Vertical Directives.

12. The Government should seek to have all of the HACCP elements negotiated within the Horizontal Directive.

13. The Government should review the application of the Meat Products (Hygiene) Regulations 1994, and the guidance issued subsequently, to clarify the position regarding which premises are intended to be covered by the regulations.

14. Pending HACCP implementation, selective licensing arrangements for premises not covered by the Meat Products (Hygiene) Regulations 1994 should be introduced by new regulations.

15. The licensing arrangements should include appropriate requirements for the documentation of hazard analysis, labelling and record-keeping to facilitate product recall and temperature control and monitoring. In relation to training, there should be a requirement for all food handlers to have undertaken at least basic food [hygiene] training and for all supervisory staff (and those who run small, one-person operations) to be trained to at least intermediate level. In addition the licence should cover matters relating to the suitability of premises, equipment and hygiene practices to a level equivalent to that required by the 1994 Regulations.

16. In relation to the physical separation requirements of licensing:
a. there should be separation, in storage, production, sale and display, between raw meat and unwrapped cooked meat/meat products and other ready to eat foods. This should include the use of separate refrigerators and production equipment, utensils and wherever possible, staff;
b. Where the use of separate staff cannot be achieved, alternative standards (such as completion and implementation by the operator of a HACCP or the provision and use of additional facilities, *e.g.* for hand washing in the serving area) might be regarded as sufficient to permit the award of a licence;
Where neither a. nor b. can be achieved, the premises concerned should not be permitted to sell both raw and unwrapped cooked meat/cooked meat products (although they may be permitted to sell pre-wrapped cooked/ready to be prepared elsewhere and brought in for that purpose).

Chapter 8 – Point of consumption

17. Food hygiene training should be provided wherever possible within the primary and secondary school curriculum.

18. Guidance and education about food handling and hygiene should be included in all food and catering education and training courses and should be reinforced through periodic advertising and awareness initiatives.

19. Steps should be taken by local authorities to encourage the adoption of HACCP principles in non-registered premises where there is catering for functions for groups of people involving the serving of more than just tea, coffee and confectionery goods.

20. Employers should ensure that food handlers, in particular those working

with vulnerable groups and/or in sensitive areas such as nursing homes and day-care centres, are aware of and implement good hygiene practice. They should be trained in food hygiene at least to the basic and preferably intermediate level.

Chapter 9 – Enforcement

21. The Government should give a clear policy lead on the need for the enforcement of food safety measures and the accelerated implementation of HACCP.

22. The Government and Local Authorities should ensure that there are suitable and adequate Environmental Health Officer skills and resources to address enforcement and education/awareness issues.

23. The Government should consider earmarking local authority funds for these purposes.

24. Local authorities should designate an environmental health officer, with appropriate training, experience and expertise, to head food safety within the authority.

Chapter 10 – Surveillance

25. The Scottish Office Department of Health should take steps to improve the implementation and monitoring of the recommendations of the Advisory Committee on the Microbiological Safety of Food (ACMSF) and this Group on laboratory testing of stool specimens.

26. In discussion with relevant professional groups, a standard case definition and a standard protocol should be agreed for testing and defining clinical cases of infection with *E. coli* O157 and their use promoted in all suspected *E. coli* O157 food poisoning investigations.

27. On completion of investigations, it should be the responsibility of the Consultant in Public Health Medicine (CPHM) to provide the Scottish Centre for Infection and Environmental Health (SCIEH) with a minimum data set (in the form of a standard proforma) for all general outbreaks of infectious intestinal disease, including food poisoning.

28. For large (or otherwise significant) outbreaks a full written report should be completed and consideration given to its publication. Copies of written reports should go to SCIEH.

29. In particular there should be written, and published, a full report of the Central Scotland outbreak.

Chapter 11 – Research

30. Any further proposals for research related to *E. coli* O157 should be subject to

normal processes for funding consideration and peer review, but with appropriate weight given to the threat the organism represents to public health.

Chapter 12 – Handling and control of outbreaks

31. Health Boards and local authorities should ensure that designated medical officers (DMOs) have adequate time and opportunity to contribute, with their environmental health officer colleagues, to the public health activities of local authorities; and they should be expected to report on their work as DMOs at least annually to both relevant Council committees and the Health Board.

32. Local Authorities and Health Boards should ensure availability of adequate numbers of personnel with appropriate skills in public health medicine and environmental health, together with the laboratory facilities and resources they will require to meet their obligations for disease control and environmental health.

[End of extract from Pennington Report]

12.2.2   *E. coli* and *E. coli* O157:H7

The strain of *E. coli* which was causative of the illness in the outbreak was *E. coli* O157:H7 phage type 2 possessing verocytotoxin gene VT2.

Strains of *E. coli* normally occur in the human gut and in the gut of many animals, and are gut commensals not normally associated with causing illness. However, some strains of *E. coli* are toxigenic and have high host specificity, colonising the gut, attaching to the gut mucosa and producing toxin normally causing diarrhoea, but sometimes more serious conditions (Table 12.1).

Most strains of *E. coli* O157 are verocytotoxic – the most serious of the various pathogenic strains of *E. coli*. There are two principal kinds of verocytotoxins, VT1 and VT2, which have similar effects but are immunologically distinguish-

Table 12.1 *Pathogenic strains of E. coli*

Enteroinvasive *E. coli*	EIEC	Causes a dysentery-like illness sometimes with bloody diarrhoea
Enteropathogenic *E. coli*	EPEC	Causes vomiting and diarrhoea in babies
Enterotoxigenic *E. coli*	ETEC	Causes watery diarrhoea and dehydration – 'travellers diarrhoea', and diarrhoea in babies
Verocytotoxin producing *E. coli* (enterohaemorrhagic *E. coli*)	VTEC (EHEC)	Illness ranges from mild to severe diarrhoea, with blood, fever, lasting kidney damage and possibly death. Mortality rate about 5%. Particularly affects young children and the elderly.

Source: DoH, 1994.

able. The main reservoir of *E. coli* O157 seems to be cattle, while many other animals, including sheep, have been shown to carry it. Infected cattle are normally symptomless, although very young calves may be diarrhoeal. The organism is shed in the faeces which provide the main route of dissemination. When animals are in close proximity with one another they may contract the organism from each other and from the general environment, and their skin and fur may become contaminated with infected faeces.

The infection disperses in the farm environment and any faecally affected materials may contain the organism – slurries, surface water, drinking troughs, streams, crops and so on. Thus farm workers are at risk. Since the infective dose of organisms is very low – as small as 100 cells – a very small slip in hygiene could cause infection through the faecal-oral route. One child, who suffered very severe illness but who fortunately recovered from the acute illness, was infected in this way through a school farm visit (*The Independent*, 2001). The Pennington Group expressed concern that farm visits by tourists and school groups should not be stopped but that on-farm hygiene should be considerably improved.

Outbreaks of illness due to the organism can occur as a result of direct person-to-person infection so the hygiene of carers (who may, as other people also may, be asymptomatic carriers and thus sources of the organism) of those who are ill is particularly important. The other important source of the organism is infected food and water.

A survey of outbreaks of VTEC in England and Wales published in February 1996 (note – at a time before the big outbreak described in this chapter) particularly noted undercooked ground beef, unpasteurised milk, and water as sources of the organism – although of course it was not ruling out foods which should have been safe because of their heat treatment but which become contaminated after that process (Wall *et al.*, 1996). Many different food types – meats, milk, cheese, meat in sandwiches, apple juice, burgers, yoghurt and vegetable have been implicated in outbreaks of illness, references to which are listed in Wall *et al.* (1996). Since the organism is heat sensitive – being readily destroyed in the temperature range 54–65 °C, and with D-values similar to those of the majority of strains of *E. coli* its presence is due to food being raw, undercooked or contaminated after cooking. Illness associated with under-cooked meat products – particularly burgers (see Riley *et al.*, 1983) – have led to the publication of cooking guidelines for frozen and unfrozen burgers and other meat products (ACMSF, 1996). *E. coli* O157 does exhibit greater acid resistance than *E. coli*, and survives for significant time periods in acid environments such as apple juice (see Table 9.2), and is able to survive in frozen foods.

The isolation and identification of *E. coli* O157 needs specific techniques. Where it might be expected to occur in high numbers – such as in samples of diarrhoea – then the standard methods of isolating *E. coli* may provide an effective first step and colonies on which more specific tests may be implemented. However, in surveillance of foods where the organism may occur in very low counts yet be significant at those levels (an infective dose of 100 or so cells might be consumed in 100 g of foods), its numbers may have to be enriched prior to identification steps. A specific ELISA test has been developed to detect *E. coli*

O157 following enrichment which allows detection in food within 24 hours (ICMSF, 1996). More specific typing methods:

- phage typing of isolated strains (there are in excess of 80 phage types) and
- DNA subtyping (*e.g.* pulsed field gel electrophoresis)

discriminate further, and provide the clear identity needed in epidemiological investigations (ACMSF, 1995; Scotter, 2000).

12.2.3   Legislative Background to Meat Hygiene (see Figure 12.4)

Meat hygiene from the farm, through the abattoir, distribution, production processes and distribution steps is controlled in the UK through a series of Regulations relating to fresh meat, meat products, and butchers' shops. Changes in the law came about partly as a result of the outbreak described in this chapter where overlap, or gaps in the regulations, or lack of clarity in how they should be implemented were identified. These issues are not discussed in this chapter but were thoroughly evaluated by the Pennington Group and many of the recommendations (see Section 12.2.1) they made relate to inadequancies in the regulatory process.
In brief:

The Fresh Meat (Hygiene and Inspection) Regulations, 1995

The Regulations (Anon, 1995a) authorise the issuing of licences to premises dealing with fresh meat and control veterinary supervision (at abattoirs), and health marking of meat and sale of fresh meat – (which may only occur if it has met the hygiene requirements). These do not apply where fresh meat will be used exclusively for the production of meat products and in a range of other situations.

The Meat Products (Hygiene) Regulations, 1994

The Regulations (Anon, 1994) relate to premises where meat products are being made and require approval by enforcement officers of the premises ensuring that they comply with specified hygienic conditions for handling meat for human consumption, that the products are marked with clear indication of their storage temperature needs and durability under those conditions, and that storage and transportation conditions protect the products from temperature abuse and contamination.

The Food Safety (General Food Hygiene) Regulations, 1995

These Regulations (Anon, 1995b), which apply equally in England, Wales and Scotland, relate to all food businesses and define how the management of food hygiene is to be approached. Regulation 4 particularly is relevant to this chapter. It imposes various obligations on the proprietor of a food business:
   4(1) Requires that specific operations are carried out in a hygienic way.

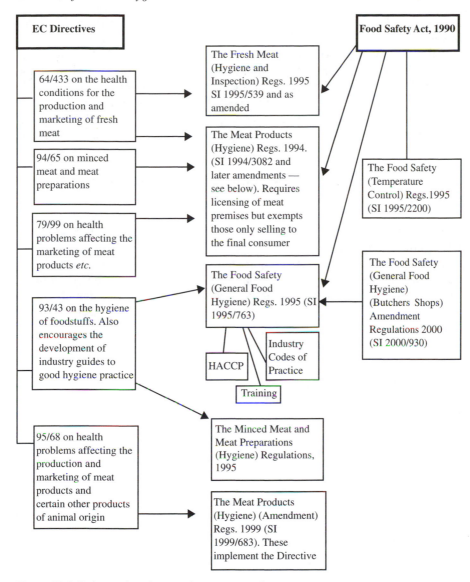

Figure 12.4 *Relevant legislation relating to meat hygiene*

4(2) Obliges the proprietor to comply with the Rules of Hygiene set out in the ten chapters of Schedule 1: Chapters I–III set out the requirements for food premises, rooms where food is prepared, treated or processed; Chapters IV–X cover transportation, requirements as to equipment, food waste, water supply and personal hygiene, and provisions applicable to food stuffs. Additionally Schedule 1 requires that "all food handlers must be trained commensurate with their work activity".

4(3) Requires a proprietor to identify and control food hazards. This is a requirement for proprietors to develop a HACCP system for their particular business.

Additionally Regulation 5 requires food handlers to report infection (actual or suspected) which could result in contamination of food to their employer ("the proprietor"). The proprietor in turn must consider excluding infected people from food handling areas, and in the case of any of a list of named infections further action (such as obtaining medical advice and/or contacting the local enforcement office with a view to pro-active control of any possible incident) is required.

The Food Safety (General Food Hygiene) (Butchers' shops) Amendment Regulations, 2000

These regulations are the result of the implementation of one the recommendations of the Pennington Group.

The Regulations (Anon, 2000) require butchers' shops which handle both raw meats and ready-to-eat cooked meats and cooked foods to be licensed. In order to obtain the licence (which needs annual renewal), the shop must demonstrate a number of qualities and actions which give confidence in their hygiene management:

* good hygienic practice;
* good temperature control of foods;
* the staff must be trained in the basic elements of food hygiene, and supervisors must have greater than basic depth of knowledge;
* HACCP procedures must be in place, and documented;
* records of the HACCP procedures and of staff training must be kept.

See also Chapter 20, which concerns the assessment of hygiene in abattoirs and the development of 'HAS' (hygiene assessment scores), in order to improve hygiene in these environments.

12.3 EXERCISES

Exercise 1. Enquiries had established that the stew, the raw meat for the soup and the pastry tops for the Wishaw Church lunch were supplied by Barr's. The Scotmid store had supplied the cooked meats for sandwiches at the Bankview Nursing Home. Putting yourself in the position of a senior investigator, and using information supplied in Sections 12.1 and 12.2 raise a series of questions designed to explore the possible ways the outbreak strain of *E. coli* O157 came to be present in stew and gravy eaten at the Wishaw Church Hall lunch and at the Bankview Nursing Home.

Exercise 2. The questions you raise in Exercise 1 should then enable you to identify reasonable precautions which could have been taken to prevent the outbreaks.

Exercise 3. Sir Graham Cox (the Sheriff who undertook the Fatal Accident Enquiry) and the Pennington Group were in agreement about the need for appropriate hygiene training at many points in the meat supply chain as exemplified by Barr's complex meat distribution arrangements. Clarify where you consider those points to be, and

- identify one or two examples of specific sorts of training and what they seek to achieve;
- consider what is needed to ensure that training knowledge is implemented;
- determine how the training aims of maintaining good hygiene can be verified.

12.4   COMMENTARY

Exercise 1. Enquiries had established that the stew, the raw meat for the soup and the pastry tops for the Wishaw Church lunch were supplied by Barr's. The Scotmid store had supplied the cooked meats for sandwiches at the Bankview Nursing Home. Putting yourself in the position of a senior investigator, and using information supplied in Sections 12.1 and 12.2 raise a series of questions designed to explore the possible ways the outbreak strain of *E. coli* O157 came to be present in stew and gravy eaten at the Wishaw Church Hall lunch, and at the Bankview Nursing Home.

Wishaw Church Hall lunch

Information in Sections 12.1.1 and 12.1.4 told you that the FAI established that the outbreak strain of *E. coli* O157 was present in the stew and gravy supplied by Barr's. Remember that a major source of *E. coli* O157 is meat, but it can be transmitted by water or through infected milk and other food materials.

Two hypotheses present themselves:

• the foods were delivered contaminated to the Church Hall;
• the foods became contaminated at the Church Hall.

In exploring hypotheses 1 and 2

To conclude that the foods were already contaminated with the outbreak strain before they were delivered to the Church Hall you need to rule out contamination after delivery (use Figure 12.2). So explore how the foods may have been handled after delivery, and through your explorations identify ways in which they could have become contaminated. Then determine how opportunity would be provided for the *E. coli* O157 to survive, grow and be mixed throughout the stew. You will notice that very little detail is provided in either Figure 12.2 or Figure 12.3 – it is for you to ask the questions about time, locations, temperatures, people, equipment use and so on.

For example, delivery to the Church Hall:

• what was delivered?
• when?
• by whom?
• how were the foods packaged?
• what was their condition?
• could one item contaminate the other items?

storage:

• how, and at what temperature were the foods stored?
• how long were they stored?

- could they/did they come into contact with each other?
- were they opened or sealed?

menu preparation:

- how was the soup from the raw meat prepared? (expect details of utensils, equipment, surfaces)
- what was the cooking method?
- when was the meat stew decanted from the gravy?
- how was it decanted?
- when and how was it re-heated?
- when were the guests served?

At the FAI the Sheriff established to his satisfaction that it was more probable that the meat stew was infected before delivery. You are not in a position to come to such a conclusion but you should consider what evidence would be needed to have confidence that the products were delivered to the Church Hall in a safe condition.

Raw meat and cooked stew (as well as pastry tops) were delivered to the Church Hall.

- prior to delivery could the raw meat (or other raw meat) have had the opportunity to contaminate the cooked stew?
- what would provide that opportunity?
- was the stew made from contaminated meat?
- how was cooking undertaken?
- how certain could the shop be that the stew was thoroughly cooked?

At the FAI the staff at Barr's were thoroughly questioned and their evidence was not always clear, and often contradictory. It became evident that there were no defined procedures, little was written down – and although staff thought they did the same thing each time they cooked meat in the boiler no account seemed to be taken of varying sizes of pieces of meat, of what the water temperature was, or core temperatures of cooked meats, or how long joints or bags of stew mixture were cooked for. There could be no confidence that the process would eliminate pathogens such as *E. coli* O157. Additionally you know (from Section 12.1.1) that the design of the premises was such that separation of raw meat from cooked meat was difficult – and therefore cross contamination likely.

On the evidence the Sheriff determined that on the balance of probabilities the cooked meat was delivered warm to the Church Hall already contaminated with *E. coli* O157. Where contamination occurred it was not possible to say. Whether infected meat was used which was inadequately cooked, or whether cooked stew was cross-contaminated after cooking, or whether the outsides of the bags were contamined with raw infected material which somehow entered the stew prior to use could not be concluded. But the inability of the Barr staff to convince the Sheriff of the reliability of the cooking process was the major factor in his view that probably the cooking was inadequate.

The outbreak strain of organism was found on various surfaces in Barr's premises as well as in the gravy, and in stools samples from infected persons affirming the probability that products from Barr's might carry the organism. Surveillance of raw meat products and butchers premises in the UK continued after this outbreak and in one survey in 1998 *E. coli* O157:H7 was isolated from 0.4% (5 out of 1400) of premises (Little and de Louvois, 1998).

Bankview Nursing Home

Figure 12.3 provides you with a flow chart for the meat sandwiches which the elderly residents at the nursing home ate. However, it does not immediately provide you with information about how that meat became contaminated with the outbreak strain. Your questions should focus on:

- was it known which types of meat sandwiches were eaten by which residents, and which staff?
- was one sort of meat or all meats incriminated?
- how was the meat handled by the purchaser?
- could it have become contaminated after purchase? Was it – for example – in contact with raw meat after purchase? If that happened would the contamination load be enough to affect all the meat eaten by the old people? Could the organism increase in the sandwiches? What is the infective dose? Was there opportunity for increase?

Then turning to the shop and exploring whether the meat could have become contaminated prior to purchase:

- how was the cooked meat displayed at the shop?
- was the cooked meat in slices when displayed, or in large pieces from which slices would be cut?
- what was used to slice the cooked meat?
- when and how was this cleaned?
- how were the types of cooked meat displayed?
- what cooked meats did the different suppliers, including Barr's, sell to the Scotmid shop?
- how were these managed?
- did the shop sell raw meats?
- who weighed the cooked meat for the purchaser?
- what were their duties?

The FAI determined that the cooked meats eaten by the Bankview residents was very likely not Barr's meat, but was meat from other suppliers which had become contaminated by contact with Barr's cooked meat (carrying the outbreak strain of *E. coli* O157). It was probable that Barr's cooked meat had been contaminated in Barr's shop or production unit. The Scotmid shop used counter techniques normal at the time of displaying meats from various sources on a single common tray, sliced on the same slicer. It was not practice, and not at that time expected by food law enforcement officers, that the slicer would be sanitised between the slicing of different types and makes of cooked meats.

Thus cross-contamination between cooked meats spread the organism.

Exercise 2. The questions you raise in Exercise 1 should then enable you to identify reasonable precautions which could have been taken to prevent the outbreaks.

Depending on your questions you will have to explore the nature of reasonable precautions to prevent the outbreak which could have been taken by the various parties you have identified. However, the principles will be simple:

- prevent occurrence of the organism;
- prevent spread by contact and other means;
- prevent growth;
- if possible kill the organism (through appropriate sanitisation schemes, or through properly controlled cooking);
- prevent re-contamination and cross-contamination.

At the FAE the Sheriff considered the food chain and the areas he identified are shown on the left-hand side of Table 12.2. You may wish to analyse each section of that Table in detail and determine what is needed in practice to prevent dissemination, survival and growth of the organism.

Exercise 3. The Sheriff and the Pennington Group were in agreement about the need for appropriate hygiene training at many points in the meat supply chain as exemplified by Barr's complex meat distribution arrangements. Clarify where you consider those points to be, identify one or two examples of specific sorts of training and their aims, consider what is needed to ensure that training knowledge is implemented, and determine how the training aims of maintaining good hygiene can be verified.

Exercises 1 and 2 should have permitted you to think about where contamination originates and how it may be minimised.

The Pennington Group identified the need for good hygiene training throughout the food chain from 'farm to fork' (Section 12.2.1).

Evidence in the FAI certainly indicated that although the Barr's shop appeared to be clean, staff had not been instructed in food hygiene, and had not been instructed in cleaning procedures appropriate to different pieces of equipment, nor in the difference between detergents and sanitisers. In fact cross-contamination was a problem, with staff crossing over between the 'raw' meat and the 'cooked' meat sides, and equipment being used for raw one minute, cooked the next with little or no sanitisation between. The fact and sheer luck that no outbreak of food poisoning associated with such practices was known to have occurred did not validate the existing practices.

Perhaps the most difficult part of this exercise is to be clear about what hygiene training seeks to achieve. Training can be focused on how to do a task or set of

Table 12.2 *Analysis from the FAE: reasonable precautions and actions which could have been taken before and after 22 November 1996*

ACTIONS POSSIBLE BEFORE *4.00pm 22 November 1996*	*ACTIONS RECOMMENDED AFTER* *4.00pm 22 November 1996*
Points at which reasonable precautions which, if taken might have prevented contaminated meat arriving in Barr's premises	**Actions – to trace and keep under competent medical surveillance those who exhibited symptoms consistent with infection by:**
The farmer	Authorities – official microbiological analysts, Lanarkshire Health Board
The abattoir	
The meat wholesaler	EH Department of N. Lanarkshire Council
Reasonable precautions which if Barr's had taken them might have prevented contaminated cooked meats products leaving their premises:	**Actions to stop members of the public eating food which authorities had reason to believe may have been unsafe by:**
Cooking methods	Authorities – official microbiological analysts, Lanarkshire Health Board
Workflows within the premises	EH Department of N. Lanarkshire Council
The cleaning of the premises	
Registration under the Meat Products (Hygiene) Regulations 1994	Staff at Barr's shop
The reasonable precautions which the enforcement agency could have taken in relation to:	
The quality and frequency of their inspections	
Enforcing compliance with the regulations appropriate to Barr's premises	

tasks, but should also include the reasons behind using those methods. At the most basic level training seeks to instruct the recipient, and a measure of the success of training could be that the trainee could retain that information at the end of the day of instruction, or a week, month or year later after being tested. But the purpose of hygiene training is not just to award certificates but to enable people to do their jobs properly and implement what they have learned. It is important that the management values the need for hygiene training and facilitates the implementation of what is learned by providing appropriate equipment, record charts, enough time and space to undertake the tasks. The type of training needed is also for management to determine. Under the Food Safety (General Food Hygiene Regulations) 1995, the implementation of HACCP or similar control system is required in all food premises. HACCP requires assessment of risks. Unless a proprietor is aware of what the microbiological hazards may be and the risk of their occurrence and takes action to reduce the risk in those premises (and has perhaps undertaken appropriate training in order to gain that awareness), outbreaks of food poisoning like the one described in this chapter may occur.

Thus:

- training is to both to instruct and to provide a greater level of understanding;
- training must be backed by systems which allow implementation of what is learned;
- management must check that implementation never waivers;
- microbiological monitoring verifies that cleaning regimes achieve their goals.

The outcome of the outbreak described in Section 12.1.1 was that many of the recommendations of the Pennington Group report were accepted by the Government. In 2000 new regulations* amending the Food Safety (General Food Hygiene) Regulations, 1995 which relate particularly to butchers' shops were brought in. These require the licensing of premises not covered by other regulations. In effect they require butchers to understand the risks associated with raw meat, and to ensure that if they sell unwrapped ready-to-eat high risk foods that these are effectively separated from all raw meat and the risk of cross-contamination from that source. Indeed the risks in the whole shop now have to be managed through an active, written HACCP system which the butcher must be able to demonstrate in order to gain the necessary licence to trade (Anon, 1999; Morrison, 1999).

12.5 SUMMARY

This case study illustrates how seriously food hygiene – at all steps in the food chain – needs to be taken. 496 people became ill, 17 of whom died as a direct result of eating cooked meat products infected with verocytotoxic *E. coli* O157:H7 and which must have entered the shop with one or more meat carcases.

Through the various investigations, particularly those of the Pennington Group, and the Fatal Accident Enquiry, many lessons were learned from it, not least that very fundamental hygiene procedures, whose value has been recognised for decades, need to be taken very seriously. Because meat is a source of food poisoning organisms cross-contamination between raw meats and cooked meats must be avoided. Implementing a whole range of food hygiene strategies is about managing and reducing that risk, among which food hygiene training for all staff is a major strand.

This outbreak also illustrates very clearly the substantial damaging conse-

*The Food Safety (General Food Hygiene) (Butchers' shops) Amendment Regulations 2000 (SI 2000/930).

quences of hygiene failure not only through the tragic deaths and illness of those infected, but also to business viability as well.

12.6 REFERENCES

ACMSF (Advisory Committee on the Microbiological Safety of Food), 1995. Report on verocytotoxin-producing *Escherichia coli*, HMSO, London, UK.

ACMSF, 1996. Advisory Committee on the Microbiological Safety of Food Report on verocytotoxin-producing *Escherichia coli.*, HMSO, London, UK.

Anon, 1994. Management of outbreaks of foodborne illness. Department of Health.

Anon, 1994. The Meat Products (Hygiene) Regulations, 1994.

Anon, 1995. The Fresh Meat (Hygiene and Inspection) Regulations, 1995.

Anon, 1995. The Food Safety (General Food Hygiene) Regulations, 1995.

Anon, 1997. *The Times* newspaper, 28 October 1997, page 8, column f.

Anon, 1999. Bradford butcher awarded 'first' HACCP accreditation, *Environmental Health News*, **16** (6), 1.

Anon, 2000. The Food Safety (General Food Hygiene) (Butchers' Shops) Amendment Regulations, 2000 (SI 2000/930).

Anon, 2001. 'Boy victim of *E. coli* is awarded £2.6 million damages', *Independent* newspaper, London, 23 January.

Cox, Sir Graham, (Sheriff Principal of Sheriffdom of South Strathclyde, Dumfries and Galloway) August 1998. Determination into the *E. coli* O157 Fatal Accident Enquiry. Available from the Sheriff Clerks Office, Sheriff Court House, Beckford Street, Hamilton, M1 6AA. Telephone 01698 282957, Price £31.

Department of Health, UK., 1994. *Management of outbreaks of food borne disease.* A booklet published by Two Ten Communications, West Yorkshire, UK.

ICMSF, 1996. *Micro-organisms in Foods.* Volume 5. *Characteristics of Microbial Pathogens.* Blackie Academic and Professional, London.

Little, C.L. and de Louvois, J., 1998. The microbiological examination of butchery products and butchers' premises in the United Kingdom. *Journal of Applied Microbiology*, **85**, 177–186.

Morrison, Y., 1999. Who cares about HACCP? *Environmental Health Journal*, **107** (3), 76–78.

Pennington Group Report, The, April 1997. Report on the circumstances leading to the 1996 outbreak of infection with *E. coli* O157 in Central Scotland. The implications for food safety and the lessons to be learned. HMSO (Edinburgh), April 1997.

Riley, L.W., Remis, R.S., Helgerson, S.D., McGee, H.B., Wells, J.G., Davis, B.R. *et al.*, 1983. Haemorrhagic colitis associated with a rare *Escherichia coli* serotype. *New England Journal of Medicine*, **308**, 681–685.

Scotter, S., Aldridge, M. and Capps, K., 2000. Validation of a method for the detection of *E. coli* O157:H7 in foods. *Food Control*, **11**, 85–95.

Wall, P.G., McDonnell, R.J., Adak, G.K., Cheasty, T., Smith, H.R. and Rowe, B., 1996. General outbreaks of verocytotoxin producing *Escherichia coli* O157 in England and Wales from 1992 to 1994. *Communicable Disease Report*, **6** (Review no. 2), R26–R33.

A Shelf Life Problem

Key issues
- Chilled desserts and bottled vegetables
- Psychrotrophic hazards and spoilage organisms
- Perception of risk
- Shelf life – safety and quality issues

Challenge
These case studies concern products made on a small scale for commercial sale.
Determining the shelf life of a product is a challenge.

13.1 THE CASE STUDIES AND BACKGROUND: PROCESSED VEGETABLES AND CHILLED DESSERTS

Moving from domestic cooking to commercial manufacture.

13.1.1 The Shelf Life of Bottled Vegetables under Ambient Storage

A food industry consultant talking:

"I once – several years ago – had occasion to visit a small food manufacturer who made North African pickled vegetables in a small business in London. The business was based on using the home recipes of the boss's mother. She had a reputation for the food she cooked and served to the family and their friends. The boss's idea was that these delicious products could be translated into commercial items, and sold to a selective, sophisticated market sector in London. The main products were pickled sweet peppers – capsicums. At home the peppers had been washed and prepared by removal of the hot seeds, and then baked in the oven. Then the cooked peppers were skinned and sliced, and packed into jars to which was then added spiced olive oil, and the filled jars baked a second time in the oven, after which the lids were quickly added and screwed down and the jars allowed to cool down.

When I visited them the company were starting to make similar bottled products, taking as great care as possible to run a hygienic process in a small production unit in London. It was a two man business – one man worked on production, the other (the boss) went out and about and generated the sales.

The entire production space was a rectangle about 8 m by 5 m, and all processes were carried out in there. In came the raw vegetables, and the 5 kg sacks of spices, the 10l drums of olive oil, the people; out went the bottled vegetables, the repacked dry spices in consumer quantities of 100 g, and the re-bottled oils, for it was the latter two products which provided basic income. A number of famous food retailers had taken a few jars of the pickled peppers (not very many because they had to be kept refrigerated), and the retailers were not at that stage prepared to dedicate much refrigeration space to them.

What the boss wanted was to convert the chilled product, with the two week shelf life he had given it, to an ambient stable product which would keep for perhaps two years. This would enable him to break into an entirely different market and not be dependent on refrigerator space, either in his factory or at the retailers. To test their shelf life at ambient he had tried keeping a few jars of the product on the shelf, and he told me 'one or two of them always fizz after a few days'. His problem was that he did not really know how to achieve the long shelf life he wanted, he only had a small business, and he had the premises he had."

Official warnings were issued (1998) about imported bottles of vegetable soup packed in glass jars and sold in Italy, and brought back to the UK by tourists following botulism in one person who had eaten those products (Anon, 1998).

13.1.2   The Shelf Life of Chilled Chocolate Dessert

In Case Study 8, in Section 8.3, Exercise 2, a cook who wanted to sell her 'home-made' chocolate dessert was described. She had a similar problem to the producer described in Section 13.1.1. She needed to predict the shelf life of her product – indeed she had decided to label each pot "Use by . . ." and "keep at 1–5 °C' because she noticed chocolate desserts in the chilled food section of her local supermarket were so labelled. She added about two weeks from the date of production which suited her distribution capability, and her developing market. She was quite confident that was OK, because the few pots that she had kept back in her refrigerator for two weeks seemed to be perfectly satisfactory.

13.2   EXERCISE

Decide for yourself how each of the two small food manufacturers should determine the microbiological shelf life of their products, and what they, or their advisor (you?) need to think about before a satisfactory shelf life can be achieved. Do you think they perceive any risks associated with storing these foods?

13.3 COMMENTARY

> **Exercise.** Decide for yourself how each of the two small food manufacturers should determine the microbiological shelf life of their products, and what they, or their advisor (you?) need to think about before a satisfactory shelf life can be achieved. Do you think they perceive any risks associated with storing these foods?

When production is moved from a normal domestic scale to a larger scale (family weddings and so on) or even to a commercial one problems can arise – sometimes due to strains on equipment and space resources (Anon, 1974), sometimes due to lack of appreciation of the technical demands of commercial production (Ryan *et al.*, 1996) and lack of perception of risk.

In thinking about the microbiological shelf life of products there are two main factors: safety, then quality.

Shelf life

Shelf life is the time period that a product remains safe, and of acceptable quality when stored under specific, defined conditions.

Ambient stable products

Ambient stable products need either to be commercially sterile, that is having received a sterilising process and/or of a composition that after processing prevents the growth of the residual flora – for example, possessing very low water activity, low pH, very high sugar content, controlling levels of preservatives (see Case Study 9). But it is not just commercial sterility in a single pot that is needed, but in the vast majority of pots, with failure rates acceptably low. This is not necessarily easy to achieve – as the potted vegetable producer said "one or two of them always fizz after a few days". That product – as currently processed – is not a product which is suitable for ambient storage.

Many of the issues relating to ambient stable vegetable products were discussed in Case Study 7. Of prime importance is the need not to kill anyone by allowing the survival and growth of *C. botulinum* in the pickled vegetables.

If the pH of the bottled vegetables is above 4.5, to be sure that the product will not contain viable spores of *C. botulinum* the process must be reliable. A small producer must not bottle pots in any manner which cannot be precisely monitored – rate of temperature rise, temperature values achieved, pressure, temperature region of slowest temperature rise in the retort, duration of exposure . . . If a small producer does not have the equipment or the knowledge then they should not be processing vegetables for ambient storage. It is too dangerous, and unreliable processing will also mean that some containers go off in storage, which is not good for trade. Risk must be perceived, understood and managed.

In a low acid vegetable product (*i.e.* above pH 4.5) when appropriate processing does destroy the contained population of heat resistant organisms, the shelf life of the product is then defined by other factors – development of undesirable sediment, colour change (fading), development of undesirable changes unrelated to microbial growth (see Man and Jones, 2000).

Chilled products

A product like chocolate dessert has the potential to support the growth of a whole range of pathogens and spoilage organisms because it is a high moisture, high water activity, near neutral pH, nutrient product. The production process will dictate how many organisms are present in it when packed, and the temperature of storage will control their rates of growth.

Of primary concern is the safety of the product. As discussed in Case Study 8, pathogens could enter in the process, and in the absence of an appropriate heating step survive, or enter as post-processing contaminants.

Thus after consideration of how the process may reduce, eliminate or control the pathogenic population which could enter or remain in the product is the need to consider the temperature control of the product. If it (meaning every pot of it) can truly be cooled to less than 5 °C within 60 minutes of pot filling, then only those pathogens which can grow between 0 and 5 °C are of concern. But even pathogens which can grow in this temperature range do not grow at their fastest rates and only become important if sufficient time passes for their numbers to rise to a point when the infective dose might be consumed. Since growth rates are greater at higher temperatures the next question is what level of assurance is there that the product will remain between 0 and 5 °C until it is eaten.

There is a difference between what someone imagines happens, and what actually happens. Take the entrepreneurial cook (Section 13.1.2 and Section 8.3, Exercise 2). She is intending to put her desserts in the boot of her car. Quite apart from it being a totally inappropriate food storage and transportation space and its lack of hygiene, this is not refrigerated space. What temperature could the desserts rise to while in there? How long are the journeys? One of the reasons she is contemplating such a delivery method is that she feels it does not matter – she says to herself "it's only a quick journey" – she does not know whether it matters of not, but one thing is certain, she has never taken any measurements of temperature. Equally she probably has no idea of how warm the desserts will get if they are just taken out of the refrigerator, and put on the work top for a while (Figure 13.1). She will have no idea how long they take to cool down in the refrigerator. She will have no inkling whether the loading of the refrigerator affects the cooling rate of the desserts or what the temperature of the air in the refrigerator is. Does she have any perception of the risk associated with organisms growing in the food? Equally when the product is delivered to the shops, how reliable are the temperatures of the retail refrigerated display cabinets? How well do they control the temperature of the goods contained in them? In a small *ad hoc* survey we did in Southwark, London in 1986,* we found that the

* Acknowledgement to Environmental Health Officers in the London Borough of Southwark for taking the measurements for me.

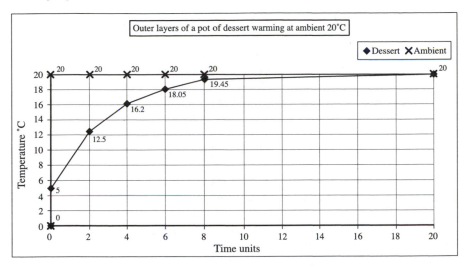

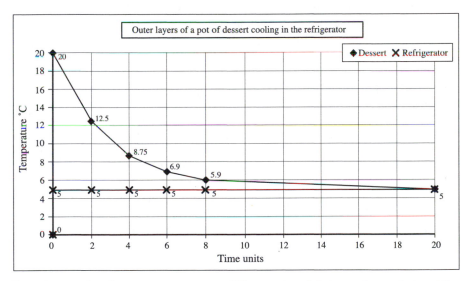

It can be seen that the greater temperature difference there is between the product and the surrounding air temperature the more rapidly the product temperature changes. It means that when products of ambient temperature are put into the refrigerator they will take a long time to reach and stabilise to the refrigerator temperature. How long that is depends on the heat transfer characteristics of the packaging and the product.

Figure 13.1 *How products warm and cool towards the surrounding temperature*

temperatures of 19 products in 19 separate refrigerators in shops had a mean temperature of 13 °C (range 5–20 °C), and the shop keepers were very surprised when told, indicating that they had not thought of measuring any temperatures and, at that time at least, refrigeration in shops was not to be relied upon. These are some examples of the variables in the storage history which will affect the actual temperature of the desserts, and thus the rate at which the contained microbial populations grows (Figure 13.2).

In terms of product safety therefore the temperature control of the product is paramount and under current food law the cold chain must be secure (Anon, 1995). Failures in temperature control give opportunity to contained pathogens to grow, so the cold chain must be well managed (see CIEH, 1997).

But assuming the product is chilled rapidly after manufacture and the chilled temperature control can be maintained with a high level of assurance – the cold chain is secure – it is then the growth of spoilage organisms which will dictate the product's shelf life.

In the commercial production of desserts very low levels of organisms which contaminate the plant can be the ones which, transferred to the product at perhaps the filling stage, grow in the product. For example *Pseudomonas* spp. are frequent plant contaminants which are psychrotrophic and can grow below 5 °C (as well as at faster rates above it).

Pseudomonads, present in high enough numbers could cause the product to change in flavour and appearance. But while Pseudomonads could be a problem in one dessert product, strains of *Bacillus* could be a problem in another product – some of them can grow under refrigeration – while moulds and yeasts contamination could be a problem a third.

What levels of these organisms can be tolerated? Work by Griffiths and Phillips (1986) showed that for some pasteurised products – such as cream – initial product counts of less than two organisms per 100 ml of product were necessary to achieve a shelf life of greater than 10 days at 6 °C. They defined end of shelf life as when the psychrotrophic bacterial count had reached $10^7 \, g^{-1}$, a point at which organoleptic change became evident. Thus in 10 days at 6 °C a population of two cells/100 ml could develop and become 10^9 cells ($10^7 \, ml^{-1}$). Clearly therefore the shelf life period is inversely related to the initial concentrations of the contaminating cells. Chocolate dessert is often sold in individual pots of about 120 ml. Were the growth characteristics to be similar in chocolate desserts and in cream (an assumption which should be viewed with caution) it would mean that no more than two contaminating organisms per pot could be tolerated. Could that level of hygiene be achieved by the cook (Section 13.1.2)? Even if she could do this, the Griffiths and Phillips (1986) data indicate the products only keep for 10 days and not the 14 days she desired – even if the temperature was controlled at 6 °C. To achieve the 14 days she either needs a much higher level of hygiene (cleaner product), together with processing which eliminates heat sensitive organisms, or temperature control at a much lower storage temperature – see Figure 13.2. Additionally in her circumstances she needs to remember that her products could be spoiled by contamination by airborne mould spores or yeasts which would grow slowly in the pots of product. Moulds would form colonies on the tops of occasional pots and could be evident during refrigerated storage. The air quality also needs to be controlled to ensure that products are not contaminated.

Thus in setting the shelf life of a product such as chocolate dessert the nature of the contaminating flora and their probable numbers need to be known, and their

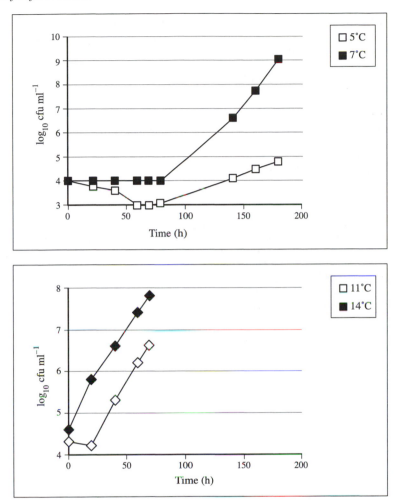

Note: with the rate of growth shown in exponential phase, together with the period of lag, if the initial population was close to 2/100 ml, at 5 °C the spoilage threshold of a population of 10^7 cfu ml^{-1} (approximately 9 log cycle increase) would be achieved in 450–500 h (21 days) of storage. At higher storage temperature the spoilage threshold would be reached sooner.

Figure 13.2 *Growth of Pseudomonas fluorescens* in chocolate dessert product*

growth rates under different temperature regimes measured, observed or predicted. At different temperatures different organisms tend to become the dominant flora. In chilled cream, free of heat sensitive organisms such pseudomonads and coliforms, the Gram positive survivors of pasteurisation – such as *Bacillus cereus* – will spoil it eventually giving rise to a bitter product (see also Case Study 21). But where present pseudomonads will tend to grow faster than bacilli and

* *Ps. fluorescens* strain 3iv – an isolate from commercial dessert product
Source: Pawsey, R.K., unpublished data.

Table 13.1 *Guideline criteria for milk, cream and dairy products*

Product examples: Liquid milk, cream, cheese, icecream, butter, dairy desserts, yoghurts and other fermented products
Storage: Frozen or chilled
Use: Ready to eat

Pathogens

Salmonella, Campylobacter, VTEC, *Listeria* and *S.aureus* will occasionally be present in raw milk. Other pathogens and parasites may occur. Vegetative pathogens will be absent in properly pasteurised milk and cream; testing for *Salmonella* and *Listeria* is not conducted routinely in these products. Low levels of spores and other thermoduric organisms may occasionally be present.

Organism	GMP	Maximum
Salmonella spp.	ND in 25 ml or g	ND in 25 ml or g
Listeria monocytogenes	ND in 25 ml or g	10^3cfu ml^{-1} or g^{-1}.
S. aureus	<20 cfu ml^{-1} or g^{-1}	10^3cfu ml^{-1} or g^{-1}
VTEC	ND in 25 ml or g	ND in 25 ml or g
(raw milk based products)		

Indicators and spoilage organisms

Criteria applicable depend on the level of heat treatment applied. Coliforms/Enterobacteriaceae and *E. coli* are most commonly used as indicators of the quality and hygienic processing of dairy products. APC may be useful for trend analysis but is not applicable to live fermented products.

Products	Organism	GMP cfu ml^{-1} or g^{-1}	Maximum cfu ml^{-1} or g^{-1}
Soft cheese (raw milk)	*E. coli*	<10^2	10^4
Processed cheese	Aerobic plate count	<10^2	10^5
	Anaerobic plate count	<10	10^5
Other cheeses	Coliforms/ Enterobacteriaceae	<10	10^4
	E. coli	<10	10^3
Pasteurised milk and cream	Coliforms/ Enterobacteriaceae	<1	10^2
Other pasteurised milk	Coliforms/ Enterobacteriaceae	<10	10^4
products	*E. coli*	<10	10^3
	Yeasts (yoghurt)	<10	10^6

Key: ND = not detectable; GMP = good manufacturing practice.

Reminder: Indiscriminate application of microbiological testing and criteria is not advocated and should be avoided. It may not be necessary to apply all tests listed in a product category and on some occasions it may be relevant to test for additional micro-organisms and/or toxins. **GMP** values are those expected immediately following production of the food under good manufacturing conditions. **Maximum** values are those regarded as the maximum acceptable at any point in the shelf life of the product. When applying microbiological tests and criteria all elements of the microbiological criterion including the sampling plan must be clearly specified.

will be the determinants of the end of shelf life. Pseudomonads and Enterobacteriaceae enter dairy products – creams, milks, desserts – from the processing plant and its environment. But other organisms can also enter product from the plant. In July 2000 *Staphylococcus aureus* caused over 12 000 people to suffer food poisoning due to consuming infected pasteurised milk products, the source of contamination being an uncleaned and colonised section of pipe in the milk processing plant followed by growth in the product (Anon, 2000).

So a chilled product like chocolate dessert, when made by a small operator with little access to technology, must be made from good quality ingredients, preferably heat processed to eliminate any heat sensitive pathogens, handled hygienically so that there is minimal post-processing contamination, chilled rapidly after being packaged, held between 0 and 5 °C and consumed within 24 hours. To establish a shelf life longer than this relies not only on the above factors being implemented within a well designed HACCP system, but also through microbiological tests which give guidance to the type of initial microbial load the product may carry and how that load, which could include psychrotrophic pathogens, could increase in storage. Guideline quality standards at the point of production have been published for such products (IFST, 1997) and also for such products at the point of sale (Table 11.10) (PHLS, 2000). It will be noted that these do not present criteria for spoilage organisms such as pseudomonads, but rather give indications of what the maxima for the aerobic colony count should be, and for organisms which indicate contamination or post-processing mishandling and/or thermal abuse – the Enterobacteriaceae and *E. coli* (see Table 13.1).

To summarise, risk must be perceived, understood and managed.

13.4 REFERENCES

Anon, 1974. *Salmonella agona* food poisoning. *British Medical Journal*, **3**, 477.

Anon, 1995. The Food Safety (Temperature Control) Regulations, 1995. HMSO, London, UK.

Anon, 1998. 'Botulism warning' in *Environmental Health News*, p. 3, issue dated 4 September 1998.

Anon, 2000. 'Bad milk leaves 12,000 sick in Japan'. Associated Press/Reuters Press release, July 6, 2000.

CIEH, 1997. *Industry Guide to Good Practice*. Chadwick House Group Ltd., London, UK.

Griffiths, M.W. and Phillips, J.D., 1986. Prediction of the shelf-life of pasteurized milk at different storage temperatures. *Journal of Applied Bacteriology*, **65**, 269–278.

Hatt, B. and Wilbey, A., 1994. Temperature control in the food chain. *Journal of the Society of Dairy Technology*, **47** (3), 77–80.

IFST, 1997. Development and use of microbiological criteria for foods. *Food Science and Technology Today*, **11** (3), 160.

Man, D.C. and Jones, A.A., 2000. *Shelf Life Evaluation of Foods*, 2nd Edition. Aspen Publishers Inc., MD, USA.

PHLS, 2000. Gilbert, R.J., de Louvois, J., Donovan, T. *et al.* – Working group of the PHLS Advisory Committee for Food and Diary Products. 'Guidelines for the microbiological quality some ready-to-eat foods sampled at point of sale'. *Communicable Disease and Public Health*, **3** (3), 163–167.

Ryan, M.J., Wall, P.G., Gilbert, R.J., Griffen, M. and Rowe, B., (1996). Risk factors for outbreaks of infectious intestinal disease linked to domestic catering. *Communicable Disease Report*, **6** (Review 13), 6 December, R179–R183.

Additional useful hygiene references

Scott, E. and Bloomfield, S.F., 1990. The survival and transfer of microbial contamination via cloths, hands and utensils. *Journal of Applied Bacteriology*, **68**, 271–278.

Tebbutt, G.M., 1991. An assessment of cleaning and sampling methods for food-contact surfaces in premises preparing and selling high risk foods. *Epidemiology and Infection*, **106**, 319–327.

CHAPTER 14

Airline Food and Control Failure

Key issues
- Cook-chill foods and airline foods
- *Staphylococcus aureus*
- Global dissemination of pathogens by air travel
- Risk factors and high and low risk foods
- HACCP

Challenge

This case study concerns the safety not only of the airline food, but also of the aircraft itself – for if the pilot and the aircrew succumb to food poisoning who will fly the plane? The challenge is therefore to consider the control systems necessary to ensure that neither this nightmare scenario nor illness in the passengers occurs.

14.1 THE CASE STUDIES – FOOD POISONING OUTBREAK ABOARD AN INTERCONTINENTAL FLIGHT

Masterton and Green (1991) showed that the numbers of air travellers thoughout the world changed from 31 million in 1949 to 955 million in 1986. In 1975 UK residents made 12 million visits abroad; in 1998 they made 50.9 million visits. The numbers of air passenger kilometres travelled within the UK alone doubled between 1985 (3.6 billion passenger kilometres) to 7.0 billion passenger kilometres in 1998 (ONS, 2001). Statistics show that since the end of World War II travel, and air travel in particular continues to rise. Air travel facilitates business, and facilitates the vast world wide tourism industry.

The majority of air flights serve at least one snack, while long haul flights – for example Australia to London, Tokyo to New York, London to San Francisco, Delhi to Moscow – may serve two or even three meals. Airline catering is big business – currently it is quoted as worth $14 billion per annum, with annual growth of 5%, directly employing 100 000 people directly, and an equal number indirectly (International Flight Caterers Association, 2001). The science and management skills of serving safe hot meals in flight have developed over the years by adjusting to experiences which have not always been good.

To give two examples:

- According to Tauxe *et al.* (1987) in August 1982, all ten crew members on a flight from Lisbon (Portugal) to Boston (USA) "were affected by staphylococcal food borne illness. [. . .] Fortunately, the crew were still able to operate the aircraft, and the plane landed safely". This outbreak, of great interest, does not appear to have been written up in the literature and Tauxe *et al.* (1987) give an unpublished memorandum as its reference.
- Earlier, in 1975, an outbreak of staphylococcal food poisoning on a flight from Tokyo for Paris affected 57% of the 344 passengers, all of whom were Japanese soft drink dealers on a charter trip to Europe, and some of the crew (Effersoe and Kjerulf, 1975). The plane left Tokyo, flew to Anchorage, Alaska and took on food and a new crew. It left and 8.75 h later made its scheduled stop in Copenhagen from where it could not continue its flight to Paris because by that time several passengers were ill, suffering general malaise and vomiting. Effersoe and Kjerulf (1975) say "before hospital admission (*i.e.* 0.5 to 5 h after [breakfast]) 141 had general malaise and vomiting, 129 had diarrhoea, and 111 reported abdominal pain." Of the passengers who became ill, most showed symptoms within 2.5 h of eating breakfast, with some becoming ill within half an hour, and a few as much as 9 h later.

The adverse outcomes of the latter outbreak were diverse, affecting the health of the passengers, their business intentions in Europe and their clients' business; affecting the airline and its image and affecting the catering company in Anchorage and its employees. Luckily, however, the pilots were not affected, otherwise . . . .

The outbreak and its investigation was described by Eisenberg *et al.* (1975): On leaving Anchorage each passenger was served a snack about one hour after departure, and 1.5 hours before reaching Copenhagen a breakfast which comprised yoghourt, roll and butter, and a cheese omelette topped with two ham slices. Food preparation was undertaken at the caterers in Anchorage where staff worked in teams preparing different food items for the affected aircraft, but also preparing food for other aircraft as well.

Figure 14.1 shows in a schematic way how the breakfast supplied to the aircraft was prepared. Team 1, comprising five people (three cooks and two catering assistants), prepared on 31 January, and on 1 February the breakfast which was served on the plane on 2 February. Team 2 (see Table 14.1) prepared the snack, and Team 3 prepared the dinner for the crew which was served to them instead of breakfast.

The food was prepared in such a way to ensure that the four galleys on the aircraft, which served the three seat price categories (first class, middle class and tourist class), had the correct numbers of passenger meals available.

A very thorough investigation followed which revealed that the food served for breakfast following the Anchorage stop was the cause. Specifically the ham served with egg omelette had been infected. The causative organisms were an enterotoxin producing strain of coagulase positive *Staphylococcus aureus* phage group III [phages 53 and 83A], and originated from inflamed open lesions on the

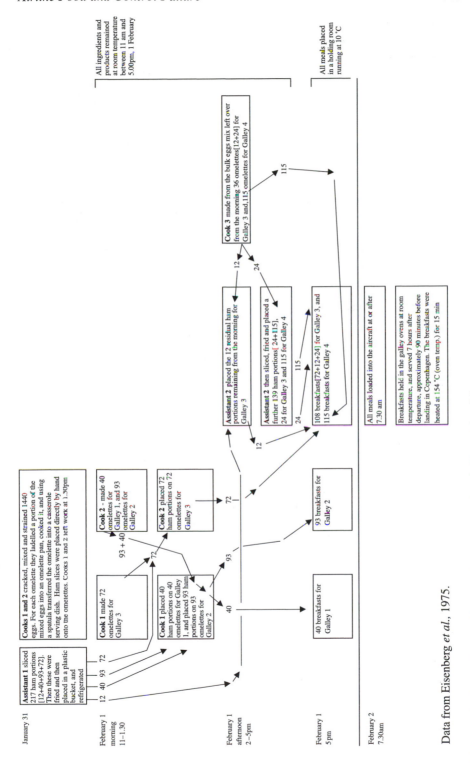

Data from Eisenberg *et al.*, 1975.

Figure 14.1 *The Anchorage–Paris food poisoning outbreak: Team 1 breakfast preparation*

Table 14.1 *Management of staff preparing meals for the flight: Anchorage–Paris via Copenhagen, 2 February, 1984*

	Team 1 *Breakfast preparation*	Team 2 *Snacks preparation*	Team 3 *Dinners for crew*
1 February, morning	See Figure 14.1	Open sandwiches of tuna, open sandwiches of beef, open sandwiches of chicken, open sandwiches of shrimp salad, chocolate cake	(What the meal comprised is not reported) served to crew instead of breakfast
	5 p.m. All breakfasts placed in holding room maintained at 10 °C	5 p.m. All sandwiches placed in holding room maintained at 10 °C	5 p.m. placed in holding room maintained at 10 °C
2 February, 7.30 a.m.	All breakfasts loaded into plane and stored at ambient in galley ovens	All sandwiches loaded onto plane cooled by dry ice (solid carbon dioxide)	
	Served about 7 hours after departure from Anchorage, *i.e.* about 1.5 hours prior to arrival in Copenhagen	Served about 1 hour after departure from Anchorage	Served to crew instead of breakfast

Based on data in Effersoe and Kjerulf, 1975.

hands of one of the cooks (Cook 1) at the ground based caterers in Anchorage. The same organisms were isolated from the stools and vomit of patients, and from three omelette samples and three ham samples. Swabs taken in the kitchen of the table tops, the meat slicer mixing bowl and the floor were negative for *Staph. aureus.* Additionally five cans of ham of the same date code as that used for the breakfast were negative both for enterotoxin and there was no evidence of *Staph. aureus* contamination.

14.2   BACKGROUND

In March 1984 a number of outbreaks of salmonellosis among long haul air travellers leaving London for the United States and for destinations in the Middle East and Africa attracted much publicity both because of the high numbers of cases, estimated at 2747 by Tauxe *et al.* (1987), and also because some passengers threatened law suits (see *The Times* newspaper, March 24, 1984 page 3 column f; and April 26, 1984, page 3 column e). Illness appeared one or two days after the passengers had disembarked the aircraft, disrupting their leisure or the work they had travelled to pursue. On investigation the cause was attributed to *Salmonella enteritidis* infected aspic glaze, which was purchased as a powder by the ground based London flight-kitchen staff and rehydrated by them for use

Table 14.2 *Food-borne outbreaks aboard commercial aircraft, 1947–1984*

Outbreak Number	Date	Agent	Implicated food	Number of flights	Attack rate Passengers (number affected/total exposed)	Crew (number affected/total exposed)	Food origin
1	August 1947	*Salmonella typhi*	Sandwiches	1	4/17	0/4	Anchorage, Alaska
2	January 1961	*Staphylococcus aureus*	Chicken	1	13/128	0/?	Vancouver, British Columbia
3	January 1967	?	Oysters	1	0/?	23/?	London
4	May 1969	Multiple	?	1	21/42	?	Hong Kong
5	July 1969	Multiple	?	1	24/59	?	Hong Kong
6	November 1970	*Clostridium perfringens*	Turkey	8	3/18	22/62	Atlanta, Georgia
7	September 1971	*Shigella*	Seafood cocktail	1	19/43	?	Bermuda
8	February 1972	*Vibrio parahaemolyticus*	Seafood hors d'oeuvres	1	12/?	3/?	Bangkok
9	November 1972	*Vibrio cholerae* O1	Hors d'oeuvres	2	47/357	0/19	Bahrain
10	June 1973	*Vibrio cholerae* non-O1	Egg salad	1	64/?	2/?	Bahrain
11	October 1973	*Staphylococcus aureus*	Custard	3	247/440	0/?	Lisbon
12	October 1973	*Salmonella thompson*	Breakfast	1	17+/117	?	Denver, Colorado
13	February 1975	*Staphylococcus aureus*	Ham	1	196/343	1/20	Anchorage, Alaska
14	February 1976	*Salmonella typhimurium*	Cold salads	11	550/2500	?	Las Palmas, Spain
15	April 1976	*Salmonella brandenburg*	Multiple	45	232/?	58/?	Paris
16	June 1976	*Staphylococcus aureus*	Eclair	1	28/185	?	Rio de Janeiro
17	October 1976	*Salmonella typhi*	Tourist class menu	1	13/225	?	New Delhi?
18	December 1976	*Vibrio parahaemolyticus*	?	1	28/?	?	Bombay
19	1978?	*Vibrio cholerae* non-O1	Sandwiches	1	23/?	?	Dubai, United Arab Republics
20	August 1982	*Staphylococcus aureus*	Custard	2	6/502	10/?	Lisbon
21	May 1983	*Salmonella enteritidis*	Swiss steak?	2	12/?	?	New York
22	October 1983	*Shigella*	?	1	42/48	?	Acapulco
23	May 1984	*Salmonella enteritidis*	Hors d'oeuvres	32	177/?	9/?	London

Reproduced from Tauxe, R.V., Tormey, M.P., Mascola, L., Hargrett-Bean, N.T. and Blake, P.A., Salmonellosis outbreak on transatlantic flights; food borne illness on aircraft: 1947–1984. *American Journal of Epidemiology*, 1987, **125** (1), 150–157. By permission of Oxford University Press.

in the presentation of some of the hors d'oeuvres. This particular outbreak was investigated by Tauxe *et al.* (1987), who in the same paper reviewed food-borne illness on aircraft between 1947 and 1984. Data on such outbreaks were collated by them and are shown in Table 14.2. Later reviews have also been published by Roberts *et al.* (1989) and by Svensson (1998).

14.3   EXERCISES

Exercise 1. Explain how the circumstances of the 1975 outbreak (Effersoe and Kjerulf, 1975; Eisenberg *et al.*, 1975), described in Section 14.1) allowed *Staph. aureus* to grow to concentrations which ensured that every infected slice of ham was capable of causing food poisoning.

Exercise 2. Taking into account both the two *Staph. aureus* outbreaks described in Section 14.1 and the other outbreaks alluded to in Table 14.2 determine the microbiological risk factors which exist in the preparation and service of airline foods.

Exercise 3. Give your assessment of how airline catering needs to be managed in order to significantly reduce microbiological risk to passengers, the pilots and the air crew.

14.4 COMMENTARY

> **Exercise 1.** Explain how the circumstances of the 1975 outbreak (Effersoe and Kjerulf, 1975; Eisenberg *et al.*, 1975), described in Section 14.1) allowed *Staph. aureus* to grow to concentrations which ensured that every infected slice of ham was capable of causing food poisoning.

Note: more detailed reading of the paper by Effersoe and Kjerulf would give many of the details below.

Staphylococcus aureus is an organism associated with the mucous membranes, the nasal passages, skin and skin lesions in man and warm-blooded animals yet can also grow in a number of food systems. It can also be isolated from fresh meats, fish, frozen meats, raw milks and prepared foods. There are a number of toxigenic strains, some of which produce enterotoxins responsible for causing food poisoning when they accumulate in foods during their growth there. It so happened in this case that one of the staff – Cook 1 – responsible for preparing the breakfasts had inflamed infected lesions on his hands. These lesions – on the palm side of the index finger and on the palm of the right hand – came into contact with the slices of ham every time Cook 1 folded the two slices of ham for each of the 133 portions of ham for 133 breakfasts to be served from Galley 1 and Galley 2. Cook 1 was not wearing disposable gloves, and the lesions were not covered with waterproof plasters. Since the ham was not further cooked the organisms passed from the inflamed lesions onto the ham and remained there for an estimated 24 h period (morning of 1 February to morning of 2 February).

Cook 1 also prepared 72 omelettes – some for Galley 3, others for Galley 4. But the egg mixture, which may have become contaminated from Cook 1, was immediately cooked by Cook 2 into omelettes, and placed by Cook 2 into individual casserole serving dishes. The number of prepared breakfasts gradually accumulated during the morning of 1 February and in the afternoon of that day, and while this happened they remained at room temperature. Thus the toxigenic strain of *Staph. aureus* was transferred from the hands of Cook 1 (where its toxins were causing inflammation) to the ham. The ham was a canned product, and it had been sliced the previous day by Assistant 1, all the slices put into one plastic bucket, and refrigerated. Cook 1 thus had to reach into the bucket to remove slices, and in that process not only contaminated the slices selected but is likely to have contaminated the bucket itself, and other slices of ham. So when Cook 2 placed ham slices on 72 portions of omelette some of those were already infected.

Tests showed that unopened canned ham of the same batch numbers as the ham used was not infected with either *Staphylococcus* enterotoxin or with viable cells of *Staph. aureus*, so it was likely that the ham slices were uninfected up to the time Cook 1 handled the slices. Thus the growth periods for the organism in the ham were between 11 a.m. on 1 February and the time the ham was eaten on 2 February. The infecting *Staph. aureus* found itself in a nutrient environment, in the absence of other competing flora, where neither the pH nor the water activity had the potential to restrain its growth significantly. Because the ham was left at room temperature (for example at 20 °C, although busy kitchens can be hotter), the organisms had the opportunity to survive and grow.

The conditions under which *Staph. aureus* grows and produces toxin are shown in Table 14.3 and from those data it can be seen that toxin is produced more rapidly at higher temperatures. Although there are at least seven recognised enterotoxins the ones most commonly associated with food poisoning are enterotoxins A and D. Different strains of the organism produce different enterotoxins. A and D are both produced in the logarithmic phase of growth – so the greater the cell population, the greater the enterotoxin concentration in the food.

Apparently none of the passengers complained of the appearance of the ham – so presumably no colonies were visible on its surface, but because the ham had been kept warm for close on 24 h the cell concentrations on the surfaces of the ham would have been approaching a point when visible change would become evident. The breakfasts were heated before serving, so that the Japanese passen-

Table 14.3 *Staphylococcus aureus – growth and enterotoxin production*

The organism:
Gram positive, coagulase positive coccus
Heat sensitive cells, non spore producing
Cells resistant to drying
Cells resist freezing and thawing
Seven antigenically different enterotoxins produced by different strains: A, B, C_1, C_2, C_3, D, E
Heat resistant enterotoxins can survive sterilisation regimes
Enterotoxins stable in frozen storage

Enterotoxin detection:
Older method: microslide Ouchterlony technique: sensitivity to 0.1 μg/100 g food
Newer method: ELISA or latex agglutination techniques 0.1 to 10 μg/ml food extract

Growth and enterotoxin production

	Temperature °C	pH	Water activity (a_w)	Atmosphere
Optimum for growth	37	6–7	0.98	Aerobic
Growth range (under optimum conditions)	7–48	4–10	0.83–>0.99 (aerobic) 0.90–>0.99 (anaerobic)	Anaerobic–aerobic
Optimum for toxin production	40–45	7–8	0.98	Aerobic (5–20% dissolved oxygen)
Toxin production range (under optimum conditions)	10–48	4.5–9.6 (aerobic)	0.87–>0.99 (aerobic) 0.92–>0.99 (anaerobic)	Anaerobic–aerobic

Source: IMCSF, 1996.

gers, who may not have been very familiar with eating ham, received hot food. The oven temperatures were reported as 154 °C, and the heating time 15 min. However, the omelettes cannot have reached a temperature much more than 65 °C (otherwise they would have lost any softness and become 'hard boiled eggs'), while the ham on the top may have achieved much higher temperatures. This exposure to heat could have killed some of the organisms – they are heat sensitive – but evidently the heating was such that not all were killed, for viable cells were isolated from vomit and faeces. But the heating would not have destroyed the enterotoxin(s) present because they are heat resistant. So the passengers consumed ham (and omelette) which contained viable cells of *Staph. aureus* and concentrations of enterotoxin sufficiently great to cause in some passengers immediate illness. Whereas vegetative cells can be destroyed by pasteurisation regimes the enterotoxins can remain toxic even after exposure to heat regimes adequate to sterilise low acid canned foods. The dosage of enterotoxin required to cause illness was estimated by Evenson (1988) and is generally accepted to lie between 0.1 and 1.0 μg toxin per kg body weight of the consumer depending on their susceptibility. Thus whether poisoning is experienced by a consumer depends on the concentration of toxin consumed, the sensitivity of the consumer and the amount of toxic food eaten.

The concentration of toxin accumulated in each piece of ham (or egg omelette) would have depended on the initial microbial load (*e.g.* the inoculum from the hands of Cook 1) and the rate of growth of those organisms as influenced by the growth medium (the salty ham, or the egg), its pH, its water activity, its temperature and the time period during which growth could occur.

Exercise 2. Taking into account both the two *Staph. aureus* outbreaks described in Section 14.1, and the other outbreaks alluded to in Table 14.2 determine the microbiological risk factors which exist in the preparation and service of airline foods.

Food-borne illness and food poisoning are caused by a wide range of micro-organisms. In outbreaks of *Staph. aureus* poisoning, for example those described in Section 14.1, the illness has rapid onset because it is an intoxication. In these two outbreaks illness began before the journey was complete. If the pilots had experienced such illness there would have been a real risk to the safe flight and landing of the aircraft. In the second outbreak virtually all the passengers were affected while on board, causing major problems for the cabin crews and the comfort of the remaining passengers. In the 1984 Salmonella outbreaks (Section 14.2) passengers became ill after disembarking but since many flights of the same airline catered for by the same airline caterers in London were affected, the ensuing publicity ensured that the airline suffered adverse publicity and damage to its reputation.

But the range of potential microbial hazards is wider than *Salmonella* and *Staph. aureus*. Some food-borne pathogens grow rapidly and produce toxins in the food, others cause infections in the consumers which show a few days later.

Some bacteria only grow in 'warm' conditions, but others grow under chilled conditions. In some the infective dose is large, in others small.

The airline caterer has got to produce food in which the risk from all of these hazards is acceptably low.

The types of meals served on planes vary widely from snacks to full meals; from low risk foods like cheese biscuits and breads, to high risk products like prawn cocktails, creams, and custards, and other high risk products such as meat salads, hot chicken casseroles and other meat dishes. Each meal can be a complex mix of cold starter items, meat or fish main courses, and cold sweet dishes. These complex meals, with a range of mixed items to be served on each plate involve the airline caterers in the need to cook, prepare and assemble the items not only so that they look attractive, but also so that the microbial quality of the food is safe when eaten. As Figure 14.1 shows there is ample scope for the mistakes of an individual (Cook 1) to affect the safety of hundreds of meals. The complexity of the assembly of meals, together with the complex logistics of providing the right numbers and types of meals for the different galleys for a flight, compounded by the need to provide more than one meal for a flight, different menus for first class passengers, compounded further by the meals provision for other flights, separate requirements for religious observations, vegetarian meals and the needs of the crew mean that there is plenty of opportunity for mistakes.

Some main risk factors in the food preparation and service are:

- the range of potential microbial hazards and their sources;
- the types of foods produced for airline meals;
- the complexity of the food preparation processes, and the complex logistics to meet the needs of the airlines;
- the time periods between preparation and consumption; [how, for example could a customer open their breakfast dish in flight and find it covered in mould as was reported (see *Which*, 4 May, 2001, p. 4)?]
- the food temperature experience and the potential to support microbial growth;
- the effective regeneration (heating) of stored foods;
- the timing of in-flight food service to customers, and the different quality meals provided to the passengers of different seat prices;
- the meals the crew and pilots need.

Exercise 3. Give your assessment of how airline catering needs to be managed in order to significantly reduce microbiological risk to passengers, the pilots and the air crew.

Having identified the main risk factors in Exercise 2 you can turn your attention to how they should be managed.

Perhaps it is fruitful to again consider the 1975 *Staph. aureus* outbreak (Section 14.1). Figure 14.1 and Table 14.1 provide you with an idea of how catering was managed in that catering company in Anchorage, and thus give you the opportunity to propose areas where improvements could have been made.

You have evidence of how contamination of the ham by *Staph. aureus* occurred. What would you now recommend should have happened to avoid that particular event?

Immediate questions arise concerning the sore on the hand of Cook 1. Should staff have been required to wear disposable gloves when handling open food such as the ham slices. Are they available? What is the cost of such gloves? Would management provide the gloves for free? Do they give the finger sensitivity required? Would barrier cream be an alternative? Why were the sores not covered by waterproof plaster? Was plaster available? Why did the sores on the hands go unnoticed? Did the manager not notice? Why didn't Cook 1 report the problem? Why was Cook 1 working? Did Cook 1 fear losing that day or week's pay? Perhaps Cook 1 was the senior manager? Why was Cook 1 oblivious of the risk of cross-contamination? Had Cook 1 (or anyone else) had any hygiene training?

Why was the food left at room temperature all the working day? Why was the store room at 10 °C and not cooler? What precautions were there to avoid contamination of other foods by other micro-organisms?

The catering company in Anchorage did, rightly, separate the staff into teams who prepared different components for the meals provided for the flight. Crew meals were prepared by a different team, but the crew meals would have had the chance to come into contact with the food prepared by Cook 1 unless specific separation procedures were adopted. If you look at Table 14.2, outbreaks 4 and 5 are reported as due to 'multiple' microbial agents. Consider how that situation might arise.

The 1975 *Staph. aureus* airline outbreak (Section 14.1) must be attributable to multiple management failures.

The principles of the production of airline foods are those of 'cook-chill' requiring good quality ingredients, good storage conditions, hygienic production and minimisation of opportunity for cross contamination, adequate cook-

---

BOX 14.1

Low risk foods are foods which are not strongly associated with the presence of food pathogens and/or foods which will have a processing step which will reduce their microbial flora to a safe level. The raw food side of the preparation of airline meals might be categorised 'low risk', but never-to-be-cooked items like cheeses and raw salad vegetables would be 'high risk'.

High risk foods are those foods which readily support the survival or growth of food pathogens and/or will not undergo any process which will make them microbiologically safer by reducing their flora. Prepared airline foods are normally 'high risk' foods in spite of the fact that they may be heated (regenerated) in flight because that process will not destroy preformed toxins and may not destroy infective micro-organisms.

ing to reduce microbial load, fast cooling, adequate cool storage to prevent microbial increase, limited shelf life to prevent microbial increase, and adequate reheating where appropriate.

The key to the successful and safe production of airline food is separation of activities, physical separation of foods (low risk from high risk: see Box 14.1), separation by time, separation of identified food preparation areas, and separation of staff, and the management of the production processes using HACCP systems with effective critical controls in place. Such systems will minimise the presence of pathogens in ingredients, minimise cross-contamination (Roberts *et al.*, 1989), optimise the cooling rates of cooked foods and minimise the opportunity for microbial increase by good temperature and time control.

Safe food for airlines also requires the provision of completely separate meals for the air crew.

The microbiological quality achieved for the food should be such that they meet the standards recommended for ready-to-eat foods (Table 11.10) (PHLS, 2000).

14.5   REFERENCES

Effersoe, P. and Kjerulf, K., 1975. Clinical aspects of outbreak of staphylococcal food poisoning during air travel. *Lancet*, **ii**, 599–600.

Eisenberg, M.S., Gaarslev, K., Brown, W., Horwitz, M. and Hill, D., 1975. Staphylococcal food poisoning aboard a commercial aircraft. *Lancet*, **ii**, 595–599.

Evenson, M.L., Hinds, M.W., Bernstein, R.S. and Bergdoll, M.S., 1988. Estimation of human dose of staphylococcal enterotoxin A from a large outbreak of staphylococcal food poisoning involving chocolate milk, *International Journal of Microbiology*, **7**, 199–205.

ICMSF, 1996. *Micro-organisms in Foods*. Volume 5. *Characteristics of Microbial Pathogens*. Blackie Academic and Professional, London.

International Flight Caterers Association (IFCA) website: http://ifca.co.uk (Surrey Place, Mill Lane, Godalming, Surrey GU7 1EY, UK. Phone +44 0 1483 419449). Site accessed April 2001. This site also has a simple food hygiene knowledge test suitable for operatives.

Masterton, R.G. and Green, A.D., 1991. Dissemination of human pathogens by air travel. *Journal of Applied Bacteriology Symposium Supplement*, **20**, 31S–38S.

ONS (Office of National Statistics) website: http://www.statistics.gov.uk

PHLS, 2000. Gilbert, R.J., de Louvois, J., Donovan, T. *et al.* – Working group of the PHLS Advisory Committee for Food and Diary Products. 'Guidelines for the microbiological quality some ready-to-eat foods sampled at point of sale'. *Communicable Disease and Public Health*, **3** (3), 163–167.

Roberts, D., Gilbert, R.J., Nicholson, R., Christopher, P., Roe, S. and Dailley, R. (1989). The microbiology of airline meals. *Environmental Health*, **97** (3), 56–62.

Svensson, C., 1998. Food poisoning on passenger aircrafts. *Svensk Veterinartidning*, **50** (16), 745–752.

Tauxe, R.V., Tormey, M.P., Mascola, L., Hargrett-Bean, N.T. and Blake, P.A., 1987. Salmonellosis outbreak on transatlantic flights; food borne illness on aircraft: 1947–1984. *American Journal of Epidemiology*, **125** (1), 150–157.

The Times newspaper, 1984, March 24, page 3 column f.

The Times newspaper, 1984, April 26, page 3 column e.

D. Sampling, Criteria and Acceptance

CHAPTER 15

Global Dissemination of Organisms and Their Control

Key issues
- Fish meal and animal feeds; low a_w materials
- Globalisation of food supply and the spread of pathogens
- *Salmonella* spp.
- Contamination and cross contamination
- Microbial survival and significance
- Import acceptance – sampling criteria

Challenge
This case study asks you consider the movement of feeds and foods in international trade and the world wide dissemination of pathogens.

15.1 THE CASE STUDIES

15.1.1 The Paragould Outbreak of *Salmonella agona*

(Based on data in McConnel Clark, G. *et al.*, 1973)

In May 1972 in Paragould, Arkansas, USA (a small town of about 10 000 residents), four people unrelated to each other were hospitalised with severe gastro-enteritis caused, it turned out, by *Salmonella agona*. This particular strain of *Salmonella* had not been reported to cause any human salmonellosis in the USA before 1967, but since that time the numbers of reportings had gradually risen for it to become one of the most frequently isolated *Salmonella* serotypes. Initial investigations in Paragould showed that the four cases only had the use of a drive-through restaurant, Restaurant A, in common. Following this finding 13 other people were also shown to be infected with the organism but were symptomless. Five of these people were identified through a mandatory screening programme for new employees in food-handling establishments and in hospitals. Two proved to be employees at Restaurant A; two others worked at another restaurant, and the fifth was a nurse and these last three often ate at Restaurant A.

Very thorough epidemiological investigation showed that the infection rate was significantly higher in persons who frequently ate in Restaurant A rather than elsewhere, and among the 67 persons in the investigation who used Restaurant A, infection was significantly higher in those who frequently ate coleslaw salad or raw onions.

When samples from Restaurant A's kitchens were examined *S. agona* was isolated from knives, tables tops, the sink, meat slicer, fresh frozen catfish, raw chicken and shredded lettuce.

The source of *S. agona* was then pursued and Figure 15.1 indicates how by identifying and investigating the supplies for Restaurant A it was traced back to the chicken producers in Mississippi who used Peruvian fish meal in their chicken feeds. *S. agona* was not actually isolated from the Peruvian fish meal, but over a prior three or four year period independent surveillance data had shown some samples of Peruvian fish meal to be contaminated with *S. agona*. The inference was then drawn that this was the probable original source of the organism.

15.1.2 *Salmonella agona* in the UK

In 1972 a short comment in the *British Medical Journal* (Anon, 1972) described *Salmonella agona* as "a new hazard". In 1969 only two cases had been reported, and both had recently returned from Spain. But in 1970 and 1971 the number of cases increased causing *S. agona* to become, by 1972, the fourth most common serotype in human infections in England and Wales, *Salmonella typhimurium* being the most frequent (Table 15.1).

In March 1970 samples from a consignment of imported fish meal had been found to be positive for *S. agona*. Additionally, McCoy (1975), writing about trends in salmonellosis in England and Wales over the period 1941–1972, stated that by June 1970 *S. agona* human infections were being reported to the PHLS related both to the consumption of cooked chicken and to contact with one day old chicks. In July 1970 a milk-borne outbreak of *S. agona* occurred, and in the following years the organism was isolated from unpasteurised egg, from milk (which caused another large outbreak) and from broiler chickens, cattle and pigs.

15.1.3 *Salmonella agona* in Peanut Snack Foods

(Based on Killalea *et al.*, 1996)

In the UK, in the period 5 December 1994 to 30 January 1995, 27 isolates of *S. agona* related to cases of gastro-enteritis, were identifed in England and Wales, compared with 12 in the same period the year before. Many of the cases were children, there was geographical clustering, and most children had Jewish surnames.

In investigative interviews four of the parents reported that their child had eaten a kosher peanut-flavoured ready-to-eat maize-based savoury snack imported from Israel. Microbiological tests had revealed that the strain causing the

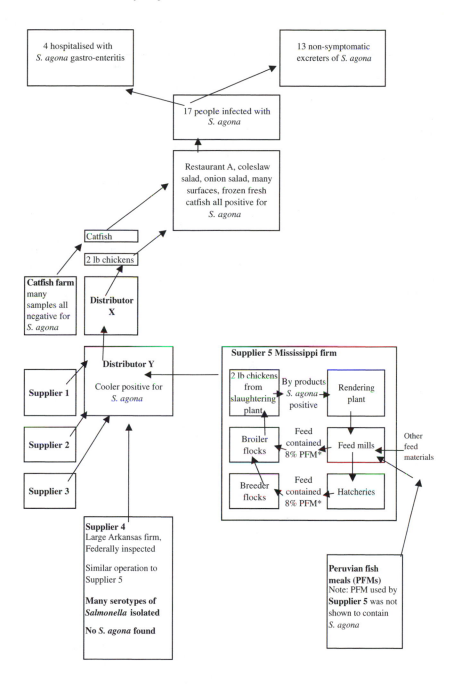

Figure 15.1 *Investigation of the source of Salmonella agona infection, Paragould, Arkansas,*
1973
(Source: McConnell-Clark, G. *et al.*, 1973)

Table 15.1 *Salmonellosis: serotypes (rank in parenthesis) isolated from human infections*

	1941/48	1949/55	1956/63	1964/68	1969/72
Salmonella meleagridis	28 (10)	31	110	15	n/a
oranienburg	125 (5)	87	70	139	n/a
montevideo	108 (6)	102 (10)	114	139	289
thompson[a]	210 (3)	667 (3)	757 (5)	89	244
newport[a]	207 (4)	392 (4)	863 (4)	182	270
bovismorbificans	43 (8)	195 (5)	229 (10)	60	n/a
brandenberg	108 (6)	19	215	684 (4)	285
dublin[a]	62 (7)	168 (6)	192	244 (9)	317
anatum	37 (9)	149 (7)	346 (8)	309 (8)	422
typhimurium[a]	2421 (1)	15238 (1)	22009 (1)	8316 (1)	10360 (1)
enteritidis[a]	261 (2)	708 (2)	973 (3)	830 (2)	3528 (2)
stanley[a]	5	140 (8)	306 (9)	585 (5)	585 (10)
heidelberg		104 (9)	1585 (2)	330 (6)	1088 (5)
bredeney	4	42	351 (7)	218 (10)	739 (9)
saintpaul	6	69	403 (6)	124	800 (7)
panama	4	4	58	808 (3)	2302 (3)
virchow	2	29	43	321 (7)	932 (6)
indiana			1	200	797 (8)
agona					1178 (4)
infantis	108 (6)	11	142	126	514
4,12:d:-					379
derby[a]	7	82	202	117	307
muenchen	2	40	127	74	265
menston		24	145	46	n/a
choleraesuis[a]	16	64	102	44	n/a
senftenberg	20	60	83	122	n/a

[a]'native' serotypes present, Great Britain 1923–39.
n/a = figures not available at the time of the report.
Reproduced from McCoy, J.H., Trends in *Salmonella* food poisoning in England and Wales, 1941–72, *Journal of Hygiene, Cambridge*, 1975, **74**, 271–282. By permission of Cambridge University Press.

illnesses was *S. agona* phage type 15. A subsequent case-control study revealed a strong association between consumption of the savoury snack and *S. agona* PT15 infection ($P = 0.0002$).

Tests on products of different dates of manufacture extending over a four month period, and conducted in several countries to which the snack was exported, showed a persistent infection of the product with the identical* strain of organism. Once a UK Government hazard warning was issued on 10 February 1995, and the distribution of batches of the product was suspended, the incidence of *S. agona* infection in the UK subsided to its previous level.

In Israel in approximately the same period (October 1994 to January 1995) more than 2200 cases of *S. agona* had been observed, representing a significant rise in its incidence but the source of infection was not identified until a link was

* Identical by phage type and by pulsed field gel electrophoresis.

Table 15.2 *International tests on the kosher peanut flavoured snack made in Israel*

Date of manufacture	Country in which samples were tested	Best before date (end of month)	Test result for Salmonella agona
4 October 1994	Canada	not stated	Contaminated packets found
6 October 1994	UK	Mar 95	3/38 samples positive for *Salmonella agona*
30 October 1994	USA	not stated	Contaminated packets found
6 November 1994	UK	Apr 95	44/53 packets positive for *S. agona* PT15, and containing an estimated 2–45 organisms per 25 g packet
unknown	UK	May 95	0/187 positive for *S. agona*
18 November 1994	Canada	not stated	Contaminated packets found
19 December 1994	Canada	not stated	Contaminated packets found
6 February 1995	Israel	not stated	Contaminated packets found

Source: Killalea *et al.*, 1996.

made to the peanut snack* through the findings in the UK and elsewhere (Table 15.2). Shohat *et al.* (1996), who investigated the connection, comment that the consumption of the peanut snack was almost universal among children in Israel, with children who ate about four packets a week significantly more exposed than those who ate about two packets.

Over the period from the end of December 1994 to February 1995 the production factory in Israel was investigated with many samples taken from production lines, ingredients, stools samples from workers and 450 packets of the snack. No specific source of the organism was found, none of the environmental samples taken in the factory were positive, nor were any of the stools samples. But two packets of snack were *Salmonella* positive – one with *S. agona* (the strain was identical to that found in the product in the UK and elsewhere), the other with *Salmonella enteritidis*. *S. enteritidis* was also found in a re-useable plastic bag in which the snack was stored for 48 hours before packaging.

Extensive controls were introduced in the factory, and following that time the national incidence of *S. agona* in Israel fell back to its previous levels prior to this outbreak.

---

* The snack contained corn grits, hydrogenated soya bean oil, antioxidants, encapsulated vitamins, salt and a peanut butter coating (Shohat, 1996).

15.2   BACKGROUND

15.2.1   Animal Feeds in the 1970s and 1980s

In the 1970s and 1980s animal feeds were made from a number of resources.

By 1980 it was estimated that between 30 and 40% of the total world catch of fish was used to manufacture animal feeds (ICMSF, 1980) including using the by-catch fish which were not the primary species sought and which were considered to be too bony, oily or otherwise unsuitable. In addition vegetable materials, maize and soy beans particularly, were used as were waste products of the meat industry – from rendered mixtures of blood, bone, brains, viscera, skins and fur. Furthermore the wastes generated by intensive rearing of animals also provided source feed materials for other animals.

In a paper entitled "The value of processed poultry waste as a feed for ruminants', El-Sabban *et al.* (1970) describe the use of poultry waste – excreta, shed feathers, wasted feed, and broken eggs – in different processed formats as feed for sheep in whom weight gain and meat characteristics were measured and compared to other feeding regimes. In comparison to vegetable based feeds, in spite of the slightly lower digestibility found, the poultry waste feeds were reported as satisfactory. However, this type of approach – the feeding of animal based products to animals which normally eat only plant materials (ruminants) – has been shown to be a flawed philosophy. Devastatingly in the 1990s the bovine spongiform encephalopathy (BSE) problem in cattle is believed to have originated through the use of cattle feeds based on materials containing the brain, nervous tissue and other body parts of sheep infected with the BSE prion (the infective agent), in which the feed manufacturing process failed to destroy it.

Furthermore it is possible that the infective agent was then transmitted to human beings through consumption of infected beef meats, causing the human illness new variant Creuztfeld Jacob disease (nvCJD) (Phillips, 2000).

Meat waste, and fish based feed materials were processed into meals or pellets for animal feeds after rendering (removal or reduction of the fat content). The rendering process exposed the animal source materials to temperatures close to 121 °C. It should therefore have destroyed heat resistant bacterial spores as well as heat sensitive organisms such as *Enterobacteria* in general and *Salmonella* in particular and resulted in material effectively micro-organism free. But of course that depended on the quality of the management of the rendering and subsequent processes. But authoritative voices (ICMSF, 1980) held the view that because the product was destined for animals' consumption, insufficient attention had been paid to sanitation principles which should have been implemented with the same rigour as for human food, and laid the blame at the feet of both the health authorities and the industry.

Stott *et al.* (1975) described the basic principles of how pelletised feed was made. First the mixed feed materials were heated up to temperatures between 63 and 85 °C and then forced though a former to create the pellets. The heating processes could be either through direct contact with steam or through raising the temperature of the meal in a steam jacketed vessel. When direct exposure to steam was used the contact time was very short – a matter of 2–3 seconds, whereas when heated in a jacketed vessel the residence time (and hence heat exposure) approximated to 2–3 minutes. The hot meal was then extruded and the pellets cooled using air drawn from the mill atmosphere, sieved to remove dust and fine particles, and then sent by conveyor to a vast storage bin. The longer residence time heating processes could reduce the microbial load of Enterobacteriaceae a hundred to a thousand fold, but the 2–3 s direct exposure to steam could only reduce population by < 1 log cycle.

The world wide trade in animal feed materials built up over the post-second world war years, with Europe and the USA importing much of their animal feed materials from South America as well as elsewhere. During that time, regulations laid down the conditions under which feed manufacture was to occur, but the same legislation often failed to specify how the process should be monitored in order to be sure that it was achieving the time/temperature exposure goals throughout all materials treated. For example in the UK in the early 1980s regulations (The Diseases of Animals (Protein Processing) Order, 1981) not only allowed unspecified human effluent to be included in feed, but under a regulatory regime which exercised very little control over the functioning of the processing plant. Furthermore import, surveillance and other testing provided evidence that animal feeds in many countries were potentially and then actually contaminated with *Salmonella* and Table 15.3 demonstrates some of this evidence.

Through the use of feed materials infected at low levels with *Salmonella*, intensively reared chickens, pigs and other animals have been shown to pick up *Salmonella* infections – but often without exhibiting symptoms, while excreting the organisms in their faeces thus providing new sources of infection and amplifying the prevalence levels of the organism. *Salmonella* infections in animal flocks can weaken or kill young animals as well as debilitating some of the older animals. Animal feeding trials with different types of feed have indicated that *Salmonella* infection rates can be reduced by different strategies including, for example, feeding pellets instead of meals, a reflection possibly of the thoroughness of the sterilisation to which the feed was subject (Edel *et al.*, 1970). In the 1970s *Salmonella* organisms were showing up frequently in the meat products destined for human beings (see Table 15.4) – the organisms originating from the infected meat animals, and being spread largely through contamination in the abattoirs and in the meat preparation stages. Some authors expressed concern relating to the spread of the organisms and Edel *et al.* (1970) suggested the institution of salmonella-free pork centres to supply children, the ill and other vulnerable individuals in hospitals. However, they also observed that this was a dream not realisable until other sources of *Salmonella* infection on the farm – such as rodents, insects, water supply and dry feeds, were also under control.

But more than 30 years on, the problems of *Salmonella* infections in feed materials, in animals and in fish products are not solved because effective hygienic practice is still lacking in the production of feeds and foods. For example, Hatha and Lakshmanaperumalsamy (1997) observe that the monsoon season correlates with the increased incidence of *Salmonella* they found in several species of market fish. Secondly, Harris *et al.* (1997), working in the USA, found

Table 15.3 *Salmonella: examples of incidence in feeds of animal origin*

Material	Country in which samples were tested and/or originated	Comments
Dog food, feather meal, fish meal, poultry by-products, tankage	USA	*Salmonella* spp. were found in 6.8–33% of samples. (Morehouse, L.G., and Wedman, E. E., 1961) *Salmonella* and other disease-producing organisms in animal by-products. A survey. *Journal of the American Veterinary Medicine Association,* **139,** 989–995.)
Fish meal	Sweden	*Salmonella* spp. found in 0.08% of samples. (Karlsson, K.-A., Rutquivst, L., and Thal, E., 1963). *Salmonella* isolated from animals and animal feeds in Sweden during 1958–1962. *Nord. Veterinaermed.,* **15,** 833–850.)
Animal by-products	USA	*Salmonella* spp. found in 0.9% of 193 samples. (Moyle, A.I., 1966). *Salmonellae* in rendering plant by-products. *Journal of the American Veterinary Medicine Association,* **149,** 1172–1176.)
Meat and bonemeal	UK	*Salmonella* spp. found in 26% of 224 samples. (Timoney, J. 1968). The sources and extent of *Salmonella* contamination in rendering plants. *Veterinary Record,* 83, 541–543.)
Meat and bonemeal	England, Wales and Denmark	*Salmonella* spp. found in 21.3% of 982 samples. (Skovgaard, N., and Nielsen, B.B., 1972). Salmonellas in pigs and animal feeding stuffs in England and Wales and Denmark. *Journal of Hygiene,* **70,** 127–140.)

Source: ICMSF, 1980

that on the 30 swine farms they evaluated a number of serotypes of *Salmonella* could be found in the animals and in their feeds. *Salmonella* are still widespread in chickens and chicken products (see Richmond, 1990, 1991; Anon, 1995) and the reduction of the incidence of *Salmonella* contamination of UK-produced chicken by 50% by April 2005 has been announced as one of the aims of the UK food control body, the Food Standards Agency (FSA, 2000).

The problem with feeds manufacture is that although the rendering and pelleting processes should provide enough heat to destroy *Salmonella,* in practice the design of the process may be such that *Salmonella* is not eliminated or that re-contamination of processed material can occur. Secondly *Salmonella* are more heat resistant in a dry environment than in a moist one, so the design of the process and the point in the process at which heating occurs, as well as the temperatures and exposure times achieved, affect the extent to which *Salmonella*

Table 15.4 *Salmonella and other organisms in sausage meat, USA, 1972*

Test	Results (Number of sets of sausages = 67)	Notes
APC counts $\leqslant 500\,000\,\text{g}^{-1}$	75% of sets	
E. coli $\leqslant 100\,\text{g}^{-1}$	88% of sets	geometric means of 10 samples
Staph. aureus $\leqslant 100\,\text{g}^{-1}$	75% of sets	geometric means of 10 samples
Salmonella	28% of 529 samples pork trimmings used for sausage	
Salmonella	28% of 560 finished sausage samples	80% of positives were only positive in 25 g samples, and negative in 1.0 g and 0.1 g

Number of plants sampled = 44.
Source: Surkiewicz *et al.*, 1972.

are destroyed. Furthermore the *Salmonella* can survive in the dry material. If the water activity of the feeds is below $0.94\,a_w$ growth will not occur, but the cells will survive. This survival period could be many weeks or even years, a period whose length is determined by the combined effects of the water activity to which the organisms are exposed, and the temperature (see Table 15.5). The surviving viable cells become injured on storage, but when the feed is eaten resuscitate in the animal body and cause infection.

15.2.2 The Survival of *Salmonella* in Low Water Activity Foods and Feeds

Around the time of the Paragould outbreak, the problem of the survival of *Salmonella* in low water activity food materials was well known since survival had been studied in, for example, Italian export quality macaroni (Ferdori and Cirilli, 1968), contaminated dried milks (LiCari and Potter, 1970) and chocolate products (Barrile *et al.*, 1970). The developing US space program was seeking to avoid the survival of Salmonellae and other organisms in the dehydrated space foods (Powers *et al.*, 1971). Microbial survival was also studied in fish protein concentrates (Goldmintz, 1970). However, this author considered that survival of vegetative pathogens would probably not constitute a problem in properly prepared and stored fish protein concentrate, but that the main problem would be mould growth permitted through water absorption in the stored product, causing the water activity to rise. Certainly bacteria do decrease gradually in low water activity products such as grains – also fed to animals (Table 15.5), but even with decline in their numbers the complete disappearance of viable cells, *e.g.* *Salmonella* spp., cannot be guaranteed at any point in the dry product's useful life.

Table 15.5 *Survival of Salmonella montevideo on wheat stored at constant relative humidity*

Temperature 25 °C	Relative humidity (RH)[a]										
	7	11	22	33	43	53	62	75	84	92	98
Initial inoculum g^{-1}	10^6	10^6	10^6	10^6	10^6	10^6	10^6	10^6	10^6	10^6	10^6
Cfu g^{-1} detected at 28 weeks storage	10^4	10^4	10^4	3.6×10^3	10^3	10^2	20	None detected after 22 weeks	None detected after 16 weeks	None detected after 26 weeks	None detected after 16 weeks

[a]Constant RH achieved by equilibrium of saturated salts solutions in sealed chambers.

Note: all counts in cfu g^{-1}.

Source: Crumrine, M.H. and Foltz, V.D., 1969.

15.3 EXERCISES

Exercise 1. Bearing in mind that fish are not natural hosts for *Salmonella*, what are the circumstances under which fish meal can become, and remain contaminated with viable cells of *Salmonella*?

Exercise 2. The outbreaks described in this chapter may be indicative of the problems of globalisation within the food chain. Is there for example any link between the Paragould outbreak in 1972 and the infection of peanut snacks in 1994/95? What is your view? Is there any lesson for today common to both?

Exercise 3. At a port of entry, where a sampling regime of a bulk consignment of fish meal fails to demonstrate the presence of *Salmonella*, does that mean the organism is absent? Would such feed be acceptable for importation and would the importer be correct in believing the consignment would not disseminate *Salmonella* to the animals fed?

15.4   COMMENTARY

Exercise 1. Bearing in mind that fish are not natural hosts for *Salmonella*, what are the circumstances under which fish meal can become, and remain contaminated with viable cells of *Salmonella*?

Logically contamination could occur before processing, during processing, in storage, in transportation and at final destination.

A witness account:

"In the early 1990s I was in Concepcion, Chile. At that time the fishing vessels, laden with mackerel fished from the clean deep waters just off the continental shelf, would come into Concepcion Bay to unload their catch.

The Bay is the remains of an ancient volcano and is an almost complete circle with a narrow entrance to the Pacific Ocean, and it is not surprising that the city of Concepcion should have grown up on the shore of this huge natural bay of protected water. But the bay exchanges its waters with the Pacific only reluctantly. When I was there the Bay took the wastes of the city – its storm water, its garbage, its sewage and its industrial wastes from its fish, and pulp and paper industries.

The fishing vessels moored two or three hundred metres off shore to unload their catch. Bay water was pumped into each hold, and water and fish were progressively sucked out and transferred by pipe to shore using centrifugal pumps, which in fact damaged the fish.

As we watched this process close to one ship the filthy blood, scum and fish offal laden waters slopped around close to our feet, and in due course our little motor boat cleaved its way through it, returning us to the shore. Here the seagulls were even more prevalent than they had been around the unloading ship, screaming and diving down grabbing whatever fish pieces their could.

A silver stream of magnificent mackerel, each larger than any I had seen in the UK, poured dripping from the filthy water replenishing a giant overhead hopper from which, in turn, the fish cascaded down a conveyor belt into the open back of a waiting lorry. At the top of the conveyor the freshness of the fish was tested using a device called a Torry meter whose scale reading of 1 to 16 determined whether the fish went for canning or for meal.

I was told one 20 tonne lorry was filled every five minutes for 16 hours each day, practically all year.

Taking twenty tonnes at a time a constant procession of these unrefrigerated, barely covered, lorries groaned their way the half kilometre or so up the hill to the processing factory which made both canned mackerel and fish meal. There, with trunk raised, the lorry tipped its load into one of three open concrete holds in the ground above which on a narrow gantry a bare chested man, smoking a cigarette and shouting into a walkie talkie, directed the drivers.

All the while the screeching seabirds took their chances.

From above, looking down on three hundred tonnes of catch, the surface of the mass of mackerel glimmered and shone and moved uneasily as deep below the invisible auger turned, moving them steadily towards the canning line, or the fish meal plant for rendering and pelleting plant for fish meal depending on the quality factors of freshness, moisture, and total volatile nitrogen. As the great concrete hold emptied, a man with a high pressure hose would bloodily sluice the last few down to the scummy hole at the bottom and the relentless auger.

This process went on day and night occupying not just this factory but others too working with the same style and urgency.

The canning process

Fresh fish selected for this process were aligned, one parallel to another, on a conveyor belt which moved them steadily to the canning machine. There a chopper whacked out the central 25 centimetres of the fish and neatly popped it into a can. Head, shoulders and gills, fins and tail all went to waste, and thence to the fish meal process. Only the central quarter of a magnificent fish went to canning. The open cans were filled with liquid – water or tomato juice – and processed at 100 °C/35 minutes, after which they were sealed and autoclaved – resting in the autoclave for 80 minutes while a temperature of 116 °C was raised. Later the pallets of tins were removed from the autoclave, and the pallets labelled and stored, the individual cans being labelled 15–20 days later.

The fish meal process

The fish were passed to a continuous cooker, and then the mass was pressed, which yielded two product streams: the solids for the meal, and the liquids from which the fats were collected by centrifuge, yielding further solids. The solids passed to a vast hot air drum dryer to remove water, then on to be air cooled, then milled, after which antioxidant was added to prevent the development of rancidity. The meal was then either filled into bags, or as bulk materials filled into lorries to be taken away and stored in bulk, and thereafter to be exported.

Pellets for salmon feed

These were made from the fish meal with added wheat flour, vitamins, agglomerates, pigments and carotenes. The separate ingredients were dosed into the mix and then made into pellets whose size determined the speed of production. The pellets were heated by water vapour, and subsequently air cooled, and fish oil was added in to the bulk-pelleted mix to achieve the right proportions of protein, lipids, fibre, ash and water content. These pellets were then bagged, labelled and stored."

You can see that this eye-witness account (my own) provides you with an opportunity to raise questions about the potential points of access of *Salmonella* into the fish meal. It also allows you to question how the processors might have replied if asked how they minimised the risk of the presence of *Salmonella* in the fish meal. If *Salmonella* were present in or on the raw fish you must question how the process ensured that all portions of the meal, in whatever form (pelletised or not), were heat treated sufficiently to destroy *Salmonella*. How could such a process be monitored? You may wish to consider how HACCP would be relevant to the fish meal manufacturing process.

Fish meal is a low water content, therefore low water activity, product. But if its water content rises – perhaps through exposure to rain water, to leakage of water onto it, through condensation of water onto it from cold metal surfaces (as might happen in a ship's hold as it travels from warmer seas to colder ones) – then its water activity could rise locally where the added water occurred. Rising water activity increases the possibility of microbial growth. Moulds are those most likely to grow, but if water activity rises further other organisms can grow as well. Thus the microbial population of the processed product can change in storage and transportation, and may increase due to inherent growth, or due to contamination from external sources – such as contaminated surfaces, feed dusts, flies and the faeces of rodents, insects and birds.

Exercise 2. The case studies described in this chapter may be indicative of the problems of globalisation within the food chain. Is there, for example, any link between the Paragould outbreak in 1972, and the infection of peanut snacks in 1994/95? What is your view? Is there any lesson for today common to both?

These questions allow you to research the variety of ways in which fish meal is actually processed and transported, what the composition of various animal feeds can be and the national and international regulations which control its processing. You could also look into what is fed to animals, where it comes from, how it is made, and research how food safety is managed in international trade (Hathaway, 1999; Kruse, 1999).

It would be presumptuous to suggest that there is a direct link between the Paragould outbreak in 1972 and one 23 years later. But it is true that *S. agona* became disseminated world wide paralleling the increasing export trade of fish meal originating in Peru. The incidence of the organism got greater as more foci of infection occurred, so that the original probable source of infection – the fish meal – became less relevant as the organism established itself in meat animal populations and elsewhere. It certainly illustrates how the import and export of food commodities has the potential to distribute pathogens and thus how the expanding global economy may also spread disease (d'Aoust, 1994). This lesson is ever more pertinent today.

In determining whether anything has been learned from these outbreaks you need to aim to evaluate what, if anything, limits the spread of zoonotic disease through the export and import of animal feeds and human food material today when the international trade in food is ever expanding.

> **Exercise 3.** At a port of entry, where a sampling regime of a bulk consignment of fish meal fails to demonstrate the presence of *Salmonella*, does that mean the organism is absent? Would such feed be acceptable for importation and would the importer be correct in believing the consignment would not disseminate *Salmonella* to the animals fed?

No, it does not mean that the organism is absent from the batch tested – it simply means that in the samples taken the testing regime did not demonstrate the organism.

The probability of detecting the presence of the organism in the whole batch increases both with increased numbers of samples being taken, and with increasing levels of contamination.

Thus for a batch of feed to be acceptable several issues have to be decided in advance:

- What [low] level of occurrence of *Salmonella* is acceptable?
- How many samples need to be taken to be sure that if *Salmonella* is present in the bulk, the actual frequency of occurrence is below the threshold level of acceptability/non-acceptability?
- What size samples should be taken, and how?
- From which locations in the bulk should they be taken?
- How should they be analysed?

These issues are not easily determined and the development of microbiological standards requires detailed knowledge of the fish meal processes and their technological limitations, what is reasonably achievable in the current state of practice, knowledge of the pathogens and the methods for their detection, the likely physiological state of the organism in the feed material, whether the organisms are evenly distributed in the material, or whether they may be concentrated in certain locations, and the availability of the proposed detection systems. It thus requires good sampling design to achieve the sensitivity of results required.

The details of the criterion of acceptability of animal feed on import are explained in Chapter 14 of ICMSF (1986) leading, for *Salmonella*, to Case 11, $n = 10$, $c = 0$, $m = 0$, where $n = $ the number of samples to be taken, $c = $ the number of samples allowed to fail the standard and $m = $ the number of *Salmonella* organisms permissible in the sample.

However, feeds will only meet this standard if they are made from materials of low initial contamination rates, are processed in a way which eliminates *Salmonella* contamination, and are stored in a manner which does not allow post-processing contamination.

Acceptability in practice is defined primarily by any legally imposed criteria for both home produced and imported product, secondly by those a company imposes on its suppliers and thirdly by the use to which the feed is to be put. Animals can acquire *Salmonella* infection from feed, and may be asymptomatic while acting as a source of infection to other animals and to the rearing environment. Thus unless the feed itself has been subjected to a sterilising process such as might be provided by irradiation, or some sorts of rendering and pelleting processes, the feed – even if it proves acceptable on its microbiological criterion – may still introduce *Salmonella* into stock at low frequency.

15.5 SUMMARY

Infected Peruvian fish meal exported to the USA and to Europe may have caused the world wide dissemination of *S. agona* through infecting the animals fed, which then acted as further points of infection. Animal feed materials, and the expanding world wide trade in processed foods, offer the potential for world wide spread of infectious pathogens. There are parallels for all foods in international trade.

15.6 REFERENCES

Anon., 1972. *Salmonella agona*: a new hazard. *British Medical Journal*, **iv**, 559.

Anon, 1995. Food Safety Bulletin of the Ministry of Agriculture, Fisheries and Food, UK.

Barrile, J.C., Cone, J.F. and Keeney, P.G., 1970. A study of salmonellae survival in milk chocolate. *Manufacturing Confectioner*, **50** (9), 34–39.

Cowden, J., 1996. Outbreaks of salmonellosis, *British Medical Journal*, **313**, 1094–1095.

Crumrine, M.H. and Foltz, V.D., 1969. Survival of *Salmonella montevideo* on wheat stored at constant relative humidity. *Applied Microbiology*, **18** (5), 911–914.

D'Aoust, J.-Y., 1994. Salmonellosis and the international food trade. *International Journal of Food Microbiology*, **24**, 11–31.

Edel, W., van Schothorst, M., Guinee, P.A.M. and Kampelmacher, E.H., 1970. The effect of pellet feeding on the prevention and sanitary control of *Salmonella* infections in fattening pigs. *Tijdschrift voor Diergeneeskunde*, **95** (5), 289–298.

El-Sabban, F.F., Bratzler, J.W., Long, T.A., Frear, D.E.H. and Gentry, R.F., 1970. Value of processed poultry waste as a feed for ruminants. *Journal of Animal Science*, **31** (1), 107–111.

Ferdori, C. and Cirilli, G., 1968. Examination of export egg-macaroni for gram-negative organisms (*Salmonella, Proteus,* coliforms *etc.*). *Tecnica Molitoria*, **19** (16), 439–440.

FSA. *Food Standards Agency News* Number 3, December 2000. Also at: www.foodstandards.gov.uk.

Goldmintz, D., 1970. Survival of microorgansms in fish protein concentrate under controlled conditions. *Developments in Industrial Microbiology*, **12**, 260–265.

Harris, I.T., Fedorka-Cray, P.J., Gray, J.T., Thomas, L.A. and Ferris, K., 1997. Prevalence of *Salmonella* organisms in swine feed. *Journal of the American Veterinary Medical Association*, **210** (3), 382–385.

Hatha, M.A.A. and Lakshmanaperumalsamy, P., 1997. Prevalence of *Salmonella* in fish and crustaceans from markets in Coimbatore, South India. *Food Microbiology*, **14** (2), 111–116.

Hathaway, S., 1999. Management of food safety in international trade. *Food Control*, **10**, 247–253.

ICMSF, 1980. *Microbial Ecology of Foods*. Volume 2. *Food Commodities*. Chapter 17, *Feeds of animal origin and pet foods*, p. 459–469. Academic Press, New York and London.

ICMSF, 1986. *Micro-organisms in Foods*. Volume 2. *Sampling for Microbiological Analysis: Principles and Specific Applications* (2nd Edition). University of Toronto Press, distributed in Europe by Blackwell Scientific Publications, Oxford, UK.

Killalea, D., Ward, L.R., Roberts, D., de Louvois, J., Sufi, F., Stuart, J.M., Wall, P.G., Susman, M., Schwieger, M., Sanderson, P.J., Fisher, I.S.T., Mead, P.S., Gill, O.N., Bartlett, C.L.R. and Rowe, B., 1996. International epidemiological and microbiological study of outbreak of *Salmonella agona* infection from a ready to eat savoury snack – I. England and Wales and the United States. *British Medical Journal*, **313**, 1105–1107.

Kruse, H., 1999. Globalization of the food supply – food safety implications. Special regional requirements: future concerns. *Food Control*, **10**, 315–320.

LiCari, J. and Potter, N.N., 1970. *Salmonella* survival differences in heated skim milk and spray drying of evaporated milk. *Journal of Dairy Science*, **53** (9), 1287–1289.

McConnell-Clark, G., Kaufmann, A.F., Gangarosa, E.J. and Thompson, M.A., 1973. Epidemiology of an international outbreak of *Salmonella agona*. *The Lancet*, September 1, 490–493.

McCoy, J.H., 1975. Trends in *Salmonella* food poisoning in England and Wales, 1941–72, *Journal of Hygiene*, Cambridge, **74**, 271–282.

Mintz, E.D., Cartter, M.I., Hadler, J.L., Wassell, J.T., Zingeser, J.A. and Tauxe, R.V., 1994. Dose response effects in an outbreak of *Salmonella enteritidis*. *Epidemiology and Infection*, **112**, 13–23.

Phillips, Lord, 2000. The BSE Inquiry: the report. The inquiry into BSE and variant CJD in the UK. HMSO, London, UK.

Powers, E.M., Ay, C., El-Bisi, H.M. and Rowley, D.B., 1971. Bacteriology of dehydrated space foods. *Applied Microbiology*, **22** (3), 441–445.

Richmond, M. 1990. *The Microbiological Safety of Food. Part I*. HMSO, London, UK.

Richmond, M. 1991. *The Microbiological Safety of Food. Part II*. HMSO, London, UK.

Shohat, T., Green, M.S., Merom, D., Gill, O.N., Reisfeld, A., Matas, A., Blau, D., Gal, N. and Slater, P.E., 1996. International epidemiological and microbiological study of outbreak of *Salmonella agona* infection from a ready to eat savoury snack – II: Israel. *British Medical Journal*, **313**, 1107–1109.

Stott, J.A., Hodgson, J.E. and Chaney, J.C., 1975. *Salmonella* in animal feed. *Journal of Applied Bacteriology*, **39** (1), 41–46.

Surkiewicz, B.F., Johnston, R.W., Elliott, R.P. and Simmons, E.R., 1972. Bacteriological survey of fresh pork sausage produced at establishments under Federal inspection. *Applied Microbiology*, **23** (3), 515–520.

Tauxe, V. and Hughes, J.M., 1996. International investigation of foodborne disease: public health responds to the globalisation of food (Editorial). *British Medical Journal*, **313**, 1093.

Extending Shelf Life – Compromising Safety?

Key issues
- Raw sea foods and sushi
- *Vibrio parahaemolyticus*
- Shelf life
- HACCP in seafood management
- Microbiological criteria

Challenge
Extension of the shelf life of fresh fish offers the opportunity of cornering the quality fresh fish market in Australia, but could this also compromise the safety of sushi – specialised preparations of raw sea-foods combined with glutinous rice?

16.1 THE CASE STUDY: EXTENDING THE SHELF LIFE OF RAW SEA-FOODS IN AUSTRALIA

In Australia in the mid-1990s the market for fish was expanding. It was suggested to local fish producers that they could secure a firm grasp on the quality market because with care and good management practices Australian fish could be sold in Australia as fresh product. Unlike the imported products which, it was argued, consumers perceived as of relatively poor quality and after freezing and thawing were prone to rapid spoilage, Australian caught fish need never be frozen. However, in order to dominate the fresh fish market the Australian fishing industry had to be fully aware of the need for excellent temperature control through the production chain, right from the moment of catch (McMeekin, 1990).

In an article in *Fishing Boat News*, a trade paper for fishermen, McMeekin explained the concept of relative spoilage rate (RSR), and clearly demonstrated how by taking 0 °C as a reference point with RSR equal to 1, storage at even one or two degrees above 0 °C would significantly accelerate the rate of spoilage and reduce the good quality life of the fresh fish. For example, at 5 °C the relative spoilage rate would be 2.25, at 10 °C 4.00 and at 20 °C 9.00. Conversely by ensuring that the temperature was as close to 0 °C as practicable the spoilage rate would be close to 1.0. Thus McMeekin held out the promise to the fishing industry that provided they rigorously controlled the temperature of their products they could expect to extend the storage life of fresh fish to as much as two weeks, which would allow more flexibility in distribution, and the opportunity to win the major share of the premium priced, fresh fish market.

It may be that the rise in the consumption of fish in Australia coincides with the demographic change Australia is experiencing in which a significant part of the population is now of Asian ethnicity originating particularly from Japan, Vietnam and Indonesia. These peoples habitually eat raw and cooked fish and other sea-foods in their diets.

There is a contrast in the aetiology of food poisoning cases in Japan, an Asian country, and in the UK, a northern European country. In Japan for the three years 1996, 1997 and 1998 *Salmonella* and *Vibrio parahaemolyticus* infections caused the greatest number of outbreaks, and in 1997 and 1998 also caused the greatest numbers of cases. Yet in England and Wales, *V. parahaemolyticus* hardly appears in the statistics, the number of incidents being so very few in comparison to those caused by *Salmonella* and *Campylobacter* (Table 16.1; see also Table 4.1).

Although raw oysters have been eaten in the UK over many centuries they also represent the only seafood normally eaten raw. However, in the period 1994–1998 the City of London saw a three-fold rise in the number of Japanese food outlets selling Japanese style sushi – complex combinations of raw fish and cooked glutinous rice (Bride, 1998). In the early 1990s these products had been almost entirely limited to a few specialised restaurants catering for the small Japanese community and visitors. But with the increase in Japanese investment in the UK the numbers of Japanese people working and visiting increased, leading to a new market for these products. Thus people in the UK have the opportunity to consume another type of raw fish product in addition to oysters, and where habitually only cooked fish is eaten. Pre-packaged sushi are now widely available in the major retail supermarket food stores. So it might be expected that cases of *V. parahaemolyticus* food poisoning would arise (West, 1989), although to date (2002) outbreaks in the UK remain uncommon.

Throughout the world and not only in Australia and in the UK, because the market for fish and fish products is changing, the safety of raw fish and raw fish products is regulated by both international and national food law (Lupin, 1999), and implemented with the help of good quality official guidance (DOH, 1994). Since sushi are very complex in form, in packaging and in the fish varieties used assessing the risk associated with them requires a detailed knowledge of the production, the constituents and the associated hazards. Thus the enforcement agencies need to put in place strategies for managing the food safety risk associated with raw sea-foods ready for consumption.

16.2 BACKGROUND

16.2.1 *Vibrio parahaemolyticus*

Vibrio parahaemolyticus causes an acute gastro-enteritis characterised by diar-

Table 16.1 *Food poisoning incidence in Japan 1996–1998*

Year	Outbreaks	Cases	Outbreaks of known aetiology	Cases of known aetiology
1996	1217 (100%)	46 327	1047 (86%)	41 300 (89%)
			Rank order: *Salmonella* (1) *V. parahaemolyticus* (2)	Rank order: *Salmonella* (1) enteropathogenic *E. coli* (2) *V. parahaemolyticus* (3)
1997	1960 (100%)	39 989	1723 (88%)	29 625 (74%)
			Rank order: *V. parahaemolyticus* (1) *Salmonella* (2)	Rank order: *Salmonella* (1) *V. parahaemolyticus* (2)
1998	3059 (100%)	44 645	2953 (97%)	43 536 (93%)
			Rank order: *V. parahaemolyticus* (1)[a] *Salmonella* (2)	Rank order: *V. parahaemolyticus* (1) *Salmonella* (2)

[a]Most outbreaks small in scale but rather frequent: 234 outbreaks.
Source: WHO, 1999.

rhoea, stomach cramps, nausea, vomiting, headache and chills. It is normally a mild self-limiting condition, which occurs about 15 hours (range 4–96 hours) after consumption of the infected food, and lasts about 2–3 days, and only rarely is hospitalisation required (ICMSF, 1996). The infective dose is about 10^5–10^7 organisms dependent on the strain and the susceptibility of the consumer.

It is a mesophilic, halophilic, motile, marine vibrio which is to be found in warm (above 15 °C) inshore marine and estuarine waters throughout the world, and is frequently isolated from crabs and shrimp, as well from molluscs and fish. The majority of those strains which are pathogenic for human beings produce a thermostable haemolysin, an enterotoxin, and are known as 'Kanagawa positive'.

This obligate halophilic organism can be differentiated from other species of vibrio, some of which are also pathogenic (*e.g. V. vulnificus*) by biochemical attributes, by serology and by ELISA and DNA probe techniques. Its growth limits, survival characteristics and heat sensitivity are summarised in Table 16.2.

It is reported that where they occur outbreaks display a seasonality, with more occurring in the warmer, summer period (Anon, 1999). *V. parahaemolyticus* being

Table 16.2 *V. parahaemolyticus: characteristics*

	Temperature (°C)	pH	Water activity (a_w)	Salt (%)	Atmosphere
Growth ranges					
Range	5–43	4.8–11	0.940–0.996	0.5–10	Aerobic to anaerobic
Optimum	37	7.8–8.6	0.981	3	Aerobic

Survival
Dies when exposed to <5–7 °C
Highest mortality at 1–5 °C
Persists in frozen sea-foods, dying only slowly over long periods

Effect of heat
Temperatures above 47 °C are destructive
Destruction is rapid above 60 °C, *e.g* in crab homogenate D_{65} °C < 1 minute

Source: ICMSF, 1996.

a marine organism and associated with marine foods, it is to be expected that one of the large outbreaks (691 cases) Japan experienced was indeed associated with the consumption of a marine food, but surely not (as happened) with boiled crabs, a cooked food, because *V. parahaemolyticus* is heat sensitive.

16.3 EXERCISES

Exercise 1. Care in the management of fresh sea-food can extend its shelf life. Do you think that the Australians were wise to attempt extending the shelf life of their raw fish, or could it have encouraged a rise in the incidence of *V. parahaemolyticus* food poisoning?

Exercise 2. What are appropriate parameters for the control of the microbiological safety of sushi, and other raw fish products especially in relation to *V. parahaemolyticus*?

Exercise 3. What role would setting criteria for *Enterobacteria* and faecal indicators play in managing the risk of *V. parahaemolyticus* in raw sea-foods, and fish products to be eaten raw?

16.4 COMMENTARY

Exercise 1. Care in the management of fresh sea-food can extend its shelf life. Do you think that the Australians were wise to attempt extending the shelf life of their raw fish, or could it have encouraged a rise in the incidence of *Vibrio parahaemolyticus* food poisoning?

There are number of issues you need to consider.

The shelf life of raw fish diminishes due to chemical changes in the tissues associated with rigor mortis and its resolution, subsequent microbial growth leading to further chemical changes and development of unpleasant odours. The rates at which both of these processes occur are affected by the storage temperature, faster quality loss occurring at higher temperatures. So stored fish has a simultaneously changing chemical and microbiological profile.

Of the population of micro-organisms present the temperature will favour growth of some against others, selecting those for which it is the most favourable. The types of organisms found on marine shellfish, fish and molluscs differ according to the ambient water temperature. Although populations are always mixed, psychrotrophic Gram negative organisms tend to dominate in cold water fish; mesophilic Gram positive ones dominate in fish taken from warmer waters. But *V. parahaemolyticus*, while frequently found on warm water inshore fish in concentrations below $10^3\,cm^{-2}$, can also be found occasionally at much higher levels, and can be found in fish from cold waters – such as the North Sea – particularly in the summer months when the inshore waters are warmer. However, it is reported that most (98%) *V. parahaemolyticus* strains found are Kanagawa negative, suggesting they are non-pathogenic (ICMSF, 1996).

To answer the question fully you will need to research:

- what sorts of temperatures different Australian waters have (remembering that the north is tropical, and Tasmania in the south is temperate in climate);
- whether *V. parahaemolyticus* has been found on Australian fish and thus
- whether it is a member of the microflora of marine products from Australian waters.

At this stage you need to consider the conditions under which *V. parahaemolyticus*, and the spoilage flora grows and survives, and whether and when they would be provided. You must then consider whether *V. parahaemolyticus* could compete with the spoilage flora and whether it would reach infective levels before or after signs of spoilage due to other causes arose. You are thus addressing some of the issues which arise in 'care in the management of sea-food' and the manipulation of parameters which could help in the extension of shelf life. Do not forget to consider the effects of packaging and modification of the atmosphere in contact with the fish.

You may wish to consider downloading some predictive software from the internet which may help you in these predictions:

http://www.arserrc.gov/mfs/pathogen.htm
http://www.df.min.dk/micro/ssp

These programs will help you to understand how the populations of spoilage organisms or pathogens theoretically will respond to changes in environmental parameters – and you will be able to identify the relative effects of changes in pH, temperature and so on. However, remember that the predictions are only as good as the data they are based on, and are only indicative. These models will not be good enough to predict the fine detail of what would happen in the very complex environment of real foods. In comparing fresh fish with frozen fish you also need to consider how *V. parahaemolyticus* responds to the processes fish imported into Australia may undergo. How does it respond to cold storage, to freezing, to thawing, to being held at 0–5 °C?

Consulting Table 16.2 may help you formulate your thoughts.

You then need to relate your analysis to the food products most likely to be associated with *V. parahaemolyticus* food poisoning. To do this perhaps you could search out statistics of cases occurring in Australia and elsewhere and determine the foods associated. You will find that raw sea-foods will be associated with outbreaks, but so will cooked sea-foods and other cooked foods. For example a large outbreak, which occurred in Japan in 1996 and in which there were 691 cases, was associated with boiled crabs (Anon, 1999), yet *V. parahaemolyticus* is heat sensitive.

Then in considering whether the practice of extending the shelf life of raw seafoods will become associated with outbreaks of *V. parahaemolyticus* enteritis you need to consider whether the available best quality fish supply (*i.e.* that which is not at all close to the predicted end of shelf life, and should therefore be of best microbiological quality) would meet the volume of market demand? Furthermore do those people making fish products have enough knowledge of the appropriate control steps, and will they implement them (see Exercise 2)? Will they implement HACCP? Do they know how?

Before coming to any conclusions in this exercise you may find it helpful to also undertake Exercises 2 and 3, and then revert to finishing Exercise 1 (then see the end of Section 16.4).

Exercise 2. What are appropriate parameters for the control of the microbiological safety of sushi, and other raw fish products especially in relation to *Vibrio parahaemolyticus*?

In Exercise 1 you have already done some of the background work necessary to develop an answer to Exercise 2. But notice that the word 'safety' is used. Fish products are prone to support the presence, and often the growth, of a variety of pathogenic agents – bacteria, viruses and parasitic worms in particular (Table 16.3). The presence/absence of this whole range of hazards depends greatly on the marine species, the waters from which the marine product originated, the

Table 16.3 *Sushi: potential bacterial and other hazards*

Hazard	Name	Example of association with food-borne disease	Ref.
Bacteria	*Aeromonas hydrophila*	Outbreak associated with raw oysters	1
	Bacillus cereus	Rice, and other foods including fish associated with outbreaks	4
	Campylobacter jejuni	Sushi caterers also store, handle and process poultry and/or red meat products *Campylobacter* normally associated with raw poultry	7
	Clostridium botulinum type E	Common in marine sediments; marine products throughout world associated with outbreaks Growth can occur in anaerobic fish flesh	6
	Clostridium perfringens	Outbreaks associated with fish have been known	4, 9
	Listeria monocytogenes	Widely distributed in the environment; has been isolated from shrimp and crabs; listeriosis has been associated with smoked mussels, shrimps and mussels	8
	Plesiomonas shigelloides	Generally considered waterborne; has been isolated from many fish and crustacea and shellfish eaten raw	7, 9
	Salmonellae	Marine and fresh water fish, molluscs and shellfish become contaminated from sewage polluted waters. Three cases of infected salmon sushi reported	7, 9
	Shigella spp.	Infection is similar to that from *Salmonella*	7, 9
	Staphylococcus aureus	Infection from human carriers in food preparation	9, 10
	Vibrio cholera 01	Inshore pollution of waters	7, 9
	Vibrio cholera non 01	Inshore pollution of waters	7, 9
	Vibrio parahaemolyticus	Inshore pollution of waters	7, 9
Viruses	Small round structured viruses (SRSV), and hepatitis A virus	Inshore pollution of waters, together with accumulation in tissues due to filter feeding habits of shellfish and molluscs	9

Parasites	Nematode worm	*Anisakis simplex* and similar species	Squid, herring, salmon, mackerel and perhaps tuna can be infected. A problem in Japan, and other raw fish eating countries	3, 5, 7
	Nematode worm	*Angiostrongylus* spp.	Associated with freshwater shrimps originating in SE Asia	2, 3
	Fluke	*Paragonimus* spp. 'Lung fluke'	Associated with raw or undercooked crab, crayfish or shrimp	2, 3, 5
	Broad fish tapeworm	*Diphyllobothrium latum* and *pacificum*	Normally associated with freshwater fish: perch, salmon, trout, pike, turbot. Increased incidence reported in USA due to increased consumption of sushi	5
Other		Scombroid fish poisoning	Associated with scombroid fish: tuna, mackerel; and some non scombroid fish	9

References relating to Table 16.3

1. Abeyta, C. Jr., Kaysner, C.A., Wekel, M.M., Sullivan, J.J., Stema, G.N. *et al.*, 1986. Recovery of *Aeromonas hydrophila* from oysters implicated in an outbreak of food borne illness. *Journal of Food Protection*, **49**, 643–646.
2. Brier, J.W., 1992. Emerging problems in seafood-borne parasitic zoonoses. *Food Control*, **6** (1), 2–7.
3. Cook, G.C., 1993. Zoonotic parasitic infections transmitted by food and drink: diagnosis and management. *Health and Hygiene*, **14**, 5–9.
4. Cowden, J.M., Wall, P.G., Adak, G., Evans, H., LeBaigue, S. and Ross, D., 1995. Outbreaks of foodborne infectious intestinal disease in England and Wales: 1992 to 1994. *Communicable Disease Reports CDR Review*, **5** (8), R105–R117.
5. Higashi, G.I., 1985. Foodborne parasites transmitted to man from fish and other aquatic foods. *Food Technology*, **39**, 69–92.
6. HMSO, 1992. Advisory Committee on the Microbiological Safety of Food. Report on vacuum packaging and associated processes. London, UK.
7. ICMSF, 1996. *Micro-organisms in Foods*, Volume 5. *Microbiological specifications of food pathogens*. Blackie Academic and Professional, London, UK.
8. McLauchlin, J. and Nicols, G.L., 1994. Listeria and seafood. *PHLS Microbiology Digest*, **11** (3), 151–154.
9. Scoging, A.C., 1991. Illness associated with seafood. *Communicable Disease Report, CDR Review*, **1** (11), R117–R122.
10. Tranter, H.S., 1991. *Foodborne staphylococcal illness* in *Foodborne Illness – A Lancet Review*. E. Arnold, London, UK.

Source: Bride, A., 1998.

hygiene of handling, and the subsequent production, transportation and retail processes.

Secondly you need to familiarise yourself with the nature of sushi. If you can, go to your local supermarket and take a look. Read the label and see the ingredients list. Otherwise you may have to rely on the internet, local knowledge or books.

Control of risk of *V. parahaemolyticus* in sushi requires management techniques which:

• minimise the use of contaminated marine products, involving selection of types of fish, and sources;
• ensure the temperature of the products during production is always below the growth threshold, *i.e.* below 7 °C;
• are aware that food composition and adjusting the pH below the optimum for growth will discourage *V. parahaemolyticus* growth;
• only permit the highest levels of hygiene in production to avoid cross-contamination at all stages of production;
• ensure strict temperature control in distribution;
• allocate a very short shelf life to the sushi, defining the storage parameters;
• in retail sale do not allow exposure to ambient temperatures for periods greater than 2 hours, and preferably shorter periods.

In managing all microbiological hazards in fish products, and thus controlling the overall safety risk, through the use of HACCP consideration has to be given to all pathogens likely to be present. For example cysts can, and must, be destroyed in raw fish and this can be achieved by freezing (and afterwards thawing) the types of fish affected with this problem. Other pathogens may actually be encouraged if insufficient thought is given to the type of packaging. For example, the survival and growth of *Clostridium botulinum* may be encouraged by vacuum packaging product. Is vacuum packaging being considered ? If so what else, if anything, controls the organism? What other hazards are there and how are they controlled (see Lupin, 1999)?

Exercise 3. What role would setting criteria for *Enterobacteria* and faecal indicators play in managing the risk of *Vibrio parahaemolyticus* in raw sea-foods, and fish products to be eaten raw?

Fish and other marine species have their own microflora, and do not naturally carry those organisms typical of the mammalian microflora, including *Escherichia coli*, the faecal coliforms, and enterococci. When human enteric organisms are found in marine foods it may be indicative either of the product being sourced from grossly faecally polluted waters, or of later contamination and mishandling in premises and with equipment which themselves are colonised with *Enterobacteria*.

Setting criteria for *Enterobacteria* in fish products provides opportunity to accept or reject products on the basis of their contamination level and would thus only provide general guidance on the safety or otherwise of the product in relation to the possible presence of human enteric pathogens (Anon, 1992). High numbers would indicate that there has been opportunity for population growth, probably due to temperature abuse.

Knowledge of the presence of unacceptable counts of *Enterobacteria* helps to manage the risk of cross-contamination, and temperature abuse, both factors of significance in the specific problem of management of risk from *Vibrio para-haemolyicus*. However *V. parahaemolyticus* is not necessarily associated with faecal pollution and its numbers correlate poorly with *Enterobacteria*, but it is likely to occur where the products are fished from warm inshore waters, or in warm seasons. Where there is known to be a possibility of its occurrence specific criteria should be set, and the organism specifically tested for.

If the presence of *V. parahaemolyticus* is identified it does not mean that pathogenic strains are present, merely that they could be. Further tests on the strains isolated would be needed to resolve that issue. Criteria for fish products are defined in ICMSF, Volume 2 (ICMSF, 1986).

Exercise 1 – conclusions

If fresh Australian fish were managed with all appropriate controls in place using a HACCP approach microbial contamination at source would be minimal. Excellent temperature control throughout the production and transportation and retail chains would retain the fish at the lowest temperature above freezing, around 1–2 °C. This would also control the rate of natural chemical biodegrada-tion. Equally retaining 1–2 °C would guarantee prevention of growth of *V. parahaemolyticus*, and provide the temperature range at which it dies most rapidly. Optimal hygiene practice is needed to ensure that high quality raw fish products do not experience cross-contamination from sources of contaminating organisms and *V. parahaemolyticus*. Thus it seems possible to provide conditions in which extending the shelf life of fresh fish would not provide opportunity for increased cases of gastro-enteritis caused by *V. parahaemolyticus*.

16.5 REFERENCES

Anon, 1992. National Advisory Committee on Microbiological Criteria for Foods – recommendations. Microbiological criteria for raw molluscan shellfish. *Journal of Food Protection*, **55** (6), 463–480.

Anon, 1999. WHO Surveillance Programme for the control of foodborne infections and intoxications in Europe. *Newsletter*, No. 62, December.

Bride, A.J., 1998. MSc Dissertation: An evaluation of microbiological risks associated with the consumption of Japanese style raw fish products (sushi and sashimi) in the City of London. South Bank University, London, UK.

DOH, 1994. EC 91/492. Shellfish hygiene Directive and EC 91/493 Fish Hygiene Direc-tive: Environmental Health Officers Guidance Package. Department of Health, Lon-don, UK.

ICMSF, 1986. *Micro-organisms in Foods*. Volume 2. *Sampling for Microbiological Analy-sis: Principles and Specific Applications*. 2nd Edition. University of Toronto

Press/Blackwell Scientific Publications, Oxford and London, UK.

ICMSF, 1996. *Micro-organisms in Foods.* Volume 5. *Microbiological Specifications of Food Pathogens.* Blackie Academic and Professional/Chapman and Hall, London, UK.

Lupin, H.M., 1999. Producing to achieve HACCP compliance of fishery and aquaculture products for export. *Food Control,* **10**, 267–275.

McMeekin, T., 1990. Temperature and sea-food spoilage. *Fishing Boat World*, February, p. 43.

West, P.A., 1989. The human pathogenic vibrios: a public health update with environmental perspectives. *Epidemiology and Infection,* **103**, 1–34.

WHO, 1999. WHO surveillance programme for control of foodborne infections and intoxications in Europe. *Newsletter*, No. 62, December 1999, 5–6.

Acceptable, Unsatisfactory and Unacceptable Concentrations of Pathogens in Ready-to-eat Food

Key issues
- Meat pâté
- *Clostridium perfringens*
- Food poisoning risk factors
- Developing microbiological criteria
- Microbiological quality guidelines

Challenge

This case study concerns commercially processed Belgian pâtés, made in 2000, retail samples of which were shown in January 2001 to have 'unsatisfactory' levels of *C. perfringens* present, resulting in the national withdrawal of the product range from the UK market. The challenge to you is to consider the circumstances under which *C. perfringens* presents an unacceptable human health risk, and how that relates to the presence of the organism in food.

17.1 THE CASE STUDY: *C. PERFRINGENS* IN BELGIAN PÂTÉS, 2000–2001

17.1.1 Outline

At the beginning of January 2001 environmental health officers employed by a local authority took some routine samples of pâté on retail sale in supermarkets and delicatessens. The microbiological results indicated that it contained levels of *C. perfringens* which were 'unsatisfactory'. As a result the company instigated a product recall, removing all the affected brands from the shops. They had independent samples taken and the preliminary results confirmed those of the local authority. The products were Belgian style pâtés, and were labelled 'use by 23 March 2001'.

In a national survey of cold ready-to-eat sliced meats from catering establishments in the UK during which 3494 samples were analysed (Gillespie, Little and Mitchell, 2000), 892 samples (26%) were classified using guidelines published by the PHLS (PHLS, 2000) as of 'unsatisfactory' quality, and among the 15 samples categorised as 'unacceptable' high levels of *Escherichia coli*, *Staphylococcus aureus*, *Listeria* species and/or *C. perfringens* were found.

In the withdrawal of the product from the shelves guidelines which had been drawn up by the Public Health Laboratory Service were used (Table 11.10; PHLS, 2000). But in the absence of such guidelines taking a decision on the potential risk associated with infected product could present an official with problems.

17.2   BACKGROUND

17.2.1   Some Recipes for Pâtés

Observation of the labelling of retail pâtés gives some indication of their composition (Table 17.1), although it does not reveal how they have been made.

Table 17.1 *Belgium pates: examples of composition and labelling*

Chilled Ardennes pâté	*Organic Ardennes pâté*	*Brussels pâté*
A decorated pâté, presented in a plastic dish and vacuum packed, 450 g	*Bottled, ambient stable, 180 g*	*Bottled, ambient stable, 250 g*
pork 51%	pork, organically farmed	pork
pork liver 31%	pork liver, from organically farmed pork	pork liver
pork stock	pork stock, from organically farmed pork	pork stock
		eggs
		cream
onions	onions, organically farmed milk powder, from organically farmed cows	onions
salt	salt	salt
starch		
dextrose	dextrose	dextrose
spices	spices and herbs, organically farmed	spices
milk proteins		
	thickeners E410, E415	thickener E412
		garlic
emulsifier E472c		
antioxidant E301	antioxidant E300	antioxidant E301
preservative E250	preservative E250	preservative E250
Keep refrigerated 0–5 °C	*Best before 08–2002*	*Best before 10–2002*
Use by 25 April 2001	*Store in cool dry place. Refrigerate after opening and consume within 3 days, and by the date shown*	*Store in cool dry place. Refrigerate after opening and consume within 3 days, and by the date shown*

Note: for your purposes 'today's' date is 9 April 2001.

17.2.2 Sodium Nitrite as Preservative

Table 17.1 shows that preservative E250 was used in the three recipes. E250 is the preservative sodium nitrite which may be added up to certain maximum levels to listed foods – defined (in the UK) in the regulations (Anon, 1995), made under the Food Safety Act, 1990 (Anon, 1990) (Table 17.2). The addition of nitrites or nitrates to foods can control the germination of spores of clostridia which having survived in the food after its processing might both compromise the safety and quality of the food.

Table 17.2 *Preservatives: sodium and potassium nitrite and nitrate*

	Preservative[a]	Food	Indicative in-going amount	Residual amount
E249 E250	Potassium nitrite Sodium nitrite	Non-heat treated, cured, dried meat products	$150\,mg\,kg^{-1}$	$50\,mg\,kg^{-1}$
		Other cured meat products, canned meat products. Foie gras, foie gras entier, blocs de fois gras	$150\,mg\,kg^{-1}$	$100\,mg\,kg^{-1}$
E251 E252	Sodium nitrate Potassium nitrate	Cured meat products; canned meat products	$300\,mg\,kg^{-1}$	$250\,mg\,kg^{-1}$ expressed as $NaNO_3$
		Hard, semi-hard and semi-soft cheese; dairy based cheese analogue	$300\,mg\,kg^{-1}$	$50\,mg\,kg^{-1}$ expressed as $NaNO_3$
		Pickled herring and sprat	$300\,mg\,kg^{-1}$	$200\,mg\,kg^{-1}$ residual amount formed from nitrite, expressed as $NaNO_2$
		Foie gras, foie gras entier, blocs de fois gras	$300\,mg\,kg^{-1}$	$50\,mg\,kg^{-1}$ expressed as $NaNO_3$

[a] In which 'preservative' means any substance which prolongs the shelf life of a food by protecting it against deterioration caused by micro-organisms.

Source: Anon, 1995. Miscellaneous Food Additives Regulations, 1995. (SI 1995/3187), HMSO, London, UK.

17.2.3   *C. perfringens* and Food Poisoning

The major source of the organism is the gastrointestinal tract of food animals, their faeces and gut contents, and thus is often associated with raw meat. In the UK outbreaks of *C. perfringens* food-borne illness occur very much less frequently than do outbreaks caused by *Salmonella* and *Campylobacter*, and have often been associated with institutional catering kitchens, rather than with products made for display and subsequent retail sale (Tables 17.3, 17.4 and 4.1). There are two significant food poisoning strains of *C. perfringens* (types A and C), with type C capable of causing severe food-borne illness in people who are significantly undernourished, and type A being the strain usually incriminated in food-borne illness in the UK.

The ICMSF (1986) classified *C. perfringens* type A as 'a moderate hazard with potential for limited spread' (ICMSF, 1986) (Table 17.5).

Table 17.3 *Food poisoning: incidence of outbreaks of C. perfringens food poisoning*

Year	General outbreaks	Family outbreaks	Sporadic cases	All cases[a]
1980	53	2	2	1056
1981	44	2	–	918
1982	65	4	–	1455
1983	63	5	–	1624
1984	64	4	–	1716
1985	58	6	–	1466
1986	49	2	8	896
1987	47	4	–	1266
1988	56	1	3	1312
1989	52	3	–	901
1990	51	2	5	1442
1991	43	1	7	733
1992	33	3	–	805
1993	37	1	1	562
1994	27	3	7	449
1995	26	–	1	342
1996	33	1	3	720
1997	35	3	1	721
1998	22	–	4	523
1999[b]	9	N/A	N/A	245

[a]Numbers positive, and/or ill.
[b]Provisional figures.

Source: www.phls.co.uk. Reproduced with permission of the PHLS Communicable Diseases Surveillance Centre. © PHLS.

Table 17.4 *General outbreaks of illness due to C. perfringens, England and Wales: weeks 50–52/99 and week 01/00*

Health Authority	Place of outbreak	Month of outbreak	Number ill	Cases positive	Suspect vehicle	Evidence
Shropshire	School	September	92	1	Coronation chicken	M
Liverpool	Residential institution	November	7	5	None	–
Manchester	Residential institution	December	15	3	None	–
Bexley and Greenwich	Prison	October	8	8	None	–
Enfield and Haringey	Restaurant	October	4	–	Chicken roti and rice	M
Sheffield	Canteen	December	5	–	Turkey	–

Key: M = microbiological: identification of an organism of the same type from cases and in the suspect vehicle or vehicle ingredient(s), or detection of toxin in faeces or food.
Source: Anon, 2000. Reproduced with permission of the PHLS Communicable Diseases Surveillance Centre. © PHLS.

17.3 EXERCISE

What should be the bases for describing the levels of *C. perfringens* found in the pâté as acceptable, unsatisfactory or unacceptable? How do these principles relate to the presence of other pathogens in foods?

Table 17.5 *Bacterial hazard categories*

	Organism	Comments
1. Organisms which represent no direct health hazard	Spoilage organisms	
2. Organisms which represent a low, indirect health hazard	Indicator organisms	
3. Moderate hazards: limited spread	Cause food-borne disease but subsequent spread is rare: *Bacillus cereus, Clostridium perfringens* type A *Vibrio cholera* non-O1 *Vibrio parahaemolyticus* *Staphylococcus aureus* (enterotoxigenic strains) *Campylobacter fetus* subsp. *jejuni* *Yersinia enterocolitica* *Yersinia paratuberculosis*	Almost ubiquitous and cause illness when the dose consumed is high High numbers in marine fish, shellfish and crustacea can occur in 'warm' conditions Accumulated enterotoxins associated with growth to high numbers in food result in emesis in the consumer
4. Moderate hazards – potentially extensive spread	Cause food-borne disease with potentially extensive spread: β-haemolytic *Streptococcus* group A, C and G pathogenic *Escherichia coli* *Salmonella typhimurium* and other *Salmonella* serovars *Shigella flexneri* *Shigella boydii* *Shigella sonnei*	Small numbers cause disease; or, secondary infections occur among contacts of infected persons; or cause serious disease among susceptible groups; initial illness can be followed by serious sequelae; can readily be spread from raw foods of animal origin to other raw or processed foods

5. Severe hazards   Cause severe disease:
Brucella melitensis
Brucella abortus
Brucella suis
Clostridium botulinum types A, B, E and F
Clostridium perfringens type C
Hepatitis A virus
Salmonella typhi
Salmonella paratyphi A, B and C
Shigella dysenteriae I
Vibrio cholerae O1

Source: ICMSF, 1986.

17.4   COMMENTARY

> **Exercise.** What should be the bases for describing the levels of *C. perfrin-gens* found in the pâté as acceptable, unsatisfactory or unacceptable? How do these principles relate to the presence of other pathogens in foods?

This exercise allows you to clarify your own understanding of the bases on which the microbiological risk associated with a pathogen in a food is judged. You should find that you raise, perhaps initially in fairly random order, a number of points such as those below.

- What would happen to a consumer if the pâté was eaten with a particular number of *C. perfringens* cells per gram; would the quantity of pâté eaten be important?
- Is there any significant difference between consuming cells or spores?
- How does *C. perfringens* cause food poisoning?
- What is the dose range which is likely to cause food poisoning; is any consumer category more vulnerable?
- What are the sources of *C. perfringens*, and how does it spread?
- How is pâté made and processed? – how could viable cells or spores of *C. perfringens* remain present? What are the D-values for cells, and for spores in meat products? Would the product fat content affect them? What role would added nitrite play? How can balance be achieved between processing to make the product with the right taste and texture requirements, with product which has a good enough microbial profile for safety and will achieve the shelf life desired, under chilled or ambient conditions? Consider the difference between bottled pâtés, which can be stored at ambient and are designed to have a shelf life of at least 18 months, and chilled, vacuum packaged product with a shelf life of perhaps 21 days – see Table 17.1. How long was the shelf life of the products withdrawn from the market (Section 17.1)?
- What can be learned from *C. perfringens* outbreaks, and data such as those shown in Tables 17.3 and 17.4?
- What levels of microbial population reduction could be achieved by differing thermal processes?
- What would allow residual *C. perfringens* spores to germinate and grow (see for example Blankenship *et al.*, 1988)? What size population of *C. perfringens* could then be achieved in a processed food? You would need to consider the conditions under which the organism would multiply in the pâté – thinking about the growth preferences of the organism – water activity, pH, tempera-ture, oxidation–reduction potential and how these things would affect popula-tion change. Would other organisms be present and/or grow?
- What are the likely handling conditions of the product after processing to the moment of consumption? Is handling abuse possible?
- On testing could other *Clostridia* be confused with *C. perfringens* and if so, what are the implications?

Have you yet any clear idea of the issues which are influential in determining the risk associated with *C. perfringens*, and is it now possible to define 'acceptable' and 'unacceptable' limits for *C. perfringens* in the cooked product, and could you suggest some values for these limits?

You may need to re-organise your thoughts and structure them on the lines below:

The food – pâté is a food designed to be 'ready-to-eat', and should thus be safe to eat at the moment of purchase. Equally it should be safe to eat after purchase, the time period depending on whether the product is chilled or ambient stable. The Sheriff, in his judgement in Case Study Case 12 said that cooked meat stew should have been delivered to the users free from the contaminating organism *E. coli* O157:H7 which caused the outbreak. But there is a difference between Case Study 12 and Case Study 17 – the Sheriff was talking in the context of a very dangerous but heat sensitive organism readily eliminated by gentle cooking, and which could really only be present if the cooked meat product was subjected to post-cooking contamination. But while *C. perfringens* type A is less dangerous than *E. coli* O157:H7 it produces heat resistant spores not destroyed by cooking processes in which the food does not reach temperatures above 100 °C and is therefore more difficult to eliminate from the food. After germination of the spores the vegetative cells can grow in the product if its storage temperature is above 10 °C. So pâté prepared in this way might contain some viable cells or spores of *C. perfringens*.

Consideration of Table 17.1 shows three pâtés of similar, but not identical, composition. Their textures are different with the Ardennes pâté being of a coarser cut than the Brussels pâté, which is very fine, smooth, and contains cream and eggs. They are also presented differently, one type being ambient stable for at least 16–18 months, the other designed for chilled storage with a residual shelf life of 12 days at 0–5 °C. (What sort of shelf life did the products in Section 17.1 have?) In Table 17.1 two products contain preservative E250 – sodium nitrite – and one does not. Thus the products vary, yet each may contain *C. perfringens* after processing. How many can be tolerated? Should processing be at a temperature/time combination to eliminate *C. perfringens* from the product, *i.e.* not detectable in the package weight, or can some organisms at low level be tolerated – not detectable in 25 g, or not detectable in 1 g product or some other standard?

The organism is an anaerobic spore bearing organism which causes food-borne gastro-enteritis through the production of enterotoxin while forming spores in the human gut. It occurs in meats and is likely therefore to be present in the raw meat pâté mixture. Its spores could survive mild heat treatment – particularly in the central regions of the pâté where heat penetration would be slowest and the lowest temperature achieved. Surviving spores could be activated by the heat, germinate in the cooling pâté and grow if the pâté remained warm centrally. Under optimal conditions it can multiply very fast indeed with doubling times close to 10 minutes (Schroder and Busta, 1971). Given the opportunity it can reach very high populations in meat products.

The consumer of the infected food suffers diarrhoea and abdominal pain which usually becomes evident within 12–18 hours of consuming it. The infective dose is normally considered to be in excess of 10^6 cells, some of which survive the acid in the stomach, pass through to the small and large intestines and multiply there. Sufferers excrete high numbers of the organism, but the faeces of healthy non-sufferers may contain low levels of *C. perfringens*.

For food poisoning to occur three factors need to come together – the food, its population of pathogenic cells, and a consumer who eats sufficient of the infected food to receive an infective dose of organisms, and who is sufficiently vulnerable to the enterotoxin to show symptoms of illness.

The risk factors are thus:

- the nature of the organism, the moderate severity of the illness caused, and the fact that the potential for the organism to spread is limited;
- the type of food;
- the presence of the organism in the food and its potential to grow there;
- underprocessing which fails to adequately reduce the population;
- post-processing cooling which is so slow that it allows population growth;
- storage abuse of the finished product, either before or after opening the sealed container, which facilitates microbial growth;
- the vulnerability of the consumer;
- the amount of the infected food the consumer eats, and the number of organisms consumed.

After considering the characteristics of the organism, the severity of the illness it causes and its infective dose range, determining *acceptable* count thresholds, below which the food would be considered safe, is approached by consideration of two factors:

- what low population could reasonably be achieved by processing the food under good manufacturing practices?
- the potential for control of that residual population in defined storage conditions of defined duration.

so that at any time when the pâté might be eaten, population density would remain below that which would cause illness.

Unacceptable counts are deemed those which could rapidly lead to populations which would cause food poisoning on ingestion.

Since a common heating process for producing chilled pâtés might be described as 'pasteurisation' rather than a 'sterilisation', on testing it would be expected that there could be some surviving viable vegetative cells as well as spores. The overall microbial quality would be indicated by the aerobic plate count and among these could be both heat sensitive and heat resistant organisms such as *Bacilli*. Equally the testing for the anaerobic population could be set up either to detect *Clostridia* in general or *C. perfringens* specifically.

However, if the products were intended to be ambient stable for 18 months (refer to Table 17.1) and without the addition of preservative, the product would have to be subjected to sterilisation and the processed product should be commercially sterile.

Table 17.6 indicates what might happen to a batch of food after it has been sampled and analysed. Interpretation of the microbial counts found should take

Table 17.6 *Microbiological criteria: the possible experiences of food after microbi-ological sampling*

1. The food can be expected to experience conditions which reduce the level of concern
2. The food can be expected to experience conditions at which the level of concern remains the same
3. The food can be expected to experience conditions which increase the level of concern

Source: ICMSF, 1986.

account of the three possibilities shown. In the case of the pâté the numerical value of the 'acceptability' threshold has to recognise that no further processing (*i.e.* cooking) is intended before the product is consumed, yet the product is at risk of thermal abuse – that is it might be exposed to a temperature and time regime in storage which allows microbial increase.

In the development of microbial criteria for foods the categorisation of the microbial hazards (Table 17.5) into five groups has been achieved by the colla-tion and consideration of what is known about the organisms and how and when they cause food-borne illness. This knowledge has been combined with the three possible post-sampling 'histories' of the food (Table 17.6) to provide 15 separate conditions for product criteria (ICMSF, 1986). So finally in the interpretation of the microbial counts found in a retail food sample it is necessary to take into account what is expected to happen to that food – whether it is to be cooked (which could reduce the microbial population), to be consumed straight away, or to experience a future period of storage which might allow opportunity for microbial increase before consumption. Thus the pâté on retail sale (Section 17.1) was ready for immediate consumption, but might not have been eaten for some days to come, either because it failed to sell until then, or it was purchased and then stored.

Standards for the acceptability of the presence of different pathogens in food also differ. For microbial hazards which are categorised as 'moderate with the potential for extensive spread' or 'severe' their presence is tolerated only at very low level – absence of *Salmonella* in 25 g is very common. But absence from a greater quantity 50 g, or 100 g might be applied if the consumer, such as a baby, is very vulnerable. But the pathogen (*C. perfringens*) is a moderate hazard with limited potential to spread so greater numbers of the organism may be tolerated, provided the food is handled properly after that moment of testing. So in a chilled pâté where the expectation is that it will be held chilled at 5 °C until it is eaten, and the quantity eaten would not normally exceed perhaps 100 g, where the pathogen cannot multiply below 10 °C (*C. perfringens*), counts per gram up to 10^2 would be acceptable, and greater numbers up to $10^4\,\mathrm{g}^{-1}$ would also be tolerable but described as 'unsatisfactory', and more than $10^4\,\mathrm{g}^{-1}$ would be 'unacceptable' (see Tables 11.10 and 11.11).

'Unsatisfactory' values are those which may indicate that some form of abuse has been experienced by the product in or after manufacture, and constitute a warning. For organisms such as *C. perfringens* if the food were to be consumed at that time the product would be unlikely to cause illness but time is running out when that could be considered to be the case.

But for other organisms (categories 4 and 5 in Table 17.5), such as *Salmonella*, exceeding the acceptable threshold immediately places the product beyond 'unsatisfactory' and straight into the 'unacceptable' category because the presence of the organism at that concentration places the consumer at an unacceptable level of risk of food poisoning.

Thus in interpreting the microbial count of a pathogen found in a food – the pâté in this case study – 'acceptability' or otherwise would be based on the factors outlined above. You may also wish to relate this case with Case Study 2, the duck pâté due to which eight people died of botulism, and to Case Study 11, where the presence of Listeria in ready-to-eat foods is discussed.

Further reading: IFST, 1997; ICMSF, 1986.

17.5 SUMMARY

C. perfringens is an anaerobic, spore-bearing organism capable of causing food poisoning. Outbreaks are frequently associated with precooked meat in which the population has developed in the interior regions at temperatures at which growth is possible. Pâté is a comminuted meat product, often subjected to a pasteurisation heat process, so control of the residual population depends on good temperature control.

Microbiological standards for any pathogen seek to ensure that, if the organism is present, its numbers will be low enough not to present an unacceptable risk to a consumer.

17.6 REFERENCES

Anon, 1990. Food Safety Act, 1990. HMSO, London, UK.

Anon, 1995. Food Safety (Miscellaneous Food Additives) Regulations, 1995. HMSO, London, UK.

Anon, 2000. General outbreaks of illness, England and Wales: weeks 50–52/99 and week 01/00. *Communicable Disease Report*, **10** (2), 10.

Blankenship, L.C., Craven, S.E., Leffler, R.G. and Custer, C., 1988. Growth of *Clostridium perfringens* in cooked chilli during cooling. *Applied and Environmental Microbiology*, **54** (5), 1104–1108.

Gillespie, I., Little, C. and Mitchell, R., 2000. Microbiological examination of cold ready-to-eat sliced meats from catering establishments in the UK. *Journal of Applied Microbiology*, **88**, 467–474.

ICMSF, 1986. *Micro-organisms in Foods*. Volume 2. *Sampling for Microbiological Analysis: Principles and Applications*. 2nd Edition. Blackwell Scientific Publications, Oxford, UK.

IFST, 1997. Development and use of microbiological criteria for foods. *Food Science and Technology Today*, **11** (3), 137–177.

PHLS, 2000. Gilbert, R.J., de Louvois, J., Donovan, T. *et al.* Working group of the PHLS Advisory Committee for Food and Diary Products. Guidelines for the microbiological quality of some ready-to-eat foods sampled at point of sale. *Communicable Disease and Public Health*, **3** (3), 163–167.

Schroder, D.J. and Busta, F.F., 1971. Growth of *Clostridium perfringens* in meat loaf with and without added soybean protein. *Journal of Milk and Food Technology*, **34**, 215–217.

E. Managing Risk

Managing Risk

Key issues
- Sandwiches
- Food poisoning organisms
- Factors affecting shelf life
- Ranking risk, and risk management
- HACCP training

Challenge
You are placed in the position of a manager who has to decide between two systems of production of sandwiches.

18.1 THE CASE STUDY: SHOULD SANDWICHES SOLD IN A HOSPITAL CAFÉ BE MADE BY VOLUNTEERS?

"Sandwich makers fight cuts"

This was the headline in a small article in the UK newspaper *The Telegraph* on 20 May 1999. © Telegraph Group Limited, 1999. The article reported

"Members of the Women's Royal Voluntary Service (WRVS) who make sandwiches at a Pembrokeshire [Wales] hospital are under threat because of a new hygiene rule.
The WRVS has said that all sandwiches must be pre-packed, and the 68 volunteers at Withybush Hospital, in Haverfordwest, are no longer needed. A spokesman for the WRVS said it wanted to protect its volunteers from possible prosecution under the Food Safety Act [1990] adding, "we can see the volunteer's point of view because they have never had an incident of food poisoning"."

A quotation from the sales literature of a sandwich filling company:

"XYZ are manufacturers of ready made sandwich fillings which are premixed, ready for use and conveniently packed in 2.5 kg and 1 kg containers.
Our sandwich fillings are manufactured in a new purpose built factory equipped to comply with EEC standards. Microbiological controls are enforced by

our technical team, whilst our own, on-site laboratory continually tests both raw materials and finished products.

The purchasing and storage of a large range of raw materials necessary to produce a good selection of sandwich fillings is hard work. [You need] Different suppliers, storage and handling facilities for meat, fish, cheese, tinned goods, frozen foods, chilled foods, fresh produce, and salads.

Stock control, waste and labour all take up valuable time – so let us do the work for you!

XYZ Foods is an approved manufacturer of ready made sandwich fillings to the British Sandwich Association.

All products are produced with up to 10 days shelf life and distributed nationally in chilled transport."

You can see the range of fillings a sandwich company might make (Table 18.1) – or you can get an idea of these for yourself by looking in many a sandwich bar in the street, or a food industry exhibition.

Table 18.1 *Sandwich fillings – examples*

Chicken with sage and onion	Roast and smoked ham and peppers
Coronation chicken	Bacon, egg and mayonnaise
Chicken tikka and mayonnaise	Sausage and egg mayonnaise
Tandoori chicken and mayonnaise	Corned beef and mayonnaise
Mexican chicken and mayonnaise	
Chinese chicken and mayonnaise	Crab sandwich filling
Chicken mayonnaise	Prawn mayonnaise
Chicken, sweetcorn mayonnaise	Prawn coleslaw
Turkey mayonnaise	Smoked salmon and soft cheese
Turkey, sweetcorn mayonnaie	Salmon and mayonnaise
Turkey, ham, mayonnaise	Smoked salmon sandwich filling
Turkey, ham, celery, mayonnaise	Smoked trout sandwich filling
	Tuna mayonnaise
Waldorf salad with tuna	Tuna, sweetcorn, mayonnaise
Tuna and egg salad	Tuna, peppers, mayonnaise
Curd cheese, celery and banana	Mackerel and mayonnaise
Curd cheese, ham and pineapple	
Cheese coleslaw	Cheese and mayonnaise
Cheese and apple salad	Cheese spring onion and mayonnaise
	Soft cheese, nuts and celery
	Egg mayonnaise

18.2   EXERCISE

Identify the risk factors associated with both the volunteer method of sandwich making and the commercial pre-packing method of sandwich making, and form your own evaluation of whether you feel that in relation to hospital sales the WRVS decision was appropriate.

18.3 BACKGROUND AND COMMENTARY

The Food Safety (General Food Hygiene) Regulations, 1995, made under the Food Safety Act (FSA) (1990) require all food producers in the production of their food to implement a safety system based on HACCP, and for food handlers to be trained 'commensurate with their work activity'. This applies to both commercial and voluntary concerns. Producers deemed to be producing food under inappropriate conditions can be closed down, if continuing to fail after guidance from enforcement officers. The issue at the heart of this case study is whether due diligence in the production of sandwiches could better be demonstrated by the commercial or voluntary organisation to a court of law should an outbreak of food poisoning arise. This defence is allowed in the UK under the FSA (1990) and permits the defendant to argue that they had taken all reasonable steps to prevent such a happening.

WRVS cafés in hospitals generally serve the needs of the visitors to patients, but patients who are able to walk can also use them. Thus the cafés supply consumers who could include very vulnerable hospital patients.

Sandwiches are made of bread and fillings. They may

- contain ingredients/fillings known to be at risk of containing pathogens (*e.g.* cooked chicken/*Salmonella*);
- be at risk of being contaminated in production, and their composition and storage conditions may support the survival or growth of those pathogens;
- they will not be subject to processes such as cooking which will destroy the pathogens.

They are thus 'high risk' foods.

To minimise the risk associated with such products requires well managed production processes which incorporate HACCP, or 'Assured Safe Catering' (ASC).* To achieve putting such systems in place managers may first need training (Mayes, 1994; Moy, 1994). This is a need in common to both the volunteer-made, and the commercially-made product, and should be followed through by the training of the staff. Thus you need to consider how such systems could be implemented in both the volunteer production system and in the commercial set-up, and within that process you need to consider the types of sandwiches produced.

In both cases the sandwiches have to be labelled, and stored under refrigerated conditions, but the commercially made ones have to be distributed from factory to point of sale in temperature controlled vehicles. In the UK The Food Safety (Temperature Control) Regulations, 1995 require ready to eat foods to be held at either $<8\,^{\circ}C$ or above $63\,^{\circ}C$.

The shelf life of the sandwiches must be defined. Because bread stales, commercially made sandwiches are normally given a shelf life of 1 or 2 days, but if non-staling bread, or bread which does not become soggy with the contents of the sandwiches, is used, longer shelf lives may be attributed. You should be aware

---

* Hazards are grouped as 'microbial growth', 'cross-contamination' and so on, rather than the very specific approach of HACCP in which each microbial hazard is individually addressed.

also that if bought-in fillings (see Table 18.1) are given by their manufacturers a shelf life of, for example, 10 days then the shelf life of the sandwiches would also have to fall within that date code.

Non-packaged sandwiches made locally by volunteers

• Are there risks associated with the type of sandwiches, and/or with the management systems which can be implemented with volunteers and their levels of knowledge and training (Ryan *et al.*, 1996)? Does non pre-packing imply more handling, more exposure to incidental cross contamination, less temperature control . . .?

Commercially made sandwiches, packaged, chilled and transported from a location distant from the hospital

• What does pre-packing imply? Longer allocated shelf life yet higher risk of pathogens growing and surviving? Production distant to the point of use and difficulties in temperature control? Reduction (or not) in handling and cross-contamination? Possibility of modified atmosphere packaging (what benefits could that provide?);
• What does commercial production imply? The possibility of implementing better (or worse) management systems; the need for control extending over a long cool chain from factory to consumer; confidence that product will be used within date code? How can you be sure?

You will need develop your list (as above) and to weigh the risks of each system and come to a view yourself and decide whether you feel that the WRVS decision could have been different and whether the system with lower risk was actually chosen. You also need to evaluate whether because no food poisoning outbreak has ever been associated with a system or company, that provides assurance that none will occur. You are thus ranking the two systems as 'more' or 'less' risk laden.

Cross refer to Case Study 12.

18.4   REFERENCES

Anon, 1990. Food Safety Act, 1990. HMSO, London, UK.
Anon, 1995. The Food Safety (Temperature Control) Regulations, 1995. HMSO, London, UK.
Mayes, T., 1994. HACCP training. *Food Control*, **5** (3), 191–195.
Moy, G., Kaferstein, F. and Motarjemi, Y., 1994. Application of HACCP to food manufacturing: some considerations on harmonisation through training. *Food Control*, **5** (3), 131–139.
Ryan, M.J., Wall, P.G., Gilbert, R.J., Griffin, M. and Rowe, B., 1996. Risk Factors for outbreaks of infectious disease. *Communicable Disease Review*, **6** (13), R179–R183. Available at http//:www.phls.co.uk

CHAPTER 19

Changing a Risk Management Strategy

Key issues
* Raw fruit and vegetable salads
* *Shigella* spp.
* Risk management

Challenge
Pre-prepared raw fruit and vegetable salads for retail sale or catering use represent processed foods for which achieving acceptably low food safety risk is difficult. The outbreak described relates to an incident of shigellosis associated with the retail sale of raw fruit salads. The challenge for you is to determine the procedures which should reduce risk to an acceptable level and give confidence in their suitability for retail sale.

19.1 THE CASE STUDY: *SHIGELLA FLEXNERI* IN FRUIT SALADS

Supermarkets in the UK now sell refrigerated fresh fruit salads either as sealed pre-packaged boxes, or from salad bars at which customers can help themselves by scooping out various quantities into 250 and 500 g cartons, thus making fruit combinations to their own particular tastes. Often there is a shop assistant who will weigh the filled containers, and attach the weight and price label to the lid.

In November 1998 newspapers quoting a newly published Government document (a confidential report from a Health Authority) reported that a large supermarket, in Haywards Heath, Sussex, UK, was being sued by some of its customers because fruit salads in one of its stores had been strongly linked with the severe shigellosis they had suffered. In the outbreak, in which there were 46 confirmed cases, five people and an 18 month old baby were admitted to hospital.

In an article dated 4 December 1998 (Anon, 1998) it was reported that the supermarket was due to contest liability in the High Court the following Friday. The cases had arisen earlier in the year in August but while investigations had failed to identify the presence of *Shigella* in any fruit samples analysed, there was a strong statistical link between people eating fruit and falling ill. A spokeswoman for the Environmental Health Authority concerned was reported as saying "In my view there is insufficient evidence to sustain a criminal prosecution under the Food Safety Act, but there is strong circumstantial evidence linking the outbreak to an asymptomatic food handler." The article also indicated that [faecal] samples had been taken from 18 food handlers and their holiday, training and sickness records examined. These enquiries were all negative. The spokeswoman was also reported as saying that the supermarket had complied with all of the council's requirements on food safety as a pre-condition of re-opening the bar.

19.2   BACKGROUND

19.2.1   *Shigella* spp. and Food-borne Disease

Different species of *Shigella* cause dysentery and acute painful enteritis in man, and are transmitted by direct person-to-person contact, and indirectly *via* foods and water. The illness appears one to six days after infection, following the consumption of as few as 100 cells (ICMSF, 1996). The illness can be fatal, but in those who recover the organisms can continue to be shed in faeces for up to 11 weeks, although normally the organisms are shed for a much a shorter period (PHLS, 2001). These people who become asymptomatic can be an important source of spread of the illness.

   Shigellas are very host specific, with man as the primary host. They enter the body *via* the mouth, and multiply in the lumen of the large intestine. There they attach to the intestinal wall, enter epithelial cells and cause epithelial lesions at which there is both the accumulation of mucus and the release of endotoxins resulting in epithelial cell death. These processes show themselves in the patient as severe diarrhoea in which blood and mucus are present. *Shigella* strains have in common a somatic lipopolysaccharide endotoxin. Additionally some strains may also produce a thermolabile soluble polypeptide toxin, the Shiga toxin, which has significant adverse effects on the human cells invaded. It kills the cells by preventing protein synthesis, and prevents the cells' normal fluid uptake from the colon which would then be passed to the body, both causing diarrhoea and severe dehydration. But the various strains of *Shigella* possess these capabilities to different degrees (Table 19.1).

   In the UK *Shigella sonnei* is the most common, and mildest, cause of shigellosis (bacillary dysentery) with illness caused by the other three species, *S. flexneri*, *S. boydii* and *S. dysenteriae*, occurring less commonly (Table 19.2).

   The most common route of infection is faecal–oral in which hands contaminated with faeces come into contact with water or foods, which in turn infect other people. Flies are also significant in the spread of the organisms, as is the contact of contaminated surfaces with foods. *Shigellae* spread rapidly among small children in nurseries and schools where managing their effective hand washing hygiene may be difficult. Similarly *Shigella* may spread in old people's homes and other institutions. The organism is associated with diarrhoea, with sewage, with infected drinking water, with contaminated insects, with poor hygiene and with infected people. Thus raw fruits and vegetables may become

Table 19.1 *Shigella: categorisation of types*

	High levels of toxin produced	*Low levels of toxin produced*
Invasive	*Shigella dysenteriae*: High levels of shiga toxin Severe infection	Other *Shigella* strains: Do not produce high levels of toxin
Non invasive	Produce verocytotoxin	

Source: ICMSF, 1996.

Table 19.2 *Shigella laboratory reports, England and Wales, faecal isolates, 1986– 2000*

	Shigella sonnei	*Shigella flexneri*	*Shigella boydii*	*Shigella dysenteriae*	*Total*
1986	3734	755	84	43	4616
1987	2596	809	85	50	3540
1988	2571	830	100	41	3542
1989	2227	749	108	48	3132
1990	2313	824	95	44	3276
1991	9840	674	101	46	10661
1992	17240	711	80	38	18069
1993	6186	603	57	54	6900
1994	5630	570	75	40	6315
1995	3566	450	65	32	4113
1996	1282	397	115	55	1849
1997	1467	404	79	49	1999
1998	878	299	75	43	1295
1999	907	243	70	42	1262
2000[a]	695	189	56	26	966

[a]Provisional figures.

Source: Laboratory Reports to CDSC; PHLS Laboratory of Enteric Pathogens. Last updated: 16 March 2001. Reproduced with permission of the PHLS Communicable Diseases Surveillance Centre. © PHLS.

contaminated in agriculture and in food processing through use of infected water, equipment, surfaces, or the hygiene failures of people working in food processing and handling. Once in the foods the organisms will probably only grow both if the environment is supportive and if the organisms are not in competition with other micro-organisms. Nevertheless growth of *Shigella* in raw fruits such as freshly cut raw cubes of papaya, jicama and water melon (Escartin *et al.*, 1989) and vegetables can occur (Davis *et al.*, 1988, quoted in Madden, 1992). Additionally ICMSF (1996) cites reports of the survival of *Shigella* in foods at a wide range of pH, temperatures and water activity values, organisms which are otherwise delicate and would be killed by heat in much the same manner as *E. coli* or *Salmonella* can be.

19.2.2   Salads and Food-borne Disease

Beuchat (1996) reviewed the pathogenic organisms associated with fresh produce – fresh vegetables, fresh fruits and salads. He demonstrated that 37 types of vegetables in 18 countries had been reported as associated with outbreaks of food-borne illness caused by the bacterial strains *Aeromonas, Bacillus cereus, Campylobacter jejuni, E. coli* O157:H7, *Listeria monocytogenes, Salmonella, Shigella, Staphylococcus, Vibrio cholerae,* and *Yersinia enterocolitica,* as well as Hepatitis A virus, Norwalk virus and the protozoan *Giardia lamblia.*

19.3   EXERCISE

If you re-read the case study in Section 19.1 you will understand that *Shigella* was not demonstrated as present in any samples taken. Clearly the supermarket would have been very concerned about such an incident. From where do the risks to the microbiological safety of fruit salads arise? What would the supermarket have checked? And what, after checking procedures, would give them the confidence to continue selling raw cut fruit salads?

Finally does an outbreak of food-borne disease found to be associated with a food provide reason to change existing food safety management practice?

19.4 COMMENTARY

Exercise. If you re-read the case study in Section 19.1 you will understand that *Shigella* was not demonstrated as present in any samples taken. Clearly the supermarket would have been very concerned about such an incident. From where do the risks to the microbiological safety of fruit salads arise? What would the supermarket have checked? And what, after checking procedures, would give them the confidence to continue selling raw cut fruit salads?

Finally does an outbreak of food-borne disease found to be associated with a food provide reason to change existing food safety management practice?

You could focus on a single organism group (*Shigella*), or you could additionally take a holistic view – which is after all what is needed before the sale of fresh packaged vegetable or fruit salad, or self-service fruits.

You might find it useful to consider a single type of product – for example fresh melon cubes, and research how pathogens could contaminate, survive and possibly grow in such a product (Golden *et al.*, 1993). Bear in mind that the surface of the fruit may be contaminated with soil as well as other materials and, unless steps are taken to decontaminate the outside, organisms present on the skin could contaminate the internal fruit when cut. Secondly organisms on the fruit (or vegetable) at the country of origin may be spread to importing countries. For example, in 1994 iceberg lettuce of Spanish origin was the cause of shigellosis in the same season in Norway, Sweden and the UK (Frost *et al.*, 1995), demonstrating that the importers must be confident in procedures at the point of origin.

Since *Shigella* is an organism strongly associated with faecal material, and is one which does not compete well with other organisms, which may also have originated from faeces, you need to evaluate the production chain up to and including the point of retail sale and identify specific points where such organisms may have entered the product and survived or even grown in it. Having determined those points, you need to consider what control measures would be effective to minimise risk.

Since the 1980s the market for packaged salad vegetables and fruits has grown, and concern for how to manage the safety of the products has grown with it. In 1987 Brocklehurst *et al.*, examined the microflora of salad products at the end of shelf life, and in these cases found them 'satisfactory' – not containing types or levels of organisms of concern. Decontamination methods for fresh fruits and vegetables have been subject to some research, particularly since concern about the occurrence of *Listeria monocytogenes* on fresh vegetables has grown in parallel with the expanding prepared salad and sandwich industries. Whereas clean water washing can usefully remove dirt and detritus, it does little to reduce microbial population, much of which adheres firmly to cellular structures. Because of the high surface tension of plant surfaces, concentrations of chlorine in the region of 200–300 μg ml^{-1} are needed to reduce microbial population by greater than two log cycles (Beuchat, 1996).

In the USA the International Fresh-cut Product Association has developed a model HACCP plan for these products – which is of course a very appropriate approach. But to evaluate how the outbreak (Section 19.1) described could have been prevented you also need to consider that any HACCP plan is only as good as its implementation, particularly in respect of monitoring and corrective actions taken. The control tools largely available in the management of the microbiological safety of cut fresh fruits and vegetables are good management practices, effective hygiene training of staff, management of staff who become ill and their exclusion from working with food both during and after illness according to national and international guidelines (PHLS, 1993), close monitoring of the continued practices of hygiene, as well as the avoidance of the use of any source of contaminated water, effective plant sanitation, appropriate fruit and vegetable sanitation, temperature controls to reduce microbial increase, and correct packaging. Finally at retail shop and other similar outlets management of the risk of product contamination by staff, equipment or by the public.

When the supermarket had re-analysed the production, distribution and retail sale procedures, they would then have decided, through following those processes, whether they needed to change their risk management strategy to maintain acceptably low risk from the products.

Postscripts:

1. It is also worth considering some comments made by Dean Cliver (Cliver, 1997) in relation to 153 cases of hepatitis A in Calhoun County, Michigan USA, and found to be associated with frozen (raw) strawberries. He speculated on the value of tightening rules, increasing inspection, and taking more samples for analysis commenting that this would increase the price of strawberries, but necessarily reduce risk. He also considered the value of immunising food workers against hepatitis A, saying that would not, of course protect food from other food-borne diseases carried by workers. Indeed he ended his comment by saying "... my 35 years of work in food safety tells me that a risk-free food supply is impossible ...".
2. Picking blackcurrants for my own use, at the local 'pick them yourself' fruit farm in the UK, I noticed a dead rat, flies buzzing round, lying on the ground just near where I was picking fruit. The week before I had picked strawberries, and eaten quite a number as I picked.
3. A useful reference (CDSC, 2000) links a national increase in multiresistant *Salmonella typhimurium* DT104 to salad vegetables.
4. A newpaper article dated 7 July 2001 (*Guardian* newspaper, London) reported 14 cases of salmonellosis linked to washed, ready-to-eat salads.

19.5 REFERENCES

Anon, 1998. 'Supermarket denies poisoning customers', *Environmental Health News*, **13** (7), 1.

Beuchat, L.R., 1996. Pathogenic organisms associated with fresh produce. *Journal of Food Protection*, **59** (2), 204–216.

Branigan, T., 2001. *Salmonella* risk in ready to eat salads. *The Guardian* newspaper, 7 July, p. 4.

Brocklehurst, T.F., Zaman-Wong, C.M. and Lund, B.M., 1987. A note on the microbiology of retail packs of prepared salad vegetables. *Journal of Applied Bacteriology*, **63**, 409–415.

CDSC, 2000. Case control study links salad vegetables to national increase in multiresistant *Salmonella typhimurium* DT104. *Communicable Disease Report*, **10** (37), 333, 336.

Cliver, D.O., 1987. Hepatitis A from strawberries: who's to blame? *Food Technology*, **51** (6), 132.

Davis, H., Taylow, J.P., Perdue, J.N., Stelma, G.N., Humphreys, J.M., Rowntree, R. and Greene, K.D., 1988. A shigellosis outbreak traced to commercially distributed shredded lettuce. *American Journal of Epidemiology*, **128**, 1312–1321.

Escartin, E.F., Castillo Ayala, A. and Saldana Lozano, J., 1989. Survival and growth of Salmonella and Shigella on sliced fresh fruit. *Journal of Food Protection*, **52**, 471–472, 483.

Frost, J.A., McEvoy, M.B., Bentley, C.A., Anderson, Y. and Rowe, B., 1995. An outbreak of *Shigella sonnei* infection associated with the consumption of iceberg lettuce. *Emerging Infectious Diseases*, **1**, 26–29.

Golden, D.A., Rhodehamel, E.J. and Kautter, D.A., 1993. Growth of *Salmonella* spp. in Cantaloupe, Watermelon, and Honeydew melons. *Journal of Food Protection*, **56** (3), 194–196.

ICMSF, 1996. *Micro-organisms in foods.* Volume 5. *Characteristics of Microbial Pathogens.* Blackwell Academic and Professional, London, UK.

Madden, J.M., 1992. Microbial pathogens in fresh produce – the regulatory perspective. *Journal of Food Protection*, **55** (10), 821–823.

PHLS Working Group on the control of *Shigella sonnei* infection, 1993. Revised guidelines for the control of *Shigella sonnei* infection and other infective diarrhoea. *Communicable Disease Review*, **5**, 23 April.

PHLS (Public Health Laboratory Service, UK), 2001. Guidelines on the management of communicable diseases in schools and nurseries. http//:www.phls.co.uk/advice. Accessed 14 July 2001.

Hygiene Improvement at Source

Key issues
- Raw meat in the abattoir
- Microbial load and enteric pathogens
- *E. coli* O157:H7
- Hygiene
- Risk based scoring system for hygiene improvement
- Risk management

Challenge

In abattoirs traditional meat inspection techniques, which are based on observation of visible defects in meat, such as tubercles and worms, do not detect the presence pathogens such as *E. coli* O157:H7 which cause no visible change. In managing the risk to public health from zoonotic disease spread through infected meat, approaches other than the traditional inspection techniques are being developed. Evaluation of hygienic practice through a risk based scoring system was introduced in the UK in 1994, and improvement encouraged through the regular publication of the scores associated with the named companies. The challenge to you is first to understand the system, and then to evaluate whether, on microbiological grounds, you agree with the system and its weightings.

20.1 THE CASE STUDY: ASSESSING AND SCORING HYGIENE DURING THE SLAUGHTER AND PREPARATION OF RED MEAT

Sections 20.1.1–20.1.4 give a summary of how a system of evaluating the hygiene of slaughter and meat preparation in UK abattoirs using a risk based scoring system was introduced.

20.1.1 Circular from MAFF

A circular letter from the UK Ministry of Agriculture, Fisheries and Food [MAFF] dated 22 October 1993 said:*

"To: Interested Parties on the attached list

* Reproduced with permission. Crown copyright.

Dear All,
Proposals for a meat hygiene performance monitoring system

I am writing to invite comments on the attached system* for assessing hygiene standards in licensed fresh meat slaughterhouses. Comments are invited both on the detail and the weighting of the scoring system, and on how this system should apply to the work of central and local government in enforcing meat hygiene controls. You may therefore find helpful a brief description of the structure and principles behind the proposed scoring system.

The assessment of risk, and the focussing of enforcement attention on key hazards, is a principle of the Government's approach to regulation which has recently been emphasised in the light of the Prime Minister's Deregulation initiative. This does not mean adding an extra tier of controls to the existing regulatory system, but rather ensuring that the enforcement effort is targeted towards the key areas of performance.

Following the introduction of the new Single Market legislation, we believe that it is important that the concept of a 'single standard' in meat production should be assessed on a more formal and transparent basis. We believe that our proposals, modified as necessary following this consultation exercise, will provide a mechanism for doing this.

The attached scoring system is intended to define key parameters relating to hygiene of production, and to weight them according to what we believe to be their relative importance. It is designed so as not to discriminate against any particular size of plant; any aspect which is not present in a plant and so cannot be judged, would be given the maximum (best) score. You will see that the scoring system gives the greatest weight to the hygiene of operation of the slaughterhall, as the aspect which has the greatest impact on the standard of the end product. The system thus puts most emphasis on the skill and competence of the slaughtermen and their attention to hygiene. The scores would not represent a 'risk rating' for each premises, but they would correlate with the performance of a given premises.

We envisage the assessment system having two main functions.

The first would be to provide local authorities with an assessment of the performance of premises, which could be taken into account in determining supervision and enforcement requirements. This would be consistent with the recommendation of the DTI [Department of Trade and Industry – a Government Department] Scrutiny Report on the enforcement and implementation of EC legislation in the UK, which recommended that supervision levels should be set on an individual basis using risk assessment techniques and the evidence of plant visits. We will wish to discuss with local authorities the extent to which the performance assessment could be used to set targets against which effectiveness of supervision could be measured.

---

* Attached to the circular letter was a set of guideline notes (relating to each point from A1 to E4 using the system shown in Table 20.1) and intended for the use of the relevant meat hygiene inspector before s/he evaluated the many aspects of hygiene of an abattoir, and eventually assigned a score.

Table 20.1 *Red meat slaughterhouses – assessment score sheet (original scheme)*

	Highest score – minimum risk a	b	c	Lowest score – greatest risk d	Total scored to be completed by assessor e	Overall weighting factor f	Total after weighting applied e×f
A. Antemortem							
A.1 Cleanliness of animals	70	45	20	0			
A.2 Antemortem arrangements	30	15	10	0			
	100	60	30	0			
Total score awarded for Section A						×0.05	Total out of 5
B. Slaughter and dressing							
B.1 Pithing and sticking	13	7	3	0			
B.2 Skinning/depilation	18	12	6	0			
B3 Evisceration	18	12	6	0			
B.4 Correlation of carcases and offal/inspection arrangements	12	6	3	0			
B.5 Cross contamination	5	3	2	0			
B.6 Handling, removal and storage of by-products	10	6	3	0			
B.7 Washing of carcases and offals	11	6	3	0			
B.8 Carcases in chillers	13	8	4	0			
	100	60	30	0			
Total score awarded for Section B						×0.40	Total out of 40
C. Personnel and practices							
C.1 Staff skill/training	18	10	4	0			
C.2 Staff medical programme	8	5	2	0			
C.3 Protective clothing	7	4	2	0			
C.4 Use of washbasins and sterilisers	18	10	5	0			

C.5	Action taken following a contamination incident	10	7	5	0
C.6	Further handling of carcases	13	8	4	0
C.7	Further handling of offal	13	8	4	0
C.8	Other practices	13	8	4	0
		100	60	30	0

Total score awarded for Section C ☐ × 0.30 Total out of 30

D. Maintenance and hygiene of premises

D.1	Rooms, equipment and facilities	21	15	7	0
D.2	Maintenance programme	21	15	7	0
D.3	Plant surrounds	6	3	2	0
D.4	Plant design to control waste water	15	9	4	0
D.5	Layout	10	5	3	0
D.6	Separation of 'clean' and 'dirty' operation	8	4	2	0
D.7	Separation of edible and inedible operations	7	3	2	0
D.8	Fly screening, vermin control *etc.*	7	3	2	0
D.9	Drainage and effluent	5	3	1	0
		100	60	30	0

Total score awarded for Section D ☐ × 0.15 Total out of 15

E. General conditions and management

E.1	Cleaning schedules	30	18	9	0
E.2	Water monitoring	20	11	6	0
E.3	Pest control	15	9	4	0
E.4	Management	35	22	11	0
		100	60	30	0

Total score awarded for Section E ☐ × 0.10 Total out of 10

Total out of 100

Source: Anon, 1993. Reproduced with permission. Crown copyright.

Table 20.2 Red meat slaughterhouses – assessment score sheet (revised scheme)

	Highest score – minimum risk a	b	c	Lowest score – greatest risk d	Total scored to be completed by assessor e	Overall weighting factor f	Total after weighting applied e × f
A. Antemortem							
A.1 Cleanliness of animals	75	56	29	0			
A.2 Antemortem arrangements	25	14	11	0			
	100	70	40	0			
Total score awarded for Section A					☐	× 0.08	Total out of 8
B. Slaughter and dressing							
B.1 Pithing and sticking	6	4	2	0			
B.2 Skinning/depilation	24	19	11	0			
B.3 Evisceration	24	19	11	0			
B.4 Correlation of carcases and offal/inspection arrangements	11	6	3	0			
B.5 Cross contamination	5	3	1	0			
B.6 Handling, removal and storage of by-products	8	5	3	0			
B.7 Washing of carcases and offals	9	5	4	0			
B.8 Carcases in chillers	13	9	5	0			
	100	70	40	0			
Total score awarded for Section B					☐	× 0.37	Total out of 37
C. Personnel and practices							
C.1 Staff skill/training	20	14	8	0			
C.2 Staff medical programme	4	3	2	0			
C.3 Protective clothing	7	5	3	0			
C.4 Use of washbasins and sterilisers	20	14	8	0			

C.5	Action taken following a contamination incident	10	7	4	0
C.6	Further handling of carcases	13	9	5	0
C.7	Further handling of offal	13	9	5	0
C.8	Other practices	13	9	5	0
	Total score awarded for Section C	100	70	40	0

□ × 0.30 Total out of 30

D. Maintenance and hygiene of premises

D.1	Rooms, equipment and facilities	21	17	9	0
D.2	Maintenance programme	21	17	9	0
D.3	Plant surrounds	6	4	3	0
D.4	Plant design to control waste water	15	10	5	0
D.5	Layout	10	6	4	0
D.6	Separation of 'clean' and 'dirty' operation	8	5	3	0
D.7	Separation of edible and inedible operations	7	4	3	0
D.8	Fly screening, vermin control *etc.*	7	4	3	0
D.9	Drainage and effluent	5	3	1	0
	Total score awarded for Section D	100	70	40	0

□ × 0.15 Total out of 15

E. General conditions and management

E.1	Cleaning schedules	30	21	12	0
E.2	Water monitoring	20	13	8	0
E.3	Pest control	15	10	5	0
E.4	Management	35	26	15	0
	Total score awarded for Section E	100	70	40	0

□ × 0.10 Total out of 10

Total out of 100

Source: Anon, 1999. Reproduced with permission. Crown copyright.

The second main function of the scoring system would be to enable the Government to publish regularly a summary of the statistics arising out of the work, to enable enforcement interests and the general public to be informed about trends in standards . . .
[some further text omitted]

Yours sincerely
Head, Meat Hygiene Division"

20.1.2   Monthly Reports from Meat Hygiene Service

As a result of consultations the scoring system was changed a little and the system in current use (current at August 2001) can be seen in Table 20.2. The Meat Hygiene Service has published monthly the individual scores of all abattoirs tested so that it is possible to track the progress of individual abattoirs and determine whether, in the view of the inspectors of the Meat Hygiene Service, the hygiene of individual abattoirs is improving, and also whether there is a trend towards better hygiene in abattoirs in general.

20.1.3   Press Release from Meat Hygiene Service

Six years after the consultations, in June 1999, the Meat Hygiene Service of the UK, issued the following press release:*

"*Press Release*
MHS 6/99
22 July 1999

Chief Executive Praises Staff as MHS Hits Annual Targets

The Meat Hygiene Service (MHS) has successfully met all its targets for the year 1998–99, Chief Executive Johnston McNeill announced today.
    Mr McNeill was speaking following the publication of the annual report of the MHS, whose inspection teams enforce hygiene and animal welfare standards in abattoirs and cutting plants across Great Britain. The report revealed that the MHS has, for the fourth year in a row, hit all of the targets which Ministers have set for it.
    During 1998–99 these included:

• To ensure that on a rolling three month average, at least 93% of licensed premises are achieving Hygiene Assessment System (HAS) scores of more than 65, and that 75% of licensed premises are achieving scores of more than 70;
• To ensure full compliance with Specified Risk Material regulations, which protect animal and public health from any risk of BSE;
• To implement strictly the MHS Clean Livestock Policy;

*Reproduced with permission. Crown copyright.

- And to provide comprehensive training in Hazard Analysis Critical Control Points (HACCP) and ISO 9002 auditing systems for all Official Veterinary Surgeons, Meat Hygiene Inspectors and Meat Technicians.

In addition to meeting its Ministerial targets, the MHS was awarded a Charter Mark award, which recognises outstanding public service, and concluded the first stage of its five year customer satisfaction survey.

Mr McNeill said: "This has been a challenging year for the MHS, but the dedication and professionalism of our staff is evidenced by the successful achievement of all our targets which were set for us by Ministers.

"MHS veterinary surgeons and inspectors play a vital role in ensuring that meat is safe to eat, and in helping the industry to drive up its hygiene standards. Currently the whole of the world is watching to see how Great Britain guarantees the safety of its meat, and the work done by MHS staff is a key component in creating the confidence that purchasers from abroad require.

"The frequency of veterinary supervision continues to be increased, in line with Ministerial instructions, with the MHS working towards full compliance with the relevant European Commission Directives. The MHS recognises that the result is an unwelcome increase in inspection costs, but increased supervision is essential to meet the European Union standards.

"In the area of hygiene, we have substantially exceeded our targets – for example, 89% of plants now have HAS scores of more than 70, compared with our target of 75%. The future will bring us many challenges, and we are determined to continue to deal with them just as successfully".

"Notes for Editors

1. The Meat Hygiene Service (MHS), an executive agency of the Ministry of Agriculture, Fisheries and Food (MAFF), was launched on 1 April 1995. It took over from some 300 local authorities responsibility for enforcing meat hygiene, inspection and animal welfare at slaughter legislation in licensed fresh meat premises in England, Scotland and Wales. Public health and animal welfare are safeguarded in plants by Official Veterinary Surgeons, Meat Hygiene Inspectors and Meat Technicians working as inspection teams.

2. HAS scores are a risk-based method of assessing hygiene standards. They are not required by law, but are one of the tools used to assess whether meat plants are operating satisfactorily under hygiene regulations. HAS scores have been undertaken by the MHS since its inception and continue to be an important guide to the hygiene performance in abattoirs and cutting plants. They are assessed over a period of time, typically a month, and are a general guide to long-term hygiene performance. There is no fixed acceptable or unacceptable score. However, plants using best practice will tend to score highly, while those complying with the minimum statutory requirements will tend to score less. The HAS score must be considered in the light of other information about the structure and operation of the premises. HAS scores are not percentages – a score of less than 100 does not necessarily mean that

the plant is not using best practice or the meat it produces is unfit for human consumption."

[End of press release]

20.1.4   Enforcements Results: Banded HAS Scores

In the UK, in February 2001 a publication from the Meat Hygiene Service, a division of the Food Standards Agency (FSA) published the following data (Table 20.3) and text:*

"Enforcement results: Banded HAS scores

In Great Britain, hygiene standards in slaughterhouses and cutting plants are regularly monitored by means of the Hygiene Assessment System (HAS). This is a risk based method of assessing hygiene standards arising from slaughtered livestock, the people working in the plant, the premises themselves and any other relevant sources.

Premises are assessed against performance criteria covering all significant aspects of products, each weighted according to their relative risk. The end product of an assessment under the system is the 'HAS score', a value of 0–100 with higher scores indicating better performance.

Assessments are carried out by the Official Veterinary Surgeon (OVS) at least once a month in full-throughput premises, and quarterly in low-throughput premises. The results enable the MHS [Meat Hygiene Service] to monitor the performance of licensed premises and identify those premises where additional supervision and enforcement action might be necessary.

- The Food Standards Agency took over food safety functions from the Ministry of Agriculture Fisheries and Food, and public health issues related to food from the Department of Health on 1 April 2000."

[End of press release]

20.2   BACKGROUND

In the 1990s consumers in the UK became very aware that meat and poultry products carried pathogenic organisms which were causing big outbreaks of food poisoning in spite of the fact that the incriminated organisms were usually heat sensitive types, and should have been destroyed by normal cooking processes. The words '*Salmonella*', '*Listeria*', '*E. coli* O157:H7' were being seen regularly in newspaper articles associated with reports of outbreaks, or investigations of the processes of animal rearing. Although the hygiene of food preparation should have been capable of preventing contamination of cooked foods, consumers were increasingly aware also that the scale of the food industry meant that single hygiene errors could lead to mass outbreaks of infection. Outbreaks of

*Reproduced with permission. Crown copyright.

Table 20.3 *Banded HAS scores: red meat slaughterhouses, Great Britain*

HAS band	0–30		31–40		41–50		51–65		66–75		76–85		86–100		Total
	No.	%	No.	%	No.	%	No.	%	No.	%	No.	%	No.	%	No.
2001 12 months to January 2001	0	0.00	6	0.20	23	0.70	191	5.70	1577	47.50	1073	32.30	452	13.60	3322

No. = Number of premises assessed.
% = Percentage of total number of premises assessed.
HAS = Hygiene Assessment System.

Source: Meat Hygiene Enforcement Report, number 46, page 9, February 2001 of the Food Standards Agency, UK.

salmonellosis from chicken meat and eggs occurred with some frequency (for example see Case Study 8), and *E. coli* O157:H7 poisonings associated with milk from dairy cattle and with beef were emerging (for example see Case Studies 10 and 12). It was well known that meat animals and meat carried pathogens such as *Salmonella* and *E. coli*. It had become clear to Government authorities that traditional inspection techniques used in abattoirs were no longer adequate for control of enteric pathogens because they did not identify infected carcases, there being no visible signs to observe.

However, the HACCP technique could, in theory, be used to manage the risk in raw meat production provided that sufficient knowledge existed to identify from where the hazards arose, and how they could be controlled or reduced to acceptable levels (Hathaway and McKenzie, 1991). This of course required the development of knowledge through research in order to identify how practices needed to change to minimise the spread of organisms in the slaughter and meat preparation processes, and to identify where the higher concentrations of organisms are likely to be located on a carcase – affecting both handling practice, and microbial sampling in the abattoir.

Working in Canada, Gill *et al.* (1996) evaluated the hygienic characteristics of a high speed beef carcase dressing process. They identified the brisket site as likely to be the most heavily contaminated site of the three sites they investigated: brisket, rump and neck. By following the carcases through the production processes these same authors identified three different distribution mechanisms of organisms on carcases: deposition, redistribution and decontamination which they explained in the following way:

- redistribution – is likely to involve the movement of bacteria away from, or into a sampled site of 100 cm², rather than redistribution within a site;
- deposition (increase in load) of bacteria is likely during skinning;
- modest decontamination (reduction in load) – may be achieved through trimming and washing, but can also result in redistribution of organisms.

Understanding these processes is important in determining the significance of microbial counts taken during processing and in turn evaluating the hygiene of production. Furthermore they suggested that to identify how enteric pathogens are disseminated during slaughter and meat dressing for HACCP purposes, enumerations of *E. coli* on the carcases were more useful than enumerations of the aerobic flora.

Earlier Lasta *et al.* (1992) had undertaken a four year survey of the microbial quality of beef carcases passing through six different abattoirs in Argentina and cautiously concluded they had some indication that psychrotrophic counts of whole carcases might correlate with the hygiene of management of the abattoirs. They also proposed a carcase sampling plan which would facilitate the acceptance or rejection of single lots of carcases based on sampling ten of them.

It is research work such as that instanced above which gradually clarifies how different processes disseminate organisms, what happens to the organisms, and how processes can be changed to reduce the dissemination of organisms.

But while microbial evaluation of carcases moving through a process can provide information about the effects on the microbial load of sampled carcases, it does not itself evaluate their total hygienic condition, nor their microbial safety (Gill *et al.*, 1996). More targeted analyses – looking for organisms such as *E. coli* indicating faecal contamination – will yield useful information about its dissemination over the carcase, and this may relate to the exact manner in which the carcase is handled, opening the possibility for modification of manner in which the detailed processes following slaughter are carried out. For example Gill *et al.* (1996) observed that they selected their particular sampling area of the brisket because it could become contaminated from the hide should the carcase roll over during side pulling of the hide. Microbiological evidence could back up recommendations for change of procedure which could avoid such a risk. But in the day to day running of abattoirs microbiological sampling may only occur for routine monitoring purposes, or may not occur at all.

Microbiological results normally are obtained some time after sampling and, even if quickly obtained, only after the event. High speed processing lines may deal with 250 beef cattle or several thousand chickens per hour. Microbiological testing cannot monitor a process on a minute by minute basis.

Thus in the drive to raise hygiene standards in abattoirs (see Sections 20.1.1–20.1.4) observation of facilities and practice, followed by rating the risk associated with them, provides a technique. Each abattoir needs to put in place well designed equipment and to train staff in procedures based on best practice and sound understanding of the effect of those practices on microbiological quality of the carcases, and which if implemented correctly should produce meat of good microbiological quality.

The HAS scoring system (in the UK) depends on the inspector who is rating the abattoir understanding the processes and judging whether each element in the assessment is being carried out well (top score), adequately, poorly, or totally unacceptably (lowest score). Of course in order for consistency across evaluations and between inspectors good guidelines have to be provided indicating, for example, how many carcases should be observed and how often; how much time should be spent in the abattoir; whether one visit is enough to form an evaluation or whether more than one visit is needed for an evaluation and so on.

The HAS scoring system has no force of law, but it provides a tool whereby each abattoir can see where it is not performing well. It is also a tool where the detail of past performance of an abattoir can be compared to the present. Furthermore through the overall scores achieved and the publication of those scores along with the names of each abattoir comparison between abattoirs is possible.

But the overall aim of the HAS abattoir scoring system is to achieve better quality, lower risk meat.

20.3 EXERCISES

Exercise 1. Study Table 20.1 (original scheme) and determine for yourself what the background microbiological issues are which relate to each factor A1 to E4. This will require considering from where the microbial hazards of concern arise, and the detail of the processes of meat dressing and packing. You could also perhaps consider what guidance you would offer an inspector to determine whether the score awarded fell into column *a*, *b*, *c* or *d*, to be recorded in column *e*.

Exercise 2. Study Table 20.2 (revised scheme) and reach an understanding of how it and Table 20.1 have been constructed, and why the figures in the columns *a*, *b*, *c* and *d* are what they are.

Exercise 3. You will have noticed that in Table 20.2 some risk rankings have been changed in relation to Table 20.1. Do you agree with the rankings in Table 20.2, and do you support the changes they represent in relation to Table 20.1?

20.4 COMMENTARY

Exercise 1. Study Table 20.1 (original scheme) and determine for yourself what the background microbiological issues are which relate to each factor A1 to E4. This will require considering from where the microbial hazards of concern arise, and the detail of the processes of meat dressing and packing. You could also perhaps consider what guidance you would offer an inspector to determine whether the score awarded fell into column *a*, *b*, *c* or *d*, to be recorded in column *e*.

You will notice that the assessment score sheet is set out in five sections A to E. Sections A and B deal with the processes of managing the animals, their slaughter and the meat dressing processes. Section C deals with how the slaughtermen perform their tasks and whether they follow good hygienic practices. The management is responsible for the premises (Section D) and for the facilities provided, the layout of the processes, and for the overall management (Section E). In microbiological terms pathogens originate from the animals, particularly from their fur and skin which will be contaminated with faeces and environmental dust and dirt. The animal intestine and faeces are primary sources of enteric pathogens, and these organisms can be disseminated by the slaughter and production processes as well as by human action, by flies and vermin. Water and air movements disseminate organisms. The management of the abattoir should have written procedures which provide guidance on how all activities should be done, and staff should be trained in those processes. The methods used should take account of the sources of organisms and the detail of methods used should minimise the spread of organisms and minimise allowing their growth.

Section A2 – ante-mortem arrangements – is particularly concerned with animal welfare and is focused on ensuring that the animals are distressed neither in their transportation to the abattoir nor in the lairage prior to slaughter.

What is particularly sought in the slaughter and post-slaughter processes is that organisms from the fur, skin and intestines of the animals do not contaminate the meat. This requires the use of techniques which are aware of this. The pithing rod should be sterilised immediately after use. Except for the initial perforation of the skin, cuts in the skin should go from inside (clean) to outside (dirty) and not the other way. Care should be taken in evisceration to avoid spilling the gut contents from the anus or from accidental perforation of the gut.

Additional information:

- The guidance sheets offered to inspectors are obtainable from the Meat Hygiene Service – see footnote at the end of Section 20.6.

> **Exercise 2.** Study Table 20.2 (revised scheme) and reach an understanding of how it (and Table 20.1 – original scheme) has been constructed, and why the figures in the columns *a*, *b*, *c* and *d* are what they are.

The overall activities are divided into five sections whose nature differs, and whose overall significance to the achievement of a satisfactory level of hygiene differs. These are:

- A. Ante-mortem procedures
- B. Slaughter and dressing
- C. Personnel and practices
- D. Maintenance and hygiene of premises
- E. General conditions and management

Within each section A–E are a number of activities. Any single activity can be done well or badly, or something in between. Taking an individual activity, for example A1, or C4, or E2 *etc.* it is possible to score the manner in which it is completed, against a scale of satisfaction from 100 to 0, with:

- 100 minimum risk – always well done;
- 60 normally satisfactory/occasionally defective;
- 30 defective;
- 0 high risk – seriously defective practice.

Scoring at the different levels is a matter of the skill and judgement of the assessor.

Taking a group of activities, that is Group A or B or C or D or E, each single activity within a group is scored as explained above. But within the group some can be considered as higher risk than others. The activities in each group, Group B for example, can be ranked. In creating the assessment form a decision had to be made to assign the highest score to the activity of greatest significance to the hygiene of the process. Which is second in importance and so on? Referring to column *a* the rank order chosen is:

- B2 skinning/depilation 24;
- B3 evisceration 24;
- B8 carcases in chillers 13;
- B4 correlation of carcases and offal inspection arrangements 11;
- B7 washing of carcases and offals 9;
- B6 handling, removal and storage of by-products 8;
- B1 pithing and sticking 6;
- B5 cross-contamination 5.

Would you have ranked these in this order? And would you have attributed these maximum scores?

Thus if B2 is done well it makes the greatest contribution within this group to risk reduction; if done badly (score 0) it creates the greatest risk. This is why if skinning is done as well as can be it gets the maximum score (24); if done fairly well it gets approximately 70% of the maximum score (19); done acceptably approximately 40% of the maximum score (11) or nothing.

Each activity in B has been treated in the same way, attributing 100%, approximately 70%, approximately 40% and 0% of the available marks at the various levels of risk.

If you check you will see that the total scores (B1 to B8) come in columns *a*, *b*, *c* and *d* respectively to 100, 70, 40 and 0 points.

You will also see that the sum of the scores for all C activities, all D activities and all E activities likewise add up to 100 in column *a*, and 70, 40, and 0 in *b*, *c* and *d* respectively.

Taking the five groups A to E together the sum of how well those activities are achieved constitutes the assessment of quality of the whole operation – the hygiene of preparation of meat in an abattoir. But one group of activities, A for example, may be less important than another – B for example. Since the whole set is a unit, then the sub parts A to E contribute fractions of the unit and hence are attributed weighting factors relative to their importance. The weighting factors which have been attributed are:

- A 0.08
- B 0.37
- C 0.30
- D 0.15
- E 0.10

which added together equal unity.

Would you have weighted these in this way?

Exercise 3. You will have noticed that in Table 20.2 (revised scheme) some risk rankings have been changed in relation to Table 20.1 (original scheme). Do you agree with the rankings in Table 20.2, and do you support the changes they represent in relation to Table 20.1?

Presumably as a result of the consultation exercise the weightings were changed. Look first at Table 20.4 in which the scorings shown in Table 20.2 have then been multiplied by their appropriate factor, then all the activities from A1 to E4 have been ranked according to the score they attract if done as well as expected (column *a*).

You can do this additionally for yourself with Table 20.1.

Table 20.5 has been created by comparing the rankings achieved in each activity of Table 20.1 – old scheme (you have now done this) and Table 20.2 (new scheme, as shown in Table 20.4) and you will see that the new scheme ranks A1 more highly than did the old scheme. It also ranks less highly than before B1, B5, B6, B7, C2 and E2. It is for you to decide why.

If you now refer back to Table 20.3 in which the scores for one year are summarised you will be able to consider the application of the scheme. The

Table 20.4 *Red meat slaughterhouses – assessment score sheet. Rank order of weightings in the revised scheme shown in Table 20.2*

		a	b	c	d
B.2	Skinning/depilation	8.88[a]	7.03	4.07	0
B.3	Evisceration	8.88	7.03	4.07	0
A.1	Cleanliness of animals	6	4.5	2.32	0
C.1	Staff skill/training	6	4.2	2.4	0
C.4	Use of washbasins and sterilisers	6	4.2	2.4	0
B.8	Carcases in chillers	4.81	3.33	1.85	0
B.4	Correlation of carcases and offal/inspection arrangements	4.07	2.22	1.11	0
C.6	Further handling of carcases	3.9	2.7	1.5	0
C.7	Further handling of offal	3.9	2.7	1.5	0
C.8	Other practices	3.9	2.7	1.5	0
E.4	Management	3.5	2.6	1.5	0
B.7	Washing of carcases and offals	3.33	1.85	1.48	0
D.1	Rooms, equipment and facilities	3.15	2.55	1.35	0
D.2	Maintenance programme	3.15	2.55	1.35	0
C.5	Action taken following a contamination incident	3	2.1	1.2	0
E.1	Cleaning schedules	3	2.1	1.2	0
B.6	Handling, removal and storage of by-products	2.96	1.85	1.11	0
D.4	Plant design to control waste water	2.25	1.5	0.75	0
B.1	Pithing and sticking	2.22	1.48	0.74	0
C.3	Protective clothing	2.1	1.5	0.9	0
A.2	Antemortem arrangements	2	1.1	0.88	0
E.2	Water monitoring	2	1.3	0.8	0
B.5	Cross contamination	1.85	1.11	0.37	0
D.5	Layout	1.5	0.9	0.6	0
E.3	Pest control	1.5	1	0.5	0
C.2	Staff medical programme	1.2	0.9	0.6	0
D.6	Separation of 'clean' and 'dirty' operation	1.2	0.75	0.45	0
D.7	Separation of edible and inedible operations	1.05	0.6	0.45	0
D.8	Fly screening, vermin control *etc.*	1.05	0.6	0.45	0
D.3	Plant surrounds	0.9	0.6	0.45	0
D.9	Drainage and effluent	0.75	0.45	0.15	0
	Total score	100	70	40	0

[a]Example: From Table 20.2, B2 [column *a*] = 24.
    Weighting factor = 0.37.
    24 × 0.37 = 8.88.
Source: Anon, 1999.

Government is able to report similarly each year to identify whether overall improvement across the whole industry has been achieved. Equally for individual abattoirs they can monitor their own performance by keeping a record of their scores as time passes. The enforcement officers can point to individual areas of concern within one abattoir, and monitor improvement numerically, and finally enforcement officers can see from the scheme which abattoirs in the

Table 20.5 *A comparison of the rank order of activities in the original red meat slaughterhouse scheme shown in Table 20.1 and the revised one shown in Tables 20.2 and 20.4*

	Original	Revised
highest risk rating	B2	B2
↓	B3	B3
		A1
	C1	C1
	C4	C4
	B1	
	B8	B8
	B4	B4
	B7	
	B6	
	C6	C6
	C7	C7
	C8	C8
	A1	
	E4	E4
		B7
	D1	D1
	D2	D2
	C5	C5
	E1	E1
		B6
	C2	
	D4	D4
		B1
	C3	C3
	B5	
	E2	
	A2	A2
		E2
		B5
	D5	D5
	E3	E3
		C2
	D6	D6
	D7	D7
	D8	D8
↓	D3	D3
lowest risk rating	D9	D9

geographical areas they control are the worst and perhaps put greater enforcement activity in the abattoirs where help is needed most.

However, you should also be able to see that the nature of the scheme and the marks awarded in each of the columns, the weighting given to activities within a group, and the actual weightings of the five groups (one against the other) affect how the scores total up. You should also be able to see that the design of the score sheet and the attribution of scores depend on informed opinion. The latter could benefit or disadvantage an abattoir at the margins of a score band. The score achieved could affect the opinion of customers for that abattoir influencing them towards or away from doing business.

But for all the areas of opinion in the scoring process the use of such a system should drive hygiene standards upwards.

Are you aware of any such process being applied elsewhere? For example do you think such a process could be applied in the hygiene of food preparation in retail shops?

20.5 SUMMARY

The introduction of this scheme of awarding a 'HAS' score to the implementation of hygiene in abattoirs in based on assessment of the risk associated with the numerous activities involved in the slaughter of animals and the dressing of the meat.

Its purpose is to facilitate raising standards in the industry in general, and in individual abattoirs. The design of the score sheet is based on knowledge of the meat industry and the microbiological changes to meat which can occur during processing, and assesses the relative risks. Although the assessment criteria use the best information to guide judgement, the awarding of scores relies on observation of practice and the professional opinion of the inspector.

Thus although the system has strengths in its practicality in comparison to taking microbiological samples, the weakness of the system lies in its subjectivity.

20.6 REFERENCES

Anon, 1993. Red meat slaughterhouse: assessment score sheet MH7 (Rev. 10/93). Produced by Ministry of Agriculture, Fisheries and Food, UK.

Anon, 1999. Red meat slaughterhouse: assessment score sheet MH9 (Rev. 10/99). Produced by Meat Hygiene Service, Ministry of Agriculture, Fisheries and Food, UK.

Anon, 2001. 'Enforcement results'. *Meat Hygiene Enforcement Report*, February, **46**, 9.

Gill, C.O., McGinnis, J.C. and Badoni, M., 1996. Use of total or *Escherichia coli* counts to assess the hygienic characteristics of a beef carcass dressing process. *International Journal of Food Microbiology*, **31**, 181–196.

Hathaway, S.C. and McKenzie, A.I. (1991). Post mortem meat inspection programs; separating science and tradition. *Journal of Food Protection*, **54** (6), 471–475.

Lasta, J.A., Rodriguez, R., Zanelli, M. and Margaria, C.A., 1992. Bacterial count from bovine carcasses as an indicator of hygiene at slaughtering places: a proposal for sampling. *Journal of Food Protection*, **55** (4), 271–278.

Ministry of Agriculture, Fisheries and Food, UK, 1993. Letter from Brown, K.A., Head of Meat Hygiene Division. 'Proposals for a meat hygiene performance monitoring system'.

Note: Copies of the current HAS scheme and its guidance notes can be obtained through:

MHS, South and West of England Regional Office, Rooms 607/609, Quantock House, Paul Street, Taunton, TA1 2NX, UK.

What Is Safe Food?

Key issues
- Milk, rice and other foods
- *Bacillus cereus*
- Risk analysis: assessing and managing risk
- Safe food

Challenge

This case study considers foods in which the food poisoning organism *Bacillus cereus* occurs and asks you to consider the risk of food poisoning occurring in extended shelf life pasteurised milk. You should then be able to define for yourself what constitutes a safe food.

21.1 THE CASE STUDY: RISK ASSOCIATED WITH PASTEURISED MILK, RESIDUAL *BACILLUS* SPP. SPORE POPULATIONS AND THEIR EFFECTS

Raw milk may contain a wide range of organisms, some of which originate from the cow's udder or from the teat duct. Organisms derived from mud, farm dirt and faecal material may enter the milk during milking, and to these may be added contaminants from the milking equipment and from storage tanks, pumps, pipes and so on. Some of the mesophilic organisms multiply in the warm milk until it is cooled by refrigeration to around 5 °C. At this temperature psychrotrophic organisms can grow and, depending on the storage time, may become significant spoilage organisms.

However, as discussed in Chapters 3, 4 and 10 pasteurisation has the potential to reduce the population of heat sensitive organisms, including pathogens such as *Mycobacterium tuberculosis*, *Brucella abortus*, *Salmonella*, *Staphylococcus aureus*, thermophilic *Campylobacter* and others, to acceptably low levels. But some organisms, notably spore bearing bacteria and some streptococci, resist the pasteurisation heat treatment and remain viable in the milk. Thus the microbiological quality of pasteurised milk depends on the initial quality of the raw milk, the numbers and types of organisms which survive pasteurisation, subsequent post-pasteurisation contamination as the milk flows through the processing plant, and the opportunities afforded to the organisms to grow in the time period following pasteurisation to the moment of consumption. Studies on the microbial quality of pasteurised milk have identified a wide range of organisms present, all or any of which may become the significant spoilage organisms with temperature and time of storage affecting which become dominant (Table 21.1).

Historically, when it was customary for milk to be purchased daily, the safety of pasteurised milk was protected by refrigeration, and the organisms of concern were mesophilic. But over the decades as retailing practice has changed the shelf life of pasteurised milk has been extended by refrigeration. However, the success of this practice is threatened by the numbers of organisms present in the milk, by the composition of the contained microbial flora and the ability of the psychrotrophic organisms to grow.

Griffiths and Philllips (1988), working in Scotland, found that the dominant organisms in spoiled milk at end of shelf life depended on:

• whether the initial quality of milk post-pasteurisation was 'poor' or 'good' (Table 21.2);
• on the temperature of storage.

Spoilage due to *Pseudomonas* could be expected in 'poor' quality milks when stored below 10 °C, and mixed population spoilage in which *Bacillus* became a more dominant proportion of the population could be expected in 'good' quality milks stored below 10 °C, although the numbers of *Pseudomonas* would probably still be predominant. They defined the end of shelf life as the time when the population was in excess of $\log_{10}$ 7.5 cfu ml^{-1}. They noted that:

"spoilage [in refrigerated milks] was mainly due to the growth of *Pseudomonas* spp. Enterobacteriaceae and Gram positive bacteria assumed greater importance in the spoilage of milks stored above 10 °C. Milks of good quality also contained *Bacillus* species and this group of bacteria were not detectable in milks with short shelf-lives".

However, these authors also observed that with increasing control over post-pasteurisation contamination generally, it could be predicted that there would be an increase in problems in pasteurised milk and cream due to the *Bacillus* group of organisms causing such effects as sweet curdling, flavour defects and a condition known as 'bitty cream' (Davies, 1975; Griffiths, Phillips and Muir, 1981).

In 1990 Griffith and Phillips published work studying the types of *Bacillus* species which could be found in samples of raw and pasteurised milk in Scotland (Table 21.3) and suggested that although on-farm hygiene and hygiene in the pasteurisation plant could reduce the numbers of *Bacillus* in milks, these techniques could not eliminate them and pasteurised milks and creams could be expected to contain some psychrotrophic *Bacillus*.

Thus of commercial concern is the balance between the desire for extending shelf life to satisfy distribution criteria and retail demands, and the need to sell products which will meet their shelf lives under consumer conditions of use and are safe to consume. The problem is that some members of the *Bacillus* group are

Table 21.1 Types of bacteria isolated from end of shelf life pasteurised milks stored at different temperatures: % isolates from pasteurised milks stored at (°C)[a]

Bacterial group:	Temperatures (°C)																				
	1.8	2.5	3.5	4.4	5.4	6.5	7.6	8.5	9.6	10.4	11.6	12.5	13.6	14.7	15.5	16.6	17.7	18.3	19.6	20.7	21.6
Pseudomonas	88.5	77.1	74.0	92.1	78.5	65.2	61.1	56.6	38.7	55.3	39.0	35.8	23.3	22.0	14.7	18.1	26.2	17.3	20.7	9.1	12.2
Enterobacteriaceae[b]	0.7	5.2	8.5	2.9	8.8	12.2	12.1	16.1	21.9	8.6	16.4	27.9	23.4	28.2	42.6	29.8	15.6	32.5	27.5	20	39.2
Other Gram negatives[c]	1.2	9.8	15.5	1.1	6.3	3.7	7.2	4.1	6.0	5.9	4.5	4.5	17.3	11.0	6.8	6.3	4.6	8.3	3.2	2.8	5.6
Gram positive spore-forming rods (*Bacillus*)	0.1	7.8	0.6	0.1	3.1	3.7	7.1	3.9	8.1	7.2	9.7	8.5	12.4	0.6	10.2	12.9	12.7	16.6	15.4	15.3	11.1
Non-sporeforming Gram positive rods[d]	4.6	0.1	0	3.5	0.4	2.9	0.6	7.8	0	4.3	2.4	8.1	3.9	3.2	6.4	5.5	4.5	3.4	5.4	5.3	2.6
Gram positive cocci[e]	4.9	0	1.4	0.3	2.9	12.3	11.9	11.5	25.3	18.7	28	15.2	19.7	35	19.3	27.4	36.4	21.9	27.8	47.5	29.3

[a]Isolates taken at the end of shelf life when bacterial counts were above $\log_{10} 7.5\,cfu\,ml^{-1}$.
[b]Including isolates of the genera *Enterobacter*, *Escherichia*, *Citrobacter*, *Klebsiella* and *Serratia*.
[c]Including isolates of the genera *Aeromonas*, *Acinetobacter*, *Alcaligenes*, *Flavobacterium* and *Moraxella*.
[d]Including isolates of the genera *Lactobacillus* and *Corynebacterium*.
[e]Including isolates of the genera *Micrococcus* and *Streptococcus*.

Method – The relative proportion of colony types on the plates used for enumeration were noted, and representative colonies of each type were subcultured to nutrient broth. On average 10% of colonies counted were identified. The isolates were Gram stained. Gram negative bacteria were further characterised using the Oxi/Ferm/Enterotube systems of Roche Products Ltd. Gram positive organisms were classified into genera according to morphology, and the primary tests described by Cowan and Steele (1974). These included motility, aerobic growth, presence of catalase and oxidase, acid production from glucose and oxidation-fermentation test. Spore forming bacteria were identified using the API50 CHB kits (API Laboratory Products).

Cowan, S.T. and Steele, K., 1974. *Manual for Identification of Medical Bacteria*. Cambridge University Press, Cambridge, UK.

Source: Reproduced from Griffiths, M.W. and Phillips, J.D. Modelling the relation between bacterial growth and storage temperature in pasteurised milks of varying hygienic quality. *Journal of the Society of Dairy Technology*, 1988, **41** (4), 96–102, by permission from Blackwell Scientific Publications Ltd.

Table 21.2 *Classification of pasteurised milks on the basis of initial psychrotrophic counts post-pasteurisation*

Range of shelf life in days[a]	Geometric mean of initial psychrotrophic count (count ml^{-1})[b]	Perception of initial quality
<5	80.2	Poor
5–6	13.8	↓
6–8	0.84	
8–10	0.11	↓
>10	0.01	Good

[a]Shelf life calculated as the time for the bacterial count in milk to reach $\log_{10}$ 7.5 at 6 °C.
[b]Determined by the most probable number technique (6–12 samples per category).
Source: Griffiths and Phillips, 1988, reproduced by permission (see Table 21.1).

Table 21.3 *Identity of psychrotrophic Bacillus spp. isolated from milk*

	% isolate of type	
	Raw milk	Pasteurised milk
Bacillus amyloliquefaciens	2.8 (1)[a]	
B. brevis		2.2 (2)
B. caratarum		2.2 (2)
B. cereus	37.3 (13)	36.7 (33)
B. cereus var. *mycoides*	8.6 (3)	1.1 (1)
B. circulans	28.6 (10)	7.8 (7)
B. firmus		2.2 (2)
B. lentus	2.8 (1)	5.6 (5)
B. mycoides	8.6 (3)	32.2 (29)
B. pasteurii/sphaericus	2.8 (1)	
B. polymyxa	2.8 (1)	1.1 (1)
B. pumilis		2.2 (2)
B. stearothermophilus (?)		1.1 (1)
B. thuringensis	5.7 (2)	5.6 (5)

[a]Figures in parentheses are the actual numbers of isolates.
Source: Reproduced from Griffiths, M.W. and Phillips, J.D., 1990. Incidence, source and some properties of psychotrophic *Bacillus* spp. found in raw and pasteurised milk. *Journal of the Society of Dairy Technology*, 1990, **43** (3), 62–66, by permission from Blackwell Scientific Publications Ltd.

known to be pathogenic – and there are many instances of food poisoning arising from this group of organisms (see Section 21.2). In the Netherlands in 1997 it was common practice for the shelf life of pasteurised milk to be set at 7 days when stored below 7 °C (Notermans *et al.*, 1997). But in the light of the knowledge of the probable presence of *Bacillus* spp. in pasteurised milk, of the ability of some strains to grow there, what is the risk of any consumer contacting food poisoning from this source, and would the public accept this risk if they knew?

21.2   BACKGROUND

21.2.1   *Bacillus* Food Poisoning Statistics and Two Outbreaks of *Bacillus cereus* Food Poisoning

Table 21.4 shows the recorded incidence of *Bacillus* food poisoning in England and Wales over the period 1980–1996. In comparison to the numbers of laboratory reports for *Salmonella*, or for thermophilic *Campylobacter* (see Table 4.1) in the same period the numbers of cases are relatively few. The organisms most commonly associated with food poisoning are strains of *Bacillus cereus* but strains of *B. subtilis*, *B. licheniformis* and *B. pumilis* may also be implicated. For comparison, some statistics from Japan are also shown in Table 21.5.

Table 21.4 *Bacillus food poisoning outbreaks, England and Wales, 1980–1996*

	General outbreaks	Family outbreaks	Sporadic cases	All cases (numbers positive and or ill)
1980	12	1	7	71
1981	9	0	8	72
1982	11	3	5	41
1983	16	1	7	134
1984	21	3	4	214
1985	12	3	7	81
1986	14	4	8	65
1987	16	5	3	137
1988	16	4	5	418
1989	23	4	14	164
1990	23	10	14	162
1991	12	2	9	95
1992	8	1	7	182
1993	8	3	6	31
1994	9	6	11	87
1995	12	0	10	87
1996[a]	6	1	5	27

[a]Data provisional.
Source: www.phls.co.uk. Reproduced with permission of the PHLS Communicable Diseases Surveillance Centre. © PHLS.

In 1993 Hobbs and Roberts described an outbreak of *B. cereus* food poisoning which exemplifies the characteristics of the illness. The outbreak was associated with beef stew in which pre-cooked vegetables were believed to have supported growth of the organism with the formation of heat stable toxin. These vegetables had been left at ambient temperature for some hours after which they were added to beef mince – the whole mix then being boiled for an hour immediately prior to serving to prisoners in a UK prison. Forty-three people were affected with vomiting (emesis) 1.5–3 hours after eating the stew, and some suffered diarrhoea and abdominal pain about 4 hours after the meal. The illness was not long lasting and all patients recovered.

Table 21.5 *B. cereus outbreaks in Japan, 1982–1986*

	Year					
	1982	1983	1984	1985	1986	Totals
Outbreaks	13	18	15	17	10	73
Cases	88	250	330	328	327	1323

Note: *B. cereus* food poisoning: 93–95% cases of the emetic type; the rest diarrhoeal.
73% of outbreaks due to some form of cooked rice; 16% outbreaks due to
some form of noodles; 11% due to other food.
All food poisoning: total outbreaks = 5141; total cases = 185 301.
Source: Shinagawa, 1990.

Table 21.6 *B. cereus, growth limits and other characteristics*

Characteristics		Reference
Limits for growth		
Temperature	$<4\,°C$	1
pH	<4.4	1
a_w	<0.91	1
Numbers of *B. cereus* found in samples of freshly pasteurised milk	<10 organisms/100 ml	2
Numbers of *B. cereus* indicative of association with food borne disease	$>10^5$ cfu g^{-1} or ml^{-1}	2
Concentrations in foods associated with *B. cereus* food poisoning	10^5 to $>10^9$ cfu ml^{-1} or g^{-1}	3
Numbers of *B. cereus* not consistent with good hygienic practice	$>10^5$ cfu g^{-1} or ml^{-1}	2
Not all strains are toxigenic		2

Sources:
1. Notermans, Zwietering and Mead, 1994.
2. Notermans *et al.*, 1997.
3. Gilbert and Taylor, 1976.

However, in the UK the first recorded cases (Anon, 1972, 1973) were associated with the serving of cooked rice which, after cooking, had been held unrefrigerated for some time allowing germination of the heat resistant spores, and subsequent vegetative cell multiplication with toxin production and accumulation. Investigations into subsequent similar cases showed concentrations of *B. cereus* of between 10^5 and 10^9 cfu g^{-1} (Gilbert and Taylor, 1976).

However, van Netten *et al.* (1990) investigated three specific outbreaks of food poisoning which occurred in the period 1986–1989 in Spain and in the Netherlands associated with vegetable pie, pasteurised milk and cooked cod fish. All these were refrigerated foods. In the milk outbreak there were 280 cases, and *B. cereus* was found present in milk samples at 0.4×10^6 cfu ml^{-1}. These strains were found able to grow at $7\,°C$ but not $43\,°C$, and had a growth temperature

Table 21.7 *The occurrence of B. cereus in retail foods often implicated in 'diarrhoeal-syndrome' food poisoning in the Netherlands*

Food products	Number of samples examined	Bacillus cereus positive (%)	Frequency of Bacillus cereus[a]				
			<2.0	2.0–3.0	3.0–4.0	4.0–5.0	>5.0
Spices	33	42	19	6	6	1	1
Egg yolk, powder, pasteurised	34	29	24	10	0	0	0
Egg yolk, liquid, pasteurised	29	24	22	7	0	0	0
Custard	48	19	39	0	4	3	2
Cream pastry	83	11	74	4	3	0	2
Milk, pasteurised[b]	483	8[c]	442	6	17	11	7
Roasted/fried meats	48	8	44	3	1	0	0
Rice meals	551	6	518	17	10	4	2
Roasted/fried poultry	38	5	36	1	1	0	0
Fish pate/mousses	25	4	24	1	0	0	0
Pea soup	69	3	67	1	1	0	0
Lasagne	35	3	34	0	0	1	0
Cheeses, soft, raw milk	53	2	52	1	0	0	0
Vegetable salads	54	2	53	1	0	0	0
Convenience meals	72	1	71	1	0	0	0
Mushrooms, blanched	45	0	45	0	0	0	0

[a]Log_{10} cfu g^{-1} on mannitol egg yolk polymixin agar at 30 °C.
[b]Examined after storage at 7 °C up to 'best before' date.
[c]Positively affected by inclusion of samples originating from a dairy factory having a bacteriological problem.
Note: all food samples were drawn from commercial outlets in the 's-Hertogenbosch area, The Netherlands. Where applicable 'best before' dates were identified and only fresh products were sampled.
Source: Reproduced from van Netten, P., van de Moosdijk, A., van Hoensel, P. and Mossel, D.A.A. (1990), Psychotrophic strains of *Bacillus cereus* producing enterotoxin. *Journal of Applied Bacteriology*, **69**, 73–79, by permission from Blackwell Scientific Publications Ltd.

range lower than that normally expected for *B. cereus* (expected range 10–50 °C), (Table 21.6), showing some strains of *B. cereus* to be psychrotrophic.

21.2.2 Foods Associated with *Bacillus* spp.

Bacillus spp. occur very widely in the environment and have often been isolated from soils, water and crop plants, particularly cereal grains. Sometimes they can be isolated from faeces from patients suffering *B. cereus* food poisoning but they represent a transient population and the group of organisms are not considered to be zoonotic. A number of surveys from across the world show that the organism can be found in many food types. van Netten and colleagues (1990) analysed over 1700 food samples in the Netherlands in order to ascertain the occurrence of *B. cereus* in retail foods which were known to often be implicated in diarrhoeal-syndrome food poisoning (Table 21.7). They showed that a wide range of foods and 8% of samples of pasteurised milk contained the organism. Earlier a survey in Taiwan (Wong *et al.*, 1988) showed that a significant percentage of milk based products were positive for *B. cereus* at levels below 10^3 cfu ml^{-1} (range from 5 to 800 cfu ml^{-1}), and Christiansson *et al.* (1989) working in Sweden obtained 136 isolates of *B. cereus* from milk and cream samples originating from 37 different Swedish dairy plants.

21.3   EXERCISES

Exercise 1. In the Netherlands in 1997 it was common practice for the shelf life of pasteurised milk to be set at 7 days when stored below 7 °C (Notermans *et al.*, 1997). The text in Sections 21.1 and 21.2 identifies a number of *B. cereus* food poisoning outbreaks, some of which were associated with milk products. Suppose the Dutch milk industry – for commercial reasons – had wanted to extend the shelf life of pasteurised milk from 7 to 14 days. In the light of the knowledge of the probable presence of *Bacillus* spp. in pasteurised milk, and of the ability of some strains to grow there:

- What would need to be considered to evaluate whether the 14 day shelf life milk would represent an increase in risk to consumers in comparison to the 7 day milk?
- What would be needed to maintain the risk at current levels?
- Would the public accept this level of risk if they knew about it?

Exercise 2. What is safe food?

You are being asked about risk, its assessment and management. You are also being asked about risk communication.

This exercise is based on work done by Notermans *et al.* (1997) and it would be instructive for you to read their paper. But before you do that it would be beneficial for you to think out the issues for yourself. Sections 21.1 and 21.2 provide you with some data, as do other parts of the whole book.

But you have to raise the questions.

21.4 COMMENTARY

The chapter has been designed in such a way that you are given some relevant data but you will have to evaluate it and also search this book, or other sources for information.

Exercise 1. In the Netherlands in 1997 it was common practice for the shelf life of pasteurised milk to be set at 7 days when stored below 7 °C (Notermans *et al.*, 1997). The text in Sections 21.1 and 21.2 identifies a number of *B. cereus* food poisoning outbreaks, some of which were associated with milk products. Suppose the Dutch milk industry – for commercial reasons – had wanted to extend the shelf life of pasteurised milk from 7 to 14 days. In the light of the knowledge of the probable presence of *Bacillus* spp. in pasteurised milk, and of the ability of some strains to grow there:
* What would need to be considered to evaluate whether the 14 day shelf life milk would represent an increase in risk to consumers in comparison to the 7 day milk?
* What would be needed to maintain the risk at current levels?
* Would the public accept this level of risk if they knew?

21.4.1 Risk Analysis

This is a three step process (Table 21.8).

Step 1. Risk assessment

In order to achieve the goal of the extension of the shelf life of the pasteurised milk without increasing the risk to the public from *B. cereus* food poisoning, a risk assessment would have to be undertaken. You are being asked to undertake some of the elements of that process. Your outcome will be qualitative – *i.e.* you will be able to say 'the risk increases' (or decreases) under certain circumstances. You will not be able to quantify the risk, for to do that numerical data are required which must then be manipulated. The techniques of both qualitative and quantitative risk assessment in food microbiology and food safety are currently being developed and the body of published material has been increasing since the mid-1990s (Notermans *et al.*, 1994; Baird-Parker, 1995; Notermans *et al.*, 1997; van Gerwen and Zwietering, 1998; Harrigan, 1998; Coleman and Marks, 1999; van Gerwen *et al.*, 2000).

A structured process has to be undertaken, but rather than telling you what to do it is better that you work out for yourself some questions which need to be asked, to determine what data need to be gathered or need to be generated. You will probably find it helpful to gather as much of these data as you reasonably can before you then consider them and determine whether they would tell you enough about the relationship between *B. cereus*, pasteurised milk, its distribution and its consumption to evaluate the current risk in 7 day shelf life pas-

Table 21.8 *Risk analysis: its component parts*

Risk assessment 'the quantitative estimation of the probability of a food borne hazard' (Notermans *et al.*, 1997)	Hazard identification	* is the organism known to cause food borne disease? * is the organism infectious or toxigenic? * is it associated with this specific food product? * is it likely to be present in this food? * what will processing do to the viability of the organism, or the activity of any toxins it may form? * if processing kills or removes it, is re-infection of the food a possibility?
	Exposure assessment	* what concentration of pathogen occurs in the food on consumption? * what factors affect growth and decline? * how much of the food is eaten by consumers, and how often?
	Dose–response	* what is the relationship between the numbers of organisms or toxin concentrations and human health
	Risk estimate	Computation of the above factors to estimate the level of risk
Risk management	Good manufacturing practice HACCP	
Risk communication	Telling the affected groups	

Drawn up from ideas in Notermans, Zwietering and Mead, 1994; van Gerwen and Zwietering, 1998; Coleman and Marks, 1999.

teurised milk. If this same milk were kept up to 14 days, what would happen?

Step 2. Risk management

You will need to consider what needs to change to ensure that when it is consumed the 14 day shelf life product does not cause *B. cereus* food poisoning. Can these changes be implemented and managed in such a way to guarantee the safety of the milk on consumption?

After you have addressed this problem yourself you may find it helpful to consult the following pages (Section 21.4.2) which give one approach to assessing the risk associated with the 7 day and 14 day milks, and managing the associated risk in such a way that the risk of *B. cereus* food poisoning occurring does not change.

Step 3. Risk communication

A point you are asked to think about is also – is the current risk associated with 7 day shelf life pasteurised milk acceptable to the public?

- Do the public know the level of risk they are exposed to?
- Should the industry describe to the public the current risk associated with pasteurised milk, and the presence in it of *B. cereus*?

Suppose the 14 day shelf life product, which might perhaps be very popular because it reduced the frequency of shopping, could only be introduced with an increase of risk. If told of the increased risk would the public accept that risk, or would they want the product but not the risk? Would or should they take any responsibility for their own safety? Public reaction to food risk is becoming more articulate but there is contention between the public, the food producers and the legislators (see for example, Shaw, 2000) about where responsibility lies.

These are matters on which you must develop your own opinion.

21.4.2 Guidance on Assessing and Managing the Risk Associated with 7 Day and 14 Day Shelf Life Milk

1. First consider the current risk associated with '7 day shelf life milk'
   . . . go to 2.
   Later consider the new risk associated with '14 day shelf life milk'
   . . . go to 6.

2. Find out the frequency of *B. cereus* food poisoning associated with pasteurised milk in your country and, if necessary, elsewhere.
   Get up to date data.

   - What data do surveillance techniques gather? (How reliable are the statistics?).
   - What data could be missing? (Is there a hidden problem?) See Figure 2.4.

   . . . go to 3.

3. The occurrence of *B. cereus* food poisoning associated with pasteurised milk demonstrates there is 'a risk'. The size of this risk relates to the number of people actually affected, to the number of people exposed, together with the overall frequency of their exposure. It thus also relates to how much pasteurised milk is consumed by individuals and their susceptibility to *B. cereus*, how much is consumed per annum and the actual loading of *B. cereus* at the time of drinking it.

• Are these data known?

. . . go to 4.

4. What circumstances allow *B. cereus* to occur in pasteurised milk, to multiply there and potentially cause food poisoning?

• Consider the organism: its sources and strains, its seasonality in milk, the levels at which it occurs in freshly pasteurised milk, its growth and factors which limit its growth. Consider when toxin is produced, and what the infective dose is.

. . . refer to Table 21.7, then go to 5.

5. When does the production-to-consumer chain facilitate growth and achievement of cell concentrations allowing an infective dose to be present?

• Does all milk support growth?
• What is/are the system(s) of distribution?
     How much time does milk spend in distribution?
     What temperature is milk held at in distribution?
     Is this consistent throughout the distribution chain?
• How is the product sold?
     Is it under controlled conditions?
• Who is the consumer?
     Are there vulnerable groups? How much milk do different groups drink per day and per annum?
• How do consumers use the product?
     How long do consumers store milk for? Is milk past its 'use-by' date consumed?
     Is consumer refrigeration practice appropriate and reliable?
• Can specific circumstances now be identified which could lead to *B. cereus* food poisoning?

6. What is known about the actual microbial quality of '7 day shelf life milk'? From laboratory based storage trials:

• What cell concentrations are reached under different time and temperature combinations?
• Can computer based predictive models help?
• Which time and temperature combinations lead to potentially dangerous levels?
• What critical limits for time and temperature exposure in distribution and use can be set?
• How do these data relate to how consumers actually use the milk?
• Would spoilage, due to other organisms be evident before or after unacceptable loads of *B. cereus* would be present?

- Do consumers use milk which has signs of spoilage present?
- What final criterion for *B. cereus* in the milk has been set in relation to the shelf life at the point of use to assure safety?
- What pre-distribution specification with respect to *B. cereus* has been set for the product?
- Are these criteria appropriate to the actual distribution and use the product receives?
- Are these criteria enforced?

7. Do the data in 2–6 give confidence that the current system for 7 day shelf life milk (from production to consumer use) is robust enough to also prevent, at the point of consumption, the presence of dangerous levels of *B. cereus* in 8–14 day old milk, manufactured under current production regimes and standards?

   - If yes – the risk (measured by probability of incidence of *B. cereus* food poisoning) to the consumer will not increase.
   - If no – the risk to the consumer will increase and further risk management strategies are needed.

   . . . go to 8.

8. Can *B. cereus* concentrations be reduced or removed from raw milk and the production environment?

   - Can farm hygiene produce cleaner milk?
   - Can better cleaning regimes reduce the incidence of the organism in processing plant?
   - Can reductions be consistently achieved which will require no other actions in the distribution-to-consumer chain? In other words can new, more stringent standards be set and implemented?
   - If yes, provided this is an accurate assessment, the whole responsibility for risk management lies with the milk producers. The risk to consumers should not increase due to the production and distribution of 14 day shelf life milk.
   - If no – are other strategies available?

   . . . go to 9.

9. Can growth opportunities for the organism be reduced, or eliminated? Refrigeration at a new, lower temperature may be essential to the safety of the milk.

   - In production can milk cooling rates be improved?
   - Can refrigeration temperatures throughout the production/distribution chain be lowered?
   - Can consumers refrigerate at this temperature? . . . see 12.

- If refrigeration temperatures cannot be lowered, risk to consumers increases.
- Are other strategies available to compensate?

. . . go to 10.

10. Can the storage time be portioned between production-to-retail period, and the consumer holding period? Who is intended to benefit from the extended shelf life of the 14 day product?

- Industry   . . . go to 11.
- Consumer   . . . go to 12.

11. Industry – with the intention of facilitating production management and longer distribution time periods. Good temperature control may be possible. But consumers would then receive an older 'fresh' product. Risk level is still dependent on consumer refrigeration.

. . . go to 9 and 12.

12. Consumers. Do the producers intend that consumers store the product for up to 14 days facilitating less frequent shopping?

- Would domestic refrigeration run at a low enough temperature, and reliably maintain it to prevent growth of *B. cereus*?
- Would consumers handle the milk in a way which prevented growth of *B. cereus* over the 14 day shelf life period?
- Could it be guaranteed that consumers would use the product within the 14 day period?
- Would consumer education be unnecessary?

Uncertain or negative answers in Section 12 increase risk to the public.

- Should consumers accept the blame if *B. cereus* outbreaks occurred after the introduction of the 14 day shelf life product?

Positive answers indicate confidence in the way the product would be handled at the point of consumption, and provided strategies outlined in 8–10 are implemented, and control achieved in distribution the new 14 day shelf life product can be produced and distributed and used with no increase in risk of *B. cereus* food poisoning arising from pasteurised milk than before.

What you have just done is undertake basic steps in risk assessment and determine strategies for its management. You have recognised that food cannot be absolutely free of pathogens, but the danger to the consumer that they represent must be managed. You may wish to compare your own ideas or those outlined here with Table 21.8 which outlines the accepted steps in risk analysis.

Notermans *et al.* (1994, 1997, 1998) studied pasteurised milk as a model for determining the risk associated with it from food poisoning associated with *B. cereus*. It was their work on which the idea for this Case Study is based, and it is recommended that you refer to their work to understand the detail needed to evaluate and quantify actual risk in a food in a real situation.

21.4.3 Safe Food

> **Exercise 2.** What is safe food?

You will by now have realised, particularly if you have undertaken many of the exercises in this book, and you have undertaken Exercise 1 in this last Case Study, that there is a difference between safe food and risk free food. Risk free food probably does not exist. Safe food is that in which there is an acceptably low risk. But what is 'acceptable' changes with time, with knowledge development, changes in technology and with public expectations. In Case Study 2 you may have noticed that it was said, after the enquiry into deaths of eight people who died of botulism, that nothing more than had already been done could have been done. In other words, at that time in 1922, following the procedures described was entirely satisfactory. But today we would expect more to be done – we would expect the production system to be controlled better and therefore the risk of the hazard occurring to be smaller. The widespread introduction in the 1990s of HACCP systems in food processing is a great leap forward because it anticipates the potential hazards, assesses the risk from them (but does not quantify it) and implements appropriate control procedures. But it has to be managed effectively. However, HACCP can only control known hazards. Even food produced under the best HACCP system cannot be guaranteed to be safe. That is why in the UK Food Safety Act (Anon, 1990) there is a defence for a food producer, whose food may have caused illness in consumers, called the 'due diligence' defence. It allows a producer to defend him/herself by producing evidence that all that could be done was done (as indeed was claimed in Case Study 2, mentioned above). When new pathogens emerge, or new shelf lives are expected, or new processes are implemented the level of risk associated with that food changes. Risk assessment determines whether the risk to the consumer has changed and whether it has become unacceptable and new controls need be implemented.

Safe food need be neither sterile, nor free of pathogens. Safe food simply has an acceptable level of risk under the expected conditions of use. What is needed is agreement between interested parties – legislatures, producers and consumers – as to what that level of acceptable risk is.

21.5 SUMMARY

Bacillus cereus is an organism which causes food poisoning through the pre-formation of emetic toxin in foods, and also by the formation of the diarrhoeal toxin in the human intestine after consumption of infected food. It is commonly but not exclusively associated with cooked rice and other starchy foods, but other foods, cooked meats and pasteurised milk have also caused food poisoning.

The organism produces heat resistant spores, which resist pasteurisation processes and is therefore liable to be present in the treated milk, but cell concentrations depend on the quality of the raw milk. Thermal abuse of the refrigerated milk will allow its growth, but psychrotrophic strains may multiply in chilled milk and may reach dangerous levels if sufficient time passes.

Extending the shelf life of pasteurised milk provides an opportunity to examine the associated risk. If this were proposed risk analysis would require both risk assessment to facilitate design of how this could be achieved and, if the project was deemed feasible, risk management in its implementation. But even if the new product were 'safe', actual risk in comparison to the previous product might increase. Determination of what is 'safe' requires the process of risk communication, understanding and agreement between the key interested parties – namely the consumer, the producer and the legislator.

Safe food is that food which has, to those three consenting parties, an acceptable level of risk.

21.6 REFERENCES

Anon, 1972. Food poisoning associated with *Bacillus cereus. British Medical Journal*, **i**, 189.

Anon, 1973. *Bacillus cereus* food poisoning. *British Medical Journal*, **iii**, 647.

Anon, 1990. Food Safety Act. HMSO, London, UK.

Baird-Parker, A.C., 1995. Development of industrial procedures to ensure the microbiological safety of food. *Food Control*, **6** (1), 29–36.

Christiansson, A., Satyanarayan Naiduu, A., Nilsson, I., Wadstrom, T. and Pettersson, H.-E., 1989. Toxin production by *Bacillus cereus* dairy isolates in milk at low temperatures. *Applied and Environmental Microbiology*, **55** (10), 2595–2600.

Coleman, M.E. and Marks, H.M., 1999. Qualitative and quantitative risk assessment. *Food Control*, **10**, 289–297.

Davies, F.L., 1975. Discussion of papers presented at a symposium on bitty cream and related problems. *Journal of the Society of Dairy Technology*, **28**, 85–90.

Gilbert, R. and Taylor, A.J., 1976. '*Bacillus cereus* food poisoning', pages 197–213 in *Microbiology in Agriculture, Fisheries and Food*, ed. Skinner, F.A. and Carr, J.G. Published for the Society for Applied Bacteriology by Academic Press, London and New York.

Griffiths, M.W., Phillips, J.D. and Muir, D.D., 1981. Development of flavour defects in pasteurised double cream during storage at 6°C and 10°C. *Journal of the Society of Dairy Technology*, **34**, 142–146.

Griffiths, M.W. and Phillips, J.D., 1988. Modelling the relation between bacterial growth and storage temperature in pasteurised milks of varying hygienic quality. *Journal of the Society of Dairy Technology*, **41** (4), 96–102.

Griffiths, M.W. and Phillips, J.D., 1990. Incidence, source and some properties of psychrotrophic *Bacillus* spp. found in raw and pasteurised milk. *Journal of the Society of Dairy Technology*, **43** (3), 62–66.

Harrigan, W.F., 1998. Incidents of food poisoning and food-borne diseases from 'new' or 'unexpected' causes: can they be prevented? *International Journal of Food Science and*

Technology, **33**, 177–189.

Hobbs, B.C. and Roberts, D., 1993. Food Poisoning and Food Hygiene, 6th Edition, p. 118. Edward Arnold, London, UK.

Notermans, S., Zwietering, M.H. and Mead, G.C., 1994. The HACCP concept: identification of potentially hazardous organisms. *Food Microbiology*, **11**, 203–214.

Notermans, S., Dufrenne, J., Teunis, P., Beumer, R., te Giffel, M. and Peeters Weem, P., 1997. A risk assessment study of *Bacillus cereus* in pasteurised milk. *Food Microbiology*, **14**, 143–151.

Notermans, S. and Batt, C.A., 1998. Risk assessment approach for food-borne *Bacillus cereus* and its toxins. *Journal of Applied Microbiology Symposium Supplement*, **84**, 51S–61S.

PHLS web site: www.phls.co.uk – data as at June 1998.

Shaw, A., 2000. Public understanding of food risks: what do the experts say? *Food Science and Technology Today*, **14** (3), 140–143.

Shinagawa, K., 1990. Analytical methods for *Bacillus cereus* and other *Bacillus* species. *International Journal of Food Microbiology*, **10**, 125–142.

van Gerwen, S.J.C. and Zwietering, M.H., 1998. Growth and inactivation models to be used in quantitative risk assessments. *Journal of Food Protection*, **61** (11), 1541–1549.

van Gerwen, S.J.C., te Giffel, M.C., van'Riet, K., Beumer, R.R. and Zwietering, M.H., 2000. Stepwise quantitative risk assessment as a tool for characterization of microbiological food safety. *Journal of Applied Microbiology*, **88**, 938–951.

van Netten, P., van de Moosdijk, A., van Hoensel, P. and Mossel, D.A.A., 1990. Psychrotrophic strains of *Bacillus cereus* producing enterotoxin. *Journal of Applied Bacteriology*, **69**, 73–79.

Wong, H.-C., Chang, M.-II. and Fan, J.-Y., 1988. Incidence and characterisation of *Bacillus cereus* isolates contaminating dairy products. *Applied and Environmental Microbiology*, **54** (3), 699–702.

Subject Index